Advances in
CHEMICAL ENGINEERING
ENGINEERING ASPECTS OF SELF-ORGANIZING MATERIALS

VOLUME **35**

ADVANCES IN
CHEMICAL ENGINEERING

Editor-in-Chief

GUY B. MARIN
Department of Chemical Engineering
Ghent University
Ghent, Belgium

Editorial Board

DAVID H. WEST
Research and Development
The Dow Chemical Company
Freeport, Texs, U.S.A.

PRATIM BISWAS
Department of Chemical and Civil Engineering
Washington University
St. Louis, Missouri, U.S.A.

JINGHAI LI
Institute of Process Engineering
Chinese Academy of Sciences
Beijing, P.R. China

SHANKAR NARASIMHAN
Department of Chemical Engineering
Indian Institute of Technology
Chennai, India

Advances in
CHEMICAL ENGINEERING
ENGINEERING ASPECTS OF SELF-ORGANIZING MATERIALS

VOLUME **35**

Edited by

RUDY J. KOOPMANS

PolyUrethane R&D, Freienbach
Dow Europe GmbH, Switzerland

Amsterdam • Boston • Heidelberg • London • New York • Oxford
Paris • San Diego • San Francisco • Singapore • Sydney • Tokyo
Academic Press is an imprint of Elsevier

Academic Press is an imprint of Elsevier
Radarweg 29, PO Box 211, 1000 AE Amsterdam, The Netherlands
32 Jamestown Road, London NW1 7BY, UK
30 Corporate Drive, Suite 400, Burlington, MA 01803, USA
525 B Street, Suite 1900, San Diego, CA 92101-4495, USA

First edition 2009

Copyright © 2009 Elsevier Inc. All rights reserved

No part of this publication may be reproduced, stored in a retrieval system or transmitted in any form or by any means electronic, mechanical, photocopying, recording or otherwise without the prior written permission of the publisher

Permissions may be sought directly from Elsevier's Science & Technology Rights Department in Oxford, UK: phone (+44) (0) 1865 843830; fax (+44) (0) 1865 853333; email: permissions@elsevier.com. Alternatively you can submit your request online by visiting the Elsevier web site at http://www.elsevier.com/locate/permissions, and selecting *Obtaining permission to use Elsevier material*

Notice
No responsibility is assumed by the publisher for any injury and/or damage to persons or property as a matter of products liability, negligence or otherwise, or from any use or operation of any methods, products, instructions or ideas contained in the material herein. Because of rapid advances in the medical sciences, in particular, independent verification of diagnoses and drug dosages should be made

Library of Congress Cataloging-in-Publication Data
A catalog record for this book is available from the Library of Congress

British Library Cataloguing in Publication Data
A catalogue record for this book is available from the British Library

ISBN: 978-0-12-374752-5
ISSN: 0065-2377

For information on all Academic Press publications visit our website at books.elsevier.com

Printed and bound in USA
09 10 11 12   10 9 8 7 6 5 4 3 2 1

Working together to grow
libraries in developing countries

www.elsevier.com | www.bookaid.org | www.sabre.org

ELSEVIER   BOOK AID International   Sabre Foundation

Contents

Contributors	vii
Preface	ix
Acknowledgments	xi

1. Engineering Materials from the Bottom Up – Overview   1
Rudy J. Koopmans and Anton P.J. Middelberg

1.	Introduction	1
2.	Historical Context	2
3.	Self-Assembling Peptide Literature	4
4.	A Future of Challenges and Opportunities	7
	References	8

2. Mechanisms and Principles of 1D Self-Assembly of Peptides into β-Sheet Tapes   11
Robert P.W. Davies, Amalia Aggeli, Neville Boden, Tom C.B. McLeish, Irena A. Nyrkova and Alexander N. Semenov

1.	Introduction	12
2.	Model of Hierarchical Self-Assembling Chiral Rods in Solution	13
3.	Responsiveness to External Triggers	25
4.	Peptide Self-Assembly on Surfaces	37
5.	Conclusions and Future Prospects	39
	References	40

3. Nucleation and Co-Operativity in Supramolecular Polymers   45
Paul van der Schoot

1.	Introduction	45
2.	Nucleated Assembly	51
3.	Kinetics of Nucleated Assembly	55
4.	Co-Operativity and Nucleation	61
5.	Coarse Graining Reversed	66
6.	Summary and Outlook	70
	Acknowledgement	72
	References	72

4. Recombinant Production of Self-Assembling Peptides   79
Michael J. McPherson, Kier James, Stuart Kyle, Stephen Parsons and Jessica Riley

1.	Introduction	80
2.	Recombinant Protein Production	80

Contents

    3. Host Organisms for Recombinant Expression — 87
    4. Repetitive and Self-Assembling Proteins and Peptides in Nature — 97
    5. Recombinant Peptide Production — 100
    6. Recombinant Expression of Self-Assembling Peptides — 108
    7. Perspective — 113
       Acknowledgments — 113
       References — 113

5. Inspiration from Natural Silks and Their Proteins — 119
Boxun Leng, Lei Huang and Zhengzhong Shao

    1. Introduction — 119
    2. Structure and Properties of Proteins and Silks — 120
    3. Artificial Spinning of Silk Fibroin — 133
    4. Bioapplication of Silk Fibroin — 142
    5. Biomineralization Regulated by Silk Proteins — 144
    6. Conclusion and Perspective — 148
       Acknowledgment — 148
       References — 148

6. Surface- and Solution-Based Assembly of Amyloid Fibrils for Biomedical and Nanotechnology Applications — 161
Sally L. Gras

    1. Introduction — 161
    2. Polypeptide Self-Assembly — 162
    3. The Role of Surfaces During and After Assembly — 167
    4. Applications — 189
    5. Conclusion — 205
       References — 206

7. Hybrid Systems Engineering: Polymer–Peptide Conjugates — 211
Conan J. Fee

    1. Introduction — 211
    2. Peptide–Polymer Conjugation — 213
    3. Peptide-Directed Formation of Gels — 214
    4. Cyclopeptide Nanotubes — 215
    5. Polyelectrolytic and Organometric Polymer–Peptide Conjugates — 216
    6. β-Sheet Suprastructures — 217
    7. Rod–Coil Conformations — 218
    8. Physical Manipulation of Nanostructural Orientation — 220
    9. Conclusions — 220
       References — 221

Subject Index — 223

Contents of Volumes in this Serial — 235

CONTRIBUTORS

Numbers in parenthesis indicate the pages on which the authors' contributions begins.

A. Aggeli, *Centre for Self-Organising Molecular Systems, School of Chemistry, University of Leeds, Leeds LS2 9JT, United Kingdom* (11)

N. Boden, *Centre for Self-Organising Molecular Systems, School of Chemistry, University of Leeds, Leeds LS2 9JT, United Kingdom* (11)

R.P.W. Davies, *Centre for Self-Organising Molecular Systems, School of Chemistry, University of Leeds, Leeds LS2 9JT, United Kingdom* (11)

Conan J. Fee, *Department of Chemical & Process Engineering, University of Canterbury, Christchurch 8020, New Zealand* (211)

Sally L. Gras, *Department of Chemical and Biomolecular Engineering and The Bio21 Molecular Science and Biotechnology Institute, The University of Melbourne, Victoria 3010, Australia* (161)

Lei Huang, *Department of Macromolecular Science, Key Laboratory of Molecular Engineering of Polymers of Ministry of Education, Advanced Materials Laboratory, Fudan University, Shanghai 200433, People's Republic of China* (119)

Kier James, *Astbury Centre for Structural Molecular Biology, University of Leeds, Leeds LS2 9JT, United Kingdom* (79)

Rudy J. Koopmans, *PolyUrethane R&D, Freienbach, Dow Europe GmbH, Switzerland* (1)

Stuart Kyle, *Astbury Centre for Structural Molecular Biology, University of Leeds, Leeds LS2 9JT, United Kingdom* (79)

Boxun Leng, *Department of Macromolecular Science, Key Laboratory of Molecular Engineering of Polymers of Ministry of Education, Advanced Materials Laboratory, Fudan University, Shanghai 200433, People's Republic of China* (119)

T.C.B. Mcleish, *Department of Physics and Astronomy; Centre for Self-Organising Molecular Systems, University of Leeds, Leeds LS2 9JT, United Kingdom* (11)

Michael J. McPherson, *Astbury Centre for Structural Molecular Biology; Centre for Plant Sciences, University of Leeds, Leeds LS2 9JT, United Kingdom* (79)

Anton P.J. Middelberg, *The University of Queensland, Centre for Biomolecular Engineering, St. Lucia, Australia* (1)

I.A. Nyrkova, *Institut Charles Sadron, Strasbourg Cedex, France* (11)

Stephen Parsons, *Centre for Plant Sciences, University of Leeds, Leeds LS2 9JT, United Kingdom* (79)

Jessica Riley, *Astbury Centre for Structural Molecular Biology, University of Leeds, Leeds LS2 9JT, United Kingdom* (79)

A.N. Semenov, *Institut Charles Sadron, Strasbourg Cedex, France* (11)

Zhengzhong Shao, *Department of Macromolecular Science, Key Laboratory of Molecular Engineering of Polymers of Ministry of Education, Advanced Materials Laboratory, Fudan University, Shanghai 200433, People's Republic of China* (119)

Paul van der Schoot, *Group Theoretical and Polymer Physics, Department of Applied Physics, Technische Universiteit Eindhoven, Postbus 513, 5600 MB Eindhoven, the Netherlands* (45)

PREFACE

About 10 years ago, I met with Professor Neville Boden at Leeds University when he had just inaugurated the Self Organizing Molecular Systems Centre (http://www.soms.leeds.ac.uk/). His ambition, as I understood it at the time, was to create a focused research effort on the physico-chemistry of short amino-acid sequences (peptides of less then 25 residues) and develop understanding on how molecules, through intermolecular interactions, may aggregate at various length and time scales. To an industrial researcher scouting for new materials and applications, the effort in place looked interesting but quite academic with very limited value if any to the chemical industry. However, the enthusiasm and motivation of Professor Boden and his team, together with some further detailed discussions, made me realize that in fact all materials at the macroscospic scale perform as a consequence of the very specific, often hierarchical organization potential of the composing atoms or molecules. Furthermore, the opportunity of learning from nature including the reuse of molecular building blocks looked like a way to innovate synthetic materials as to chemistry, application performance, and sustainability. Exploring such a vast space, particularly for an industrial setting, requires focus in a way that while working toward a target application, knowledge and other opportunities would be revealed. Since those days and across the world many more research groups – mainly in academia – have started working on different aspects of self-organization of nature's building blocks. The list of publications is expansive, which makes it opportune to bring together a number of key papers that cover the state of the art as well as the thinking about the main technical challenges.

With a decade of experience, I have brought together seven papers with a scope limited to self-assembly of peptides, but covering key issues for advancing materials research into product development. The patent literature activity is typically an indicator of commercial interest and application space. Chapter 1 covers in some detail, but without being comprehensive, what self-assembled peptides may bring in terms of material applications. Chapters 2 and 3 explain the underlying principles of peptide self-assembly and how experiment can be used to model the hierarchy of structure formation. Typically, the work on peptides requires a tailored molecular configuration and Chapter 4 covers the options available for preparing them in sufficient quantities. The remaining chapters focus on the technology and use of the self-assembly mechanism to create specific applications such as "silk fibers," coatings, and scaffolds to name a few. Either an "all" peptide composition can be used or combinations of peptide with synthetic oligomers or polymers – hybrid systems.

The idea to bring this topical volume together was suggested by David West, member of the editorial board but also a very good friend and long time colleague at The Dow Chemical Company. The volume provided a platform to write up most of the things that is known in this field of research, for which I am very grateful as I believe that peptides will become the new materials of the future in view of their versatility

Rudy Koopmans
Horgen, February 2009

ACKNOWLEDGMENTS

Funding under the Sixth European Framework (project NMP3-CT-2005-516961 – Bio-based functional materials from engineered self-assembled peptides (BASE)) together with a number of bilateral conections allowed not only advancing the science and technology of self-assembly peptides but also brought together many creative people on many occasions. It was in fact a symposium in 2007 organised by Prof. Anton Middelberg at University of Queensland, Australia, that connected many of the authors and set the frame for the structure of this volume. I thank all the contributing authors for their time and efforts – a nice piece of work! Writing about science and technology is nothing without the people that make the publication possible. In particular Professor Guy Marin provided continued encouragement to stick to the timeline. The Elsevier staff was always helpful to make things happen in order to turn the words into print. Obviously, there are those many invisible hands and minds that contributed to this process and whom I also want to thank. Finally, I want to thank you the reader for taking up interest in this fascinating world of peptide self-assembly. Enjoy the reading!

CHAPTER 1

Engineering Materials from the Bottom Up – Overview

Rudy J. Koopmans[1,*] and **Anton P.J. Middelberg**[2]

Contents		
	1. Introduction	1
	2. Historical Context	2
	3. Self-Assembling Peptide Literature	4
	4. A Future of Challenges and Opportunities	7
	References	8

1. INTRODUCTION

Self-assembly is a concept receiving significant attention in a variety of research fields ranging from biology to cybernetics to social sciences. Despite its seemingly self-explanatory simplicity, implying a level of spontaneous assembly or ordering beyond the individual, composing molecules, this process is difficult to formally express in mathematical terms. Still the phenomenon in the natural sciences where atoms and molecules aggregate into higher order structures at various time and length scales is an observable reality. It offers significant potential for designing highly functional and diverse materials once the underlying mechanisms are understood.

Self-organization is often considered as synonymous to self-assembly and for many practical purposes indistinct (Anderson, 2002). However, self-assembly may be taken as the simple collection and aggregation of components into a confined entity, while self-organization can be considered as spontaneous but information-directed generation of organized functional

---

[1] PolyUrethane R&D, Freienbach, Dow Europe GmbH., Switzerland
[2] The University of Queensland, Centre for Biomolecular Engineering, St Lucia, Australia

[*] Corresponding author.
E-mail address: rjkoopmans@dow.com

structures in equilibrium conditions (Lehn, 2002). For the purpose of these papers, self-assembly will be used to mean any form of organization that comes about through forces directing hierarchical structure formation from the molecular level up.

At the macroscale, such organized functional structures manifest as physical substances, that is, materials that form the basis for engineer-specified products and applications. For example, wool is a material that can be taken to make yarn for use in cloth making. At the molecular level, wool is a keratin composed of protein molecules with very specific amino acid sequences that have been spontaneously ordered into fibrous structures allowing the mechanical operation of yarn forming (Block, 1939; Corfield and Robson, 1955). However, in contrast to conventional use of either natural or synthetic materials, where a transformation process is needed to shape the product and application, for peptides and proteins the boundary conditions (e.g. temperature, pH, and solvent) direct the aggregation of such molecules into organized structures at various length and time scales, and with differing orders of complexity. These structures are the functional intermediates or potential end products that can be applied in multiple applications for different markets.

The following papers focus mainly on various aspects associated with the self-assembly of peptides. Peptides are relatively short sequences of amino acids, typically less than 50. The limited number of residues brings simplicity but still allows for sufficient differentiation to study self-assembly in its various details. The compositional freedom of the primary molecule allows for a sufficiently rich hierarchical structure creation through aggregation of individual peptides into supramolecular constructs resulting in interesting materials. This chapter looks into relevant patent literature as a reflection of the state of the art of technology in peptide self-assembly.

2. HISTORICAL CONTEXT

Materials define the face of society. Initially, since prehistoric times – and to this day – materials were selected amongst those available in nature. These included, besides stones and metals, basic ingredients obtained from plants, crops, and animals in the form of, for example, wood, flax, wool, and leather. Materials use was a skills-based activity perfected by artists and guild-members handed from one generation to the next.

Not until the late 18th and beginning of the 19th century, commencing with the Industrial Revolution and with an increasing need for natural products, did the search for more and other materials begin. Empire building, commerce, and population growth stimulated investigations into the use of more novel natural products such as rubber and cellulose. Entrepreneurialism combined with scientific understanding and discovery gave rise to new materials, although initially still based on nature's feedstock such as modified cellulose and coal tar. The prevailing scientific paradigm was analysis, knowing that organic substances obtained from plants and animals could not be created in the laboratory; they could only be isolated and examined, and perhaps broken down into simpler substances. The reverse, the production of the complex from the simple, was, at a scientific level, beyond human competence. It was the work of a creator, or of a life force operating within living systems (Service, 2005). A little more than a half-century later, Archibald Scott Couper (1858) determined the tetra-valence of the carbon atom, and Friedrich August Kekulé (1865) published his paper on the hexagonal ring of the benzene molecule. By now a language for nanoscale objects, that is, atoms and molecules, had been developed that assigned a separate name and compositional diagram to every inorganic and organic compound including composition of natural materials. Suddenly, the science of chemistry became almost unique among the sciences in creating much of the world of materials.

By the late 19th and early 20th century, chemists discovered and developed a new class of synthetically made materials now known as plastics. In the second half of the 20th century, the perceived abundance of petroleum as cheap feedstock led to an exponential growth through process engineering scale up, establishing plastics as an unsurpassed multipurpose material.

Over these past 150 years, natural and synthetic materials use became much more founded on knowledge of the underlying physical principles. The evermore sophisticated analytical tools that could probe the molecular world provided insight into the hierarchies of organization from the atoms and molecules up to the macroscopic scale of materials. It became clear that there is more to materials than just innovative molecular synthesis.

Although order and the associated structure in a material have long been recognized, it is only in the last few decades that the process of ordering, that is, organizing components and the associated hierarchical structure formation, has received much attention. It is acknowledged that structure can be built up from the nanometer (10^{-9} m) to the macroscopic scale of everyday experience, that is, "from the bottom up" (Whitesides, 2003; Whitesides and Boncheva, 2002). Materials and devices can be built from molecular components that *organize themselves* chemically using principles of molecular recognition. Far more precision and functionality can be achieved than with a "top down" assembly approach where nano-objects are constructed from larger entities without atomic-level control.

The essential realization in this spontaneous ordering process is the importance of noncovalent bonding interaction between molecules, that is, supramolecular chemistry. These conformation-specific interactions are governed by weak forces including hydrogen bonding, metal coordination, van der Waals forces, pi–pi interactions, and electrostatic Coulombic effects. The cooperative action of multiple noncovalent interaction forces is precisely the path nature takes to produce shape and form.

Already in the early 20th century, Nobel Laureate Emil Fischer recognized (Kunz, 2002) in work associated with enzyme-substrate interactions and peptides the importance of supramolecular chemistry. Today, science still tries to understand the fundamental rules of structure and functionality formation in order for technology to apply them. The challenges are legion requiring multidisciplinary approaches as their complexity is substantial. Furthermore, a reductionist approach to materials discovery and invention, as simply defined by the nature of the elemental components, does not apply. A more holistic approach is required for recognizing the importance of environment and boundary conditions on intramolecular and intermolecular forces (Hyde et al., 1997) for structure and functionality development that shape material applications. It indicates the need for appreciating what mathematics, physics, and chemistry have to offer biochemistry, biology, and life processes for understanding and ultimately engineering novel materials and their use.

The search for connection between shape, structure, and function was posed by D'Arcy Thompson in his book *On Growth and Form* first published in 1917 (Thompson, 1992). His book lets one reflect that complex forms or shapes in nature are not solely a consequence of Darwinian natural selection. They can be purely explained on the basis of geometry, physics, mathematics, and engineering and are guided by underlying physicochemical principles that drive organization of molecules to higher order structures (Ball, 1999, 2004).

These principles are best recognized when studying relatively simple molecular systems that have an ability to exploit weak interactions to create structure. Among many, peptides are the perfect choice for such studies considering their versatility in make up given the 20+ natural and synthetic amino acid residues, and their functional diversity. In addition, the amino acid sequence of the primary structure combined with the ability of forming secondary β-sheet or α-helix structures provide substantial room for the creation of hierarchical structures based on weak intermolecular forces, mainly hydrogen bonds. A limited sequence of residues also prevents additional complication from tertiary and quaternary structures as seen with proteins.

In the following papers specifically for peptides, and without claiming comprehensiveness, several research aspects are presented in the drive for innovative and sustainable natural materials.

3. SELF-ASSEMBLING PEPTIDE LITERATURE

Quickly browsing Internet resources with the keyword self-organization (or self-assembly, including the s and z versions) provides at least five hundred thousand links of varied relevance and quality. Searching the scientific literature for the same topics still provides about a hundred thousands "hits." Specifying further to peptides and proteins reduces the number of papers to a significant but more manageable number of a few thousand. Irrespective of the content of these papers, it is clear there continues to be a significant number of publications on the subject.

As much of the more relevant scientific literature will be referenced in the subsequent papers, it is worthwhile examining the associated patent literature on peptide self-assembly. Patents are pursued to protect intellectual capital with the aim of developing science and technology into a business proposition. In the last 5 years, about 100 patents have been granted in the peptide self-assembly field with focus on applications in the medical and pharmaceutical markets.

High on the list of applications claimed are methods and methodologies to prepare and use specific peptide compositions for tissue engineering, drug delivery either via vesicles or hydrogels, and films and membranes in various medical and electronic applications. The novelty of the inventions is closely linked to the self-assembly capacity of the peptides, being a reversible mechanism, controlled by the primary structure and the triggers used for assembly and disassembly. In addition, the primary peptide structure composition and functionality are claimed as options for innovative and highly specific usage in the medical, pharmaceutical, and cosmetic fields.

Professor Sam Stupp and coworkers at NorthWestern University (USA) hold an important patent portfolio aiming at exploiting self-assembling peptides for various medicinal purposes. The approach takes advantage of peptides composed of a hydrophobic alkyl tail and a β-sheet forming hydrophilic head that assembles into a cylindrical fiber with the hydrophilic peptide facing the solvent. The peptide amphiphiles (e.g., palmitoyl—AAAAGGGEIKVAV—COOH or branched versions) can form nanofibers for uses as scaffolds for tissue growth or drug delivery (Hsu and Stupp, 2008; Mata and Stupp, 2008; Stupp and Guler, 2005; Stupp et al., 2003a, b, c, d). These systems can be turned into 2D and 3D structures and possibly cross-linked, forming microtextures with specific function of enhancing neuron growth whether or not combined with actives delivery (Stupp and Kessler, 2006; Stupp et al., 2004).

Although no self-assembly attributes are claimed, similar peptide structures were developed by Sederma SA (France) and Procter & Gamble Inc. (USA). For example, palmitoyl-pentapeptides (pal-KTTKS) are used in cosmetics as skin rejuvenation ingredients claiming no skin irritation as compared to retinol-based ones (Osborne et al., 2005).

Other research groups aim at the same applications of tissue engineering and drug delivery using somewhat different approaches, changing the nature of the primary structure, the methodology of self-assembly, the tissue types targeted, or the actives delivered (Ellis-Behnke et al., 2005; Genove et al., 2005a, b; Horii et al., 2008).

The primary peptide structure brings options for sequence tailoring that generate specific functionalities and self-assembly characteristics. Examples are: amphiphilic dendritic dipeptides making helical pores (Percec, 2006);

cyclic homodetic peptides with repeating D–L chirality for assembling and disassembling molecular tubes that can act as channels to transport ions or glucose across lipid bilayers (Ghadiri, 2003); self-assembling β-sheet – barrel channel – forming peptides for actives delivery (Aggeli et al., 2003).

The primary peptide structure offers sufficient functional groups from selected amino acids that can be linked to actives, which in combination with the self-assembly capacity provide various routes to deliver them to specific receptors or locations difficult to reach (Bhatia et al., 2007; Joyce, 2005; Krafft et al., 2007; Ludwig et al., 2007; Michal et al., 2007; Stupp et al., 2005b, 2006; Zhang and Vauthey, 2003; Zhao and Kessler, 2005).

A very interesting approach exploiting peptide self-assembly is to take advantage of its reversible nature and the various self-assembly triggering mechanisms to form hydrogels. This allows easy transportation of the peptide in solution to a desired place often difficult to reach mechanically and then triggers self-assembly once in place. The hydrogel can function as a structural supporting medium either for cell growth, localizing actives, or inducing templated mineralization (Boden et al., 2004; Lynn et al., 2003, 2006; Narmoneva et al., 2003; Schneider and Pochan, 2006; Semino et al., 2002, 2004; Stupp et al., 2005a).

Alternative to fibers and 2D and 3D woven or nonwoven networks thereof formed as either self-supporting structures or as a hydrogel, it is possible to self-assemble peptides into thin self-assembled monolayers (SAMS) or multilayer structures. Such structures have been reported to act as membranes for controlled diffusion of ions and controlled movement of body fluids and contaminants (Ellis-Behnke et al., 2006, 2007; Holmes et al., 1999). Alternatively, various techniques have been put in place to provide coatings on various substrates ranging from tissue to metals and inorganics, for example, mica (Boden et al., 2002; Haynie, 2005, 2007; Haynie and Zhi, 2007; Yoo et al., 2008).

The use of self-assembling peptides is also explored beyond the medical, pharmaceutical, or cosmetics industry. Areas of interest are among others functional foods, electronics, functional coatings, and catalysis (but different from enzyme research). As an example, peptides can be designed to switch from a random coil-like primary structure organization into an α-helix or β-sheet secondary structure with unique properties. Short peptides align to form β-sheet tapes with different functionalities, for example, hydrophilic and hydrophobic on either side of the tape to form monolayer coatings (Boden et al., 1996).

While the aforementioned technologies focus either on bulk or on solid-phase assembly, a more recent innovation in peptide self-assembly has targeted the soft or fluid–fluid interface. Designer peptides of 7 amino acid "heptad repeats" can be reversibly triggered to fold into an α-helix conformation to organize both hydrophilic and hydrophobic faces, thus inducing surfactant-like structure. These peptide surfactants, however, interact with each other in the interfacial plane, through reversible metal–ion coordination at the fluid–fluid interface, to give a gel-like film that has switchable rheology. This switchable character enables rapid and reversible "on demand" phase coalescence, thus differentiating the performance of these "Pepfactants" from nonswitchable conventional surfactants (Dexter and Middelberg, 2007; Malcolm et al., 2006). Importantly, the inherent design capability embedded in the amino acid code allows the engineering of application-specific function. It includes the ability to turn-off bulk self-assembly, thus maximizing interfacial availability and mass transfer rates, as well as allowing the interfacial activity and trigger conditions to be tuned for specific applications targeting phase separation and foam stability.

4. A FUTURE OF CHALLENGES AND OPPORTUNITIES

In subsequent papers, a nonexhaustive discussion is presented that covers several topics associated with the use of peptides and proteins as component molecules to develop structure and novel materials through self-assembly. First, a paper is devoted to the underlying scientific principles of peptide self-assembly. It is important to understand which amino acid sequences will trigger self-assembly under which conditions of temperature, concentration, pH, and solvent type. Also, it is essential to know what kind of structures can be expected in order for a rational molecular design to target specific applications. The experimental data should form a good basis for testing the validity of theoretical models discussed in the second paper. The state of the art in capturing the self-assembly process into a mathematical framework that accurately simulates hierarchical structure formation is an extremely challenging subject of research. Still for peptides to be useful, it is one thing to be able to have tailored species with well-defined structures created in the laboratory at mg scale but it is another, particularly an engineering feat, to produce peptides in sufficient quantities of sufficient purity at a reasonable cost. Therefore, a third paper considers what the options and challenges are of producing peptides. Subsequently there are three papers devoted to the potential applications. One explores the space of natural and artificial silk, another examines biomedical applications associated with tissue engineering. A third paper considers alternative applications where peptides are combined with polymers to form hybrid A-B block copolymers, to steer novel structure formation guided by the self-assembly of the peptide.

This series of papers aim to demonstrate the fascinating science and technology that self-assembling peptides bring. Self-assembly can be considered as the "how" to build novel materials for the future using peptides as building blocks. With nature as a reference, being a source of inspiration, the type of materials that potentially can be shaped seems infinite. Today, the research into this field has barely begun with few but very active academic groups from around the world exploring the possibilities. Medical, pharmaceutical, cosmetic, and personal care companies are the most likely early adopters. Self-assembling peptides are very feasible solutions to important health and wellness challenges particularly in view of aspects such as biocompatibility with living organisms, highly functionally specific and controlled action, scar-free tissue engineering and repair, and agent delivery options. The key challenge here is successful clinical trial results to forge a market position of sufficient size that justifies the cost of the required research and development. In other sectors of industry, there may be additional hurdles related to regulatory legislation and socioeconomic acceptance. A major breakthrough will be needed in terms of economically providing designers peptide that perform under real-life conditions.

How long all this will take depends on the science and technology breakthroughs and the associated meme. Therefore, writing on the engineering perspectives of self-assembling peptides is by default limited to the imagination of today's scientists as reported in literature. It just remains a nonexhaustive attempt to inspire researchers, engineers, and any other interested party to discover a path of learning and applying the lessons of nature. Still the research efforts are fast paced and growing as can be gathered from the subsequent contributions. Accordingly, engineering innovative products using self-assembling peptides will become a discipline intimately linked to scientific understanding of multiple research fields, with applications in areas and markets we have just begun to imagine.

REFERENCES

Aggeli, A., Boden, N., Hunter, M., and Knowles, C., Self-assembling beta-barrel channel-forming peptides for wound dressing and other pharmaceutical uses, 2002-GB3212 2003006494 (2003).
Anderson, C. *Biol. Bull.* **202**, 3, 247–255 (2002).
Ball, P. "The Self-Made Tapestry – Pattern Formation in Nature". Oxford University Press, Oxford (1999).
Ball, P. "Critical Mass – How One Thing Leads to Another". W. Heinemann, London (2004).
Bhatia, S.N., Harris, T., and Von Maltzahn, G. Triggered Self-Assembly Conjugates (TSACs) and Nanosystems for Use as Diagnostic Agents and Targeted Drug Delivery Systems, 2007-US6141 2007106415 (2007).
Block, R.J. *J. Biol. Chem.* **128**(1), 181–186 (1939).
Boden, N., Aggeli, A., and McLeish, T.C.B. Betasheet Peptides and Gels Made Thereof, WO 96/31528 (1996).
Boden, N., Aggeli, A., Fishwick, C., Knobler, C., Fang, J.Y., and Henderson, J. Coatings, WO 02/081104A2 (2002).
Boden, N., Aggeli, A., Ingham, E., and Kirkham, J. Supramolecular Networks Made by Beta-Sheet Self-Assembly of Rationally Designed Peptides, and Their Uses as Industrial Fluids, Personal Care Products, Tissue Engineering Scaffolds and Drug Delivery Systems, 2003-GB3016 WO 2004007532 (2004).
Corfield, M.C., and Robson, A. *Biochem. J.* 59(1), 62–68 (1955).
Couper, A.S. *Annales de chemie et de physique* Série 3, **53** (1858), 488–489, and *Philosophical Magazine* **16**, 104–116 (1858).

Dexter, A.F., and Middelberg, A.P.J. *J. Phys. Chem C* **111**, 10484–10492 (2007).
Ellis-Behnke, R., Liang, Y.-X., Schneider, G., So, K.-F., and Tay, D. Compositions and Methods for Affecting Movement of Contaminants, Body Fluids or Other Entities and/or Affecting Other Physiological Conditions, 2007-US10041 2007142757 (2007).
Ellis-Behnke, R., Schneider, G., and Zhang, S. Self-Assembling Peptides for Regeneration and Repair of Neural Tissue, 2004-968790 2005287186 (2005).
Ellis-Behnke, R., Zhang, S., Schneider, G., So, K.-F., Tay, D., and Liang, Y.-X. Compositions and Methods for Promoting Hemostasis and Other Physiological Activities, 2006-US15850 2006116524 (2006).
Genove, E., Semino, C., and Zhang, S. Self-Assembling Peptides Incorporating Modifications, Method for Preparation and Use as Scaffolds in Tissue Engineering, 2004-US20549 2005014615 (2005a).
Genove, E., Zhang, S., and Semino, C. Self-Assembling Peptides Derived from Laminin-1 and Use for Wound Healing, 2004-877068 2005181973 (2005b).
Ghadiri, R.M. Cyclic Homodetic Peptides with Repeating D-L Chirality, Employable for Assembling and Disassembling Molecular Tubes, 96-632444 6613875 (2003).
Haynie, D.T. Method for Designing Polypeptides for the Nanofabrication of Thin Films, Coatings, and Microcapsules by Electrostatic Layer-by-Layer Self Assembly for Use in Medicine, 2003-652364 2005069950 (2005).
Haynie, D.T. Multilayer Films, Coatings, and Microcapsules Comprising Polypeptides, 2006-586329 2007077275 (2007).
Haynie, D.T., and Zhi, Z.-I. Biodegradable Polypeptide Films and Microcapsules, 2006-559175 2007207212 (2007).
Holmes, T., Zhang, S., Rich, A., Dipersio, C.M., and Lockshin, C. Stable Macroscopic Membranes Formed by Self-Assembly of Amphiphilic Peptides and Uses Therefor, 94-293284 5955343 (1999).
Horii, A., Zhang, S., Wang, X., and Gelain, F. Modified Self-Assembling Peptides for Cell Culture and Tissue Engineering, 2007-US20754 2008039483 (2008).
Hsu, L., and Stupp, S.I. Self-Assembling Peptide Amphiphiles, 2007-US84223 2008067145 (2008).
Hyde, S., Anderson, S., Larsson, K., Blum, Z., Landh, T., Lidin, S., and Ninham, B.W. The Language of Shape. Elsevier, Amsterdam (1997).
Joyce, T.H. Self Assembling Activation Agents Targeted Using Active Drug Release with Organic Nanotube or Alpha-DL Peptide Enclosed in Liposomes, 2004-807835 2005214356 (2005).
Kekulé, F.A. *Annalen der Chemie* 137, 129–196 (1865).
Krafft, G.A., Klein, W.L., Viola, K.L., Lambert, M.P., Pray, T.R., and Lowe, R. Neurotoxic Soluble Diffusible Non-Fibrillar Amyloid Beta Peptide Assembles and Their Metal Complexes Useful in Drug Screening and Vaccines, 2007-686570 2007213512 (2007).
Kunz, H. "Emil Fischer – Unequalled Classicist, Master of Organic Chemistry Research, and Inspired Trailblazer of Biological Chemistry". *Angew. Chem. Int. Ed.* **41**, 4439–4451 (2002).
Lehn, J.-M. *Proc. Natl. Acad. Sci. U.S.A.* **99**(8), 4763–4768 (2002).
Ludwig, F.N., Pacetti, S.D., Hossainy, S.F.A., and Davalian, D. Nanoshells Comprising Self-Assembled Material for Drug Delivery, 2006-454813 2007292495 (2007).
Lynn, D., Conticello, V., Morgan, D.A., and Dong, J. Self-Assembling-Beta-Amyloid Peptide-Based Structures and Control of Their Self-Assembly by Changes in Metal Ions Concentration and Other Environmental Parameters, 2003-US9229 2003082900 (2003).
Lynn, D., Conticello, V., Morgan, D.A., and Dong, J. Self-Assembling-Peptide-Based Structures and Processes for Controlling the Self-Assembly of Such Structures, 2004-945133 2006063919 (2006).
Malcolm, A.S., Dexter, A.F., Middelberg, A.P.J. *Soft Matter* 2, 1057–1066 (2006).
Mata, A., and Stupp, S.I. Self-Assembling Peptide Amphiphiles for Tissue Engineering, 2007-US84278 2008061014 (2008).

Michal, E., Basu, S., and Kuo, H.-C. Methods and Compositions for Treating Post-Myocardial Infarction Damage, 2006-447340 2007218118 (2007).
Narmoneva, D., Zhang, S., Kamm, R.D., and Lee, R.T. Angiogenesis and Cardiac Tissue Engineering with Peptide Hydrogels and Related Compositions and Methods of Use Thereof, 2003-US14092 2003096972 (2003).
Osborne, R., Robinson, L.R., and Tanner, P.R. Skin Care Composition Containing Dehydroacetic Acid and Skin Care Actives, 2005044219 (2005).
Percec, V. Amphiphilic Dendritic Dipeptides and Their Self-Assembly into Helical Pores, 2005-171494 2006088499 (2006).
Schneider, J.P., and Pochan, D.J. Novel Hydrogels and Uses Thereof, 2004-900344 2006025524 (2006).
Semino, C.E., Shen, C., Sherley, J., and Zhang, S. Cellular Reprogramming in Peptide Hydrogel and Uses Thereof, 2003-US21981 2004007683 (2004).
Semino, C.E., Sherley, J., and Zhang, S. Cellular Reprogramming in Peptide Hydrogel and Uses Thereof, 2002-US3607 2002062969 (2002).
Service, R.F. *Science* **95**, 309 (2005).
Stupp, S.I., Beniash, E., and Hartgerink, J.D. Compositions for Self-Assembly and Mineralization of Peptide Amphiphiles, 2003-US35902 2005003292 (2005a).
Stupp, S.I., Donners, J.J.J.M., Silva, G.A., and Behanna, H.A. Anthony, S.G. Self-Assembling Peptide Amphiphiles and Related Methods for Growth Factor Delivery, 2004-US40550 2005056039 (2005b).
Stupp, S.I., and Guler, M.O. Branched Peptide Amphiphiles, Related Epitope Compounds and Self Assembled Structures Thereof, 2004-US40546 2005056576 (2005).
Stupp, S.I., Hartgerink, J.D., and Beniash, E. Peptide Amphiphile Solutions and Self Assembled Peptide Nanofiber Networks, 2003-US10051 2003084980 (2003a).
Stupp, S.I., Hartgerink, J.D., and Beniash, E. Self-Assembly of Peptide-Amphiphile Nanofibers Under Physiological Conditions for Biomedical Applications, 2003-US4779 2003070749 (2003b).
Stupp, S.I., Hartgerink, J.D., and Beniash, E. Self-Assembly and Mineralization of Peptide-Amphiphile Nanofibers, 2002-US36486 2003054146 (2003c).
Stupp, S.I., Hartgerink, J.D., and Niece, K.L. Self-Assembling Peptide-Amphiphiles and Self-Assembled Peptide Nanofiber Networks for Tissue Engineering, 2003-US29581 2004106359 (2004).
Stupp, S.I., Hulvat, J.F., and Rajangam, K. Angiogenic Heparin-Binding Epitopes, Peptide Amphiphiles, Self-Assembled Compositions and Related Methods of Use, 2006-US7864 2006096614 (2006).
Stupp, S.I., and Kessler, J.A. Self-Assembling Peptide Amphiphiles Generating Nanofiber Scaffolds for Encapsulation, Growth and Differentiation of Neurons for Therapeutic Uses, 2006-US2354 2006079036 (2006).
Stupp, S.I., Messmsore, B.W., Arnold, M.S., and Zubarev, E.R. Encapsulation of Nanotubes Via Self-Assembled Nanostructures, 2003-US12111 2003090255 (2003d).
Thompson, D'Arcy W. "On Growth and Form". Dover Publication, New York (1992).
Whitesides, G.M. *Nat. Biotechnol.* **21**(10), 1161–1165 (2003).
Whitesides, G.M., and Boncheva, M. *Proc. Natl. Acad. Sci. U.S.A.* **98**(8), 4769–4774 (2002).
Yoo, P.J., Nam, K.T., Qi, J., Lee, S.-S., Park, J., Belcher, A.M., and Hammond, P.T. Self-Assembly of Macromolecules on Multilayered Polymer Surfaces, 2007-US2914 2008057127 (2008).
Zhang, S., and Vauthey, S. Surfactant Peptide Nanostructures for Drug Delivery, 2002-US21757 2003006043 (2003).
Zhao, L.-R., and Kessler, J.A. Compositions and Methods for Controlling Stem Cell and Tumor Cell Differentiation, Growth, and Formation, 2004-18622 2005214257 (2005).

CHAPTER 2

Mechanisms and Principles of 1D Self-Assembly of Peptides into β-Sheet Tapes

Robert P.W. Davies[1,2], **Amalia Aggeli**[1,2,*], **Neville Boden**[1,2], **Tom C.B. McLeish**[1,3], **Irena A. Nyrkova**[4], and **Alexander N. Semenov**[4]

Contents		
1.	Introduction	12
2.	Model of Hierarchical Self-Assembling Chiral Rods in Solution	13
	2.1 Theoretical rational of hierarchical self-assembling chiral rods in solution	13
	2.2 Experimental evidence of hierarchical peptide self-assembly in solution	17
3.	Responsiveness to External Triggers	25
	3.1 pH	25
	3.2 Ionic strength	28
	3.3 Other	34
	3.4 Heteroaggregates formed by complementary peptides	35
4.	Peptide Self-Assembly on Surfaces	37
5.	Conclusions and Future Prospects	39
	References	40

[1] Centre for Self-Organising Molecular Systems, University of Leeds, Leeds LS2 9JT, United Kingdom
[2] School of Chemistry, University of Leeds, Leeds LS2 9JT, United Kingdom
[3] Department of Physics and Astronomy, University of Leeds, Leeds LS2 9JT, United Kingdom
[4] Institut Charles Sadron, Strasbourg Cedex, France

[*] Corresponding author.
E-mail address: a.aggeli@leeds.ac.uk

1. INTRODUCTION

Molecular self-assembly has attracted growing international research efforts and interest due to its central importance in biology and its role in the understanding of the molecular origin of a wide range of diseases (Burkoth et al., 1998; Lashuel et al., 2000; Yamada et al., 1998). With the onset of world-wide activities in nanoscale science and nanotechnology, molecular self-assembly has also provided inspiration for innovation (Wilson et al., 2002) and new product development in the fields of new nanostructured and biologically inspired materials, for example, for drug delivery, wound healing and tissue engineering, and novel processing routes (Collier et al., 2001; de Loos, 2001; Gronwald et al., 2002; Hanabusa et al., 1996; Marini et al., 2002; Qu et al., 2000; Terech and Weiss, 1997). Molecular self-assembly is an attractive route to nanostructured materials, which can have a number of key performance properties such as massive surface area and thus increased functional properties such as adsorption or binding. Nanomaterials can be light-weight and thus appropriate for miniaturization, they can be controllable, via external on/off triggering, and thus are injectable. Furthermore, molecular self-assembly is a spontaneous phenomenon, that is, thermodynamically driven; one implication is that self-assembling nanostructures are self-healing; they further provide a cheap, easy, and potentially fast methodology for bulk production of complex functional structures. The simple processes involved in the production of self-assembling structures using conventional techniques under mild conditions, without the need for expensive, sophisticated instruments, or harsh settings, are particularly suited to large-scale industrial applications.

Bioinspired protein-like self-assembly is one of the most fascinating and fast growing areas in molecular self-assembly. Proteins and peptides are the most versatile biological building blocks in nature in terms of chemistry, conformation, and functionality. They offer routes to sustainable, large-scale production since they can be produced not only by chemical means but also through genetic engineering. Another advantage is that they are environmentally friendly, "green" polymers. Also via precise control of the protein molecule, it is possible to exert precise control of the nanostructure, intermolecular interactions, binding, and (bio-)activity. One of the main drawbacks is that the immense chemical and conformational complexities make proteins difficult to understand quantitative and thus predict and control accurately and reliably.

In order to appreciate the fundamental physical and chemical principles that govern protein-like self-assembly, and thus learn how to exploit it, it is advantageous to start by using simple model peptide systems. Peptides can be easily synthesized in large numbers of systematic variations to build structure–function relationships, they can be designed in order to minimize the chemical and conformational complexity of biological proteins, and they can evolve to progressively more complicated building blocks. Rationally designed peptides that self-assemble in 1-D into long β-sheet tapes offer one of the simplest and best understood systems to study hierarchical protein-like self-assembly (Aggeli et al., 1997a, b, 2001a, b, 2003a, b; Nyrkova et al., 2000a, b). The tapes are stabilized by precise intermolecular interactions, the aggregate structure is well-studied and unambiguously established, and the peptide building blocks offer versatility of the chemical and structural properties. Here we shall review the theoretical and experimental advances achieved so far in order to decipher the mechanisms and principles that describe and predict hierarchical β-tape self-assembly in a quantitative and reliable manner. We shall demonstrate these principles using predominantly the P_{11}-X family of rationally designed peptides. We shall summarize the conclusions of the studies with respect to the elucidation of the intrinsic self-assembling properties of the pure β-sheet structure, and the opportunities arising for functional peptide design.

2. MODEL OF HIERARCHICAL SELF-ASSEMBLING CHIRAL RODS IN SOLUTION

2.1. Theoretical rational of hierarchical self-assembling chiral rods in solution

A peptide in a β-strand conformation can be considered as a chiral rod-like unit, with complementary donor and acceptor groups aligned on opposing sides and having chemically different upper and lower surfaces (Figure 1a). This is a single step of coarse-graining from atomic detail to the nanoscale. The chiral unit is able to undergo 1D self-assembly in solution and to form the hierarchical structures in Figure 1 at concentrations depending on the values of a small set of coarse-grained interaction energies ε_j (Aggeli et al., 2001b; Nyrkova et al., 2000a, b; Weiss and Terech, 2006). An isolated monomer in solution will tend to be in a different conformation (Figure 1b), with lower free energy than in the rod-like state, giving rise to a conformational free energy change ε_{trans}.

The rod-like "monomers" self-assemble via multiple intermolecular interactions to form long twisted tapes (Figure 1c) with an association free energy change ε_{tape} per intermonomer bond. The tape is chiral due to the chirality of the monomers, which gives rise to a left-handed twist around the long axis of the tape (Figure 1c). The differences in the chemical properties of the two faces of the tape give rise to a cylindrical curvature, causing

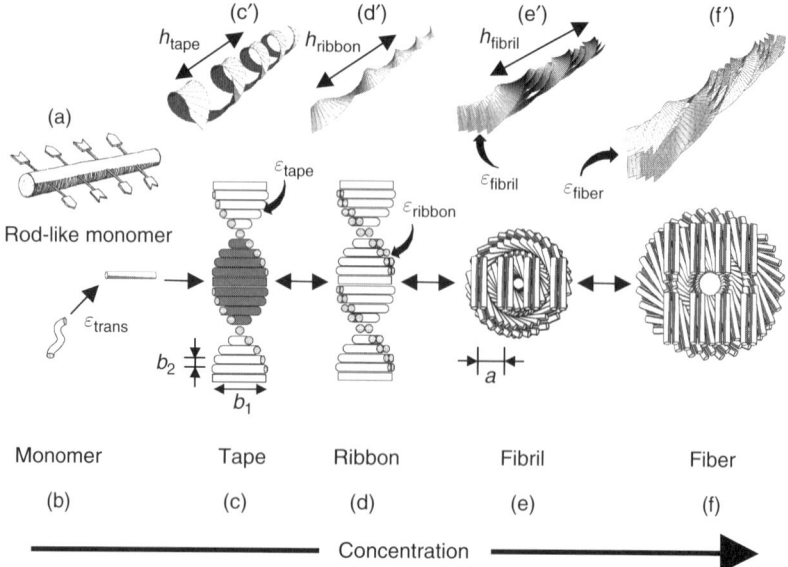

Figure 1 Model of hierarchical self-assembly of chiral rod-like monomer units (b): Local arrangements (c–f) and the corresponding global equilibrium conformations (c′–f′) for the hierarchical self-assembling structures formed in solutions of chiral molecules (a), which have complementary donor and acceptor groups, shown by arrows, via which they interact and align to form tapes (c). The black and the white surfaces of the rod (a) are reflected in the sides of the helical tape (c) which is chosen to curl toward the black side (c′). (e) and (f) show the front views of the edges of fibrils and fibers, respectively. Geometrical sizes for P_{11}-1 and P_{11}-2 peptides: inter-rod separation in a tape b_2 ($b_2 = 0.47$ nm); tape width, equal to the length of a rod, b_1 ($b_1 \approx 4$ nm); interribbon distance in the fibril, a ($a \approx 1.6$–2 nm for P_{11}-1, and $a \approx 2$–2.4 nm for P_{11}-2) (Aggeli et al., 2001b).

the tape to curl into a helical configuration (Figure 1c′), with helical pitch h_{tape} and radius r_{tape}:

$$h_{tape} = b_2 \left(\frac{2\pi}{\gamma_\theta} \right) \left(1 + \left(\frac{\gamma_\nu}{\gamma_\theta} \right)^2 \right)^{-1} \quad (1)$$

$$r_{tape} = b_2 \left(\frac{\gamma_\nu}{\gamma_\theta^2} \right) \left(1 + \left(\frac{\gamma_\nu}{\gamma_\theta} \right)^2 \right)^{-1} \quad (2)$$

where γ_ν and γ_θ are the tape bend and twist angles (in radians) per monomer rod along the tape and b_2 is the distance between adjacent rods in the tape.

One face of the tape is expected to be less soluble than the other (i.e., black is more hydrophobic if the solvent is water in Figure 1c). This difference results in intertape attraction and hence in double-tape (ribbon, Figure 1d) formation, with an associated energy $\varepsilon_{\text{ribbon}}^{\text{attr}}$ per peptide. Both faces of the ribbon are identical (white in Figure 1d) and are characterized by a saddle curvature. Hence, the ribbon does not bend, and its axis is straight at equilibrium (Figure 1d'). The white sides of the ribbons can also be mutually attractive with an associated energy $\varepsilon_{\text{fibril}}^{\text{attr}}$ per pair of interacting peptides, leading to stacking of ribbons into fibrils (Figure 1e). Furthermore, the ends of the rods at the edges of the fibrils can also be mutually attractive, causing fibrils to entwine into fibers (Figure 1f), stabilized by attraction energy $\varepsilon_{\text{fiber}}^{\text{attr}}$.

The whole set of the self-assembling structures in Figure 1 are left-handed twisted due to the chirality of the monomer. If the ribbons were not twisted, an unlimited growth of stacks of them would be expected. Instead, when twisted ribbons aggregate into stacks, fibrils with well-defined widths are formed. Fibers are formed in a similar way from twisted fibrils, but again to well-defined widths. In order to aggregate, twisted objects must bend and adjust their twist in response to the packing constraints imposed by its twisted neighbors. Hence, there is an elastic energy cost $\varepsilon_{\text{elast}}$, which must be compensated for, by the gain in attraction energy (coming from $\varepsilon_{\text{ribbon}}^{\text{attr}}$, $\varepsilon_{\text{fibril}}^{\text{attr}}$, and $\varepsilon_{\text{fiber}}^{\text{attr}}$) upon stacking. The distortion energy $\varepsilon_{\text{elast}}$ is higher for thicker stacks. This serves to stabilize the widths of fibrils and fibers. Thus, the fibril width is determined by a balance between the gain in attraction energy (coming from $\varepsilon_{\text{fibril}}^{\text{attr}}$) associated with ribbon stacking, and the elastic cost on the ribbons associated with fibril formation. If the ribbon contour length is fixed and the deformations are weak, from symmetry arguments we find that this cost is

$$\varepsilon_{\text{elast}} = \frac{1}{2} k_{\text{bend}} (\nu - \nu_0)^2 + \frac{1}{2} k_{\text{twist}} (\theta - \theta_0)^2 \qquad (3)$$

per unit length of each ribbon in the fibril, where ν and θ are the local curvature and the local twist strength of the ribbon within a fibril, $\theta_0 = 2\pi/h_{\text{ribbon}}$ is the equilibrium value of twist strength of an isolated ribbon, while its bend strength is zero ($\nu_0 = 0$), and k_{bend} and k_{twist} are the ribbon elastic constants (Nyrkova et al., 2000a, b). For a ribbon at a distance ρ from the central axis of a fibril, it can be shown that $\nu = \gamma^2 \rho/(1 + \gamma^2 \rho^2)$ and $\theta = \gamma/(1 + \gamma^2 \rho^2)$, where $\gamma = 2\pi/h_{\text{fibril}}$ (h_{fibril} is the fibril's helical pitch). The thicker the fibril is, the larger the typical ρ are, and hence the higher the cost $\varepsilon_{\text{elast}}$. Thus, the net energy gain $\varepsilon_{\text{fibril}}$ per peptide in a fibril

$$\varepsilon_{\text{fibril}} = \frac{p-1}{2p} \varepsilon_{\text{fibril}}^{\text{attr}} - \varepsilon_{\text{fibril}}^{\text{elast}} \qquad (4)$$

has a maximum at some p (p is the number of ribbons in the fibril). Hence, a well-defined width of fibrils arises, corresponding to this optimal p.

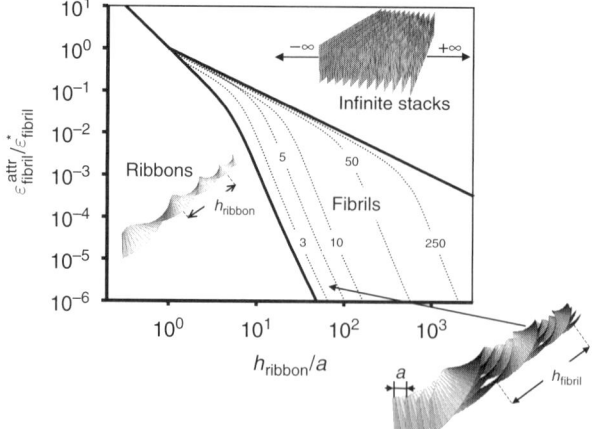

Figure 2 Phase diagram of a solution of twisted ribbons which form fibrils, where the relative helix pitch of isolated ribbons h_{ribbon}/a is plotted against the relative side-by-side attraction energy between ribbons $\varepsilon_{fibril}^{attr}/\varepsilon_{fibril}^{*}(\varepsilon_{fibril}^{*} \equiv (2\pi^2 b_2/a^2)k_{twist})$. The areas divided by the thick lines reveal the conditions where ribbons, fibrils, and infinite stacks of completely untwisted ribbons are stable. The dotted lines are lines of stability for fibrils containing p ribbons (p are written on the lines); $k_{bend}/k_{twist} = 0.1$ (Aggeli et al., 2001b).

The diagram of possible aggregate structures calculated by using this model, and seeking its structure of minimum free energy in each case, is shown in Figure 2. Fibrils with finite diameter are seen to be stable for a wide range of values of $\varepsilon_{fibril}^{attr}$ provided that the intrinsic pitch h_{ribbon} of the lone ribbon strongly exceeds the interribbon gap, a, in the fibril. For low $\varepsilon_{fibril}^{attr}$, the ribbons do not stack into fibrils. For high $\varepsilon_{fibril}^{attr}$, the ribbons form infinite aggregates (sheet-like crystallites) in which the ribbons are completely untwisted. The optimum number p of stacked ribbons per fibril, and hence the fibril diameter, increases with h_{ribbon} and $\varepsilon_{fibril}^{attr}$. This is usually accompanied by an increase in h_{fibril}.

The concentration ranges over which the various self-assembled structures are observable, their contour lengths, and abruptness of interstructure transformations with concentration are determined by the energy parameters ε_j. For example, if ε_{trans} is high enough ($\varepsilon_{trans} > 4$, all energies here are measured in $k_B T$ units) and ε_{ribbon} is small (≤ 1), the single tapes emerge abruptly at

$$c_{cr}^{tape} \cong \nu_{tape}^{-1} \exp(-\varepsilon_{tape} + \varepsilon_{trans}) \quad (5)$$

and their typical aggregation number is

$$\langle m_{tape} \rangle \cong \left[\left(\frac{c}{c_{cr}^{tape}}\right) - 1\right]^{\frac{1}{2}} \exp\left(\frac{\varepsilon_{trans}}{2}\right) \quad (6)$$

if $c_{cr}^{tape} < c < c_{cr}^{ribbon}$ (c is the total peptide concentration and ν_{tape} is the "freedom" volume of the bonds forming the tape). Next, given the tape bend and twist are not very high, that is, ε_{elast} (cf. Equation (3)) is small enough, the net ribbon energy

$$\varepsilon_{ribbon} = \frac{1}{2}\varepsilon_{ribbon}^{attr} - \varepsilon_{ribbon}^{elast} \qquad (7)$$

is positive. Hence, at concentration

$$c_{cr}^{ribbon} \cong c_{cr}^{tape} + c_{tape}^{max}, \quad c_{tape}^{max} \cong \nu_{tape}^{-1}\varepsilon_{ribbon}^{-2}\exp(-\varepsilon_{tape}) \qquad (8)$$

the ribbons emerge; above c_{cr}^{ribbon}, the population of peptide in single tapes saturates at c_{tape}^{max} and all extra peptide goes into ribbons; simultaneously the average aggregation number of ribbons grows as

$$\langle m_{ribbon}\rangle \approx \left[\left(\frac{c}{c_{cr}^{ribbon}}\right) - 1\right]^{\frac{1}{2}} \varepsilon_{ribbon}^2 \exp\left(\frac{\varepsilon_{trans} + \varepsilon_{tape}}{2}\right) \qquad (9)$$

whereas the length of tapes saturates at

$$\langle m_{tape}\rangle \cong \varepsilon_{ribbon}^{-1} \qquad (10)$$

The formulae (Equations (5), (6), (8)–(10)) are asymptotic. To realize sequentially the entire hierarchy of structures in Figure 1, with increasing monomer concentration, it is essential that $\varepsilon_{tape} \gg k_BT \gg \varepsilon_{ribbon} \gg \varepsilon_{fibril} \gg \varepsilon_{fiber}$, otherwise, some structures may not appear. These are the net energies gained per one peptide inside the corresponding structures as compared to a peptide inside the structure of the previous level.

2.2. Experimental evidence of hierarchical peptide self-assembly in solution

2.2.1. Tape-forming peptides

Peptide P$_{11}$-1 (CH$_3$COQQRQQQQQEQQNH$_2$) has been designed de novo with a sequence of glutamine (Q, Gln) residues, whose side chains are believed to interact strongly in water (Perutz et al., 1994) via hydrophobic and complementary hydrogen-bonding interactions. Arginine (R, Arg) and glutamate (E, Glu) residues were placed in positions 3 and 9 to provide molecular recognition between adjacent antiparrallel β-strand peptides in tape-like aggregates and to prevent random peptide association. These favorable intermolecular side-chain interactions, together with the cooperative intermolecular hydrogen bonding between peptide backbones, will result in high scission energy $\varepsilon_{\text{tape}}$, thus promoting β-sheet tape formation (Figure 1c). One side ("black") of the tape will be lined by the CONH$_2$ groups of the Gln residues, whilst its other side ("white") will be lined by Gln, Arg, and Glu. At low pH, there will also be a net positive charge per peptide; thus the high hydrophilicity of both surfaces of the tape, combined with the electrostatic repulsion between positively charged surfaces, will result in very small $\varepsilon_{\text{ribbon}}^{\text{attr}}$ and $\varepsilon_{\text{fibril}}^{\text{attr}}$ energies compared to $k_B T$, thus promoting predominantly single tape formation for low enough peptide concentration in acidic solutions.

Solutions of P$_{11}$-1 at very low concentrations are found to consist predominantly of monomeric random coil conformation (Figures 1b and 3a), whereas at higher concentrations $c \geq 0.01$ mM, they contain semiflexible tapes with a width $W \approx 4$ nm, equal to the expected length of an 11-residue peptide in a β-strand conformation, and persistence length $\tilde{l} < 0.3$ μm. The different chemical nature of the two sides of the tape seems to cause it to bend and twist simultaneously, resulting in curly tapes with a left-handed twist, a helical pitch $h_{\text{tape}} \approx (30 \pm 15)$ nm, and a radius $r_{\text{tape}} \approx 5$ nm. At $c \geq 1$ mM, loose ribbons are also observed, with $\tilde{l} \sim 0.3 - 1$ μm and $h_{\text{ribbon}} \approx (50 \pm 20)$ nm. These experimental data were treated with the theoretical model in order to derive the magnitudes of the bend $\gamma_\nu = 3°$ and twist $\gamma_\theta = 3°$ angles for the single tapes and the ribbons (Table 1).

Aqueous solutions of P$_{11}$-1 tapes produce FTIR spectra with absorption maxima in amide I' at 1,630 and 1,690 cm^{-1}, demonstrative of a predominantly antiparallel β-sheet structure. They also exhibit characteristic β-sheet

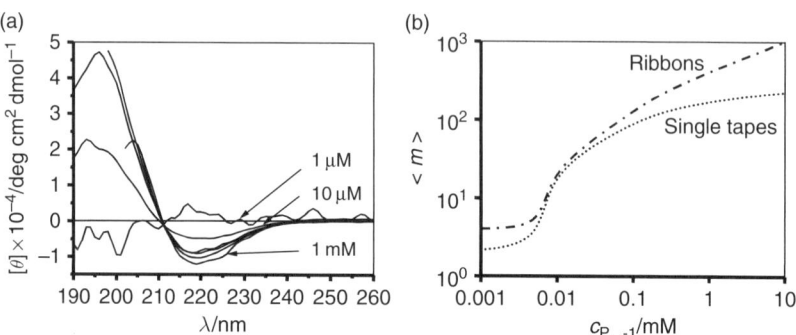

Figure 3 Self-assembling properties of P$_{11}$-1. (a) Far-UV CD spectra as a function of peptide concentration in water at pH = 2. (b) Theoretical concentration dependence of the average number $\langle m \rangle$ of peptides per single tape (dotted line) and in ribbons (dash-dot line). Minimum number of peptides in tapes is two and in ribbons is four. The predicted lengths of tapes and ribbons are in agreement with the observed lengths in the TEM pictures for the same peptide concentration (Aggeli et al., 2001b).

Table 1 Magnitudes of molecular parameters and macroscopic properties of the aqueous solutions of the two de novo-designed self-assembling β-tape forming peptides P$_{11}$-1 and P$_{11}$-2 (Aggeli et al., 2001b)

	P$_{11}$-1			P$_{11}$-2		
	β-Tapes	Ribbons	Fibrils	β-Tapes	Ribbons	Fibrils
$c_{cr}/\mu M$	8	1,000	$c > 25{,}000$	<90	90	700 ± 200
ε_{trans}/k_BT	6.5 ± 1.5			3 ± 1		
ε_{tape}/k_BT	31.0 ± 1.5			24.5 ± 1.0		
$\varepsilon_{ribbon}/k_BT$		$(3.5 \pm 1.5)10^{-3}$			0.6 ± 0.3	
$\varepsilon_{fibril}/k_BT$			$<10^{-3}$			$(2.0 \pm 0.3)10^{-4}$
Pitch h/nm	30 ± 15	50 ± 20			160 ± 40	160 ± 40
Twist angle $\gamma_\theta/°$	3	3		1	1	1
Bend angle $\gamma_v/°$	3	0			0	0
$\tilde{l}/\mu m$	<0.3	0.6 ± 0.2			1.0 ± 0.3	20–70
$L/\mu m,(c = 6\,mM)$	10^{-1}	10^{-1}–10^0		10^{-3}	10^0	10^{17}
Properties of aqueous solution	Isotropic fluid ($c < 13\,mM$) Nematic fluid/gel ($c > 13\,mM$)			Isotropic fluid ($c < 0.9\,mM$) Nematic fluid ($c \approx 0.9$–$6\,mM$) Nematic gel ($c > 6\,mM$)		

CD spectra (Manning et al., 1988) with minimum and maximum ellipticities at 218 nm and 195 nm, respectively (Figure 3a). The fraction of the peptide in β-sheet tapes starts to grow abruptly at a critical concentration $c_{cr}^{tape} \approx 0.008$ mM. The two-state transition from random coil to β-sheet with increasing concentration has an isodichroic point at 211 nm (Figure 3a). ε_{trans} and ε_{tape} were treated as fitting parameters and it was thus possible to describe well the growth of the β-sheet CD band with concentration. The best-fit energy values obtained are in Table 1. The ε_{trans} energy results in the nucleated growth of tapes, manifested by a "sudden" onset of β-sheet tape formation at c_{cr}^{tape}. By using these values of energetic parameters, this single tape model predicts a mean tape contour length for a given peptide concentration, which agrees well with the observed range of contour lengths in the TEM images for the same concentration. At $c_{cr}^{ribbon} \approx 1$ mM, loose ribbons start appearing, implying a weak attraction between tapes. This attraction may be mediated by multiple, cooperative, complementary hydrogen bonding and van der Waals interactions between the —$CONH_2$ groups of glutamine side chains, which line completely one of the two polar sides of the tapes. From the value of c_{cr}^{ribbon}, it can be estimated that the ribbons are stabilized by $\varepsilon_{ribbon} = (0.0035 \pm 0.0015)$ k_BT. Fibrils (Figure 1e′) are not observed up to $c = 25$ mM, hence using the theoretical model, it can be estimated that $\varepsilon_{fibril} < 10^{-3}$ k_BT and $\varepsilon_{fibril}^{attr} \leq 0.1$ k_BT.

2.2.2. Fibril-forming peptides

How can a tape-forming peptide be modified in a rational manner so that it becomes a fibril-forming peptide at μM–mM concentration? According to the theoretical model, in order to design a peptide with an increased tendency to associate into ribbons, the magnitude of ε_{ribbon} must be increased either by decreasing $\varepsilon_{ribbon}^{elast}$ or by increasing $\varepsilon_{ribbon}^{attr}$. The latter can be achieved by addition of salts or of appropriate cosolvents, or by changing the peptide primary structure. In particular, glutamines at positions 4, 6, and 8 of P$_{11}$-1 were replaced by phenylalanine, tryptophan, and phenylalanine, respectively. This new peptide, P$_{11}$-2 (CH$_3$CO—QQRFQWQFEQQ—NH$_2$), will form β-sheet tapes with a hydrophobic "adhesive" stripe running along one side of the tape, which will promote their association into ribbons in water. At $c \geq 0.1$ mM in water, P$_{11}$-2 is found to form long, stable semiflexible β-sheet ribbons with a width of 2–4 nm, which fits with the expected cross section of $\approx 2 \times 4$ nm^2 of these ribbons and a persistence length $\tilde{l} \approx 1$ μm (Figure 4a). At $c \geq 0.6$ mM, a second transition from ribbons to fairly rigid fibrils is observed (Figure 4b and c). The fibrils have a well-defined screw-like structure with typical minimum and maximum widths $W_1 \approx 4$ nm and $W_2 \approx 8$ nm, respectively. At even higher concentrations still, a third structural transition takes place and fibers are detected,

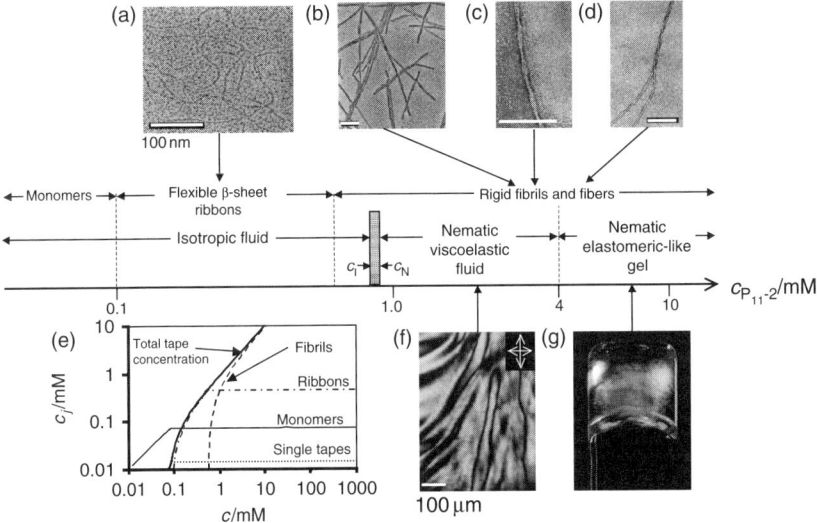

Figure 4 Self-assembling structures and liquid crystalline phase behavior observed in solutions of P₁₁-2 in water with increasing peptide concentration c (log scale). Electron micrographs (a) of ribbons ($c = 0.2$ mM), (b) and (c) of fibrils ($c = 6.2$ mM), and (d) fibers. The curves in (e) were calculated with the generalized model described in the text (see also Figure 5d). The polarizing optical micrograph (f) shows the thick thread-like texture observed for a solution with $c = 3.7$ mM in a 0.2 mM pathlength microslide; (g) shows a self-supporting birefringent gel ($c = 6.2$ mM) in an inverted 10 mM o.d. glass tube, viewed between crossed polarizers. The scale bars in (a–d) correspond to 100 nm, in (f) to 100 μm (Aggeli et al., 2001b).

typically comprised of two entwined fibrils (Figure 4d). The sequence of these structural transitions is also supported by distinctive far- and near-UV CD spectra, corresponding to P₁₁-2 monomers, ribbons, and fibrils (Figure 5a and b).

P₁₁-2 is predominantly in the monomeric random coil conformation at low concentrations (Figure 1b), whereas the fraction of peptide in β-sheet structures starts to grow abruptly at $c \approx 0.07$ mM (Figure 5a and c). Assuming only the presence of tapes in the experimental solutions, ε_{trans} and ε_{tape}, were treated as fitting parameters, it was possible to describe well the experimentally observed β-sheet growth with concentration. However, this single tape model yields a mean tape length of about 20 nm at $c = 0.2$ mM (Figure 5d), much shorter than the observed length ≥ 500 nm (Figure 4a). It is possible, however, to describe the experimentally observed growth of aggregate (solid line in Figure 5c) and simultaneously to predict the occurrence of these long aggregates (Figure 4a) by assuming the presence of tapes and ribbons in the solution. In this way, the experimental data was fitted with three parameters: ε_{trans}, ε_{tape}, and ε_{ribbon} (Figure 1d).

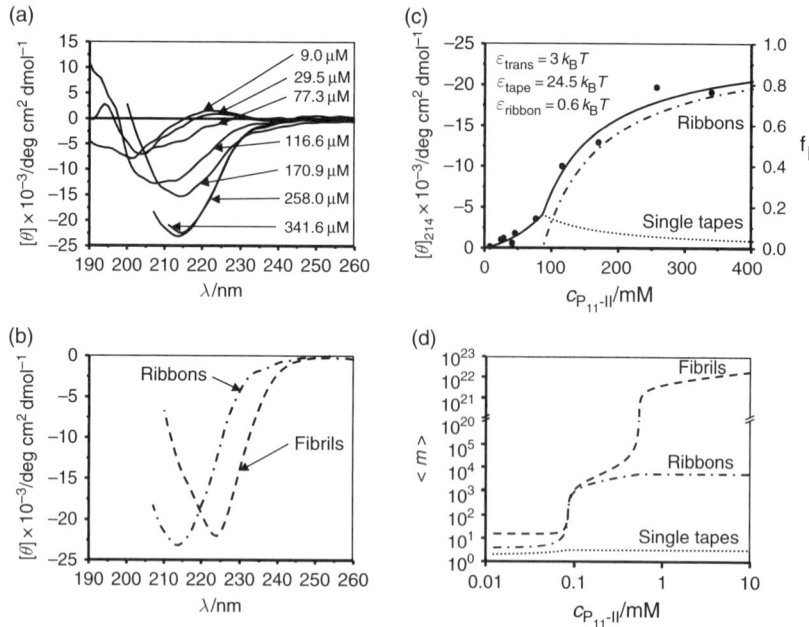

Figure 5 Self-assembling behavior of P$_{11}$-2. (a) and (b) Far-UV CD spectra in water at 20° C. (b) Comparison of the CD spectra of isotropic solutions of P$_{11}$-2 ribbons and of fibrils at $c = 0.3$ mM. The fibril spectrum reveals a red-shifted negative band (centered at 224 nm, compared to 214 nm for ribbons), possibly arising from the superposition of a strong aromatic CD band on the classical β-sheet CD spectrum. (c) Fit of the theoretical model (solid line) for the self-assembly of peptides into single and double β-sheet tapes, to the measured concentration dependence of the mean residue ellipticity [θ] of the negative CD band at $\lambda = 214$ nm. [θ]$_{214}$ is taken to be a linear function of the fraction f_β of peptides in β-sheet tapes. The fractions of peptides involved in single and double tapes are represented with dotted and dash-dot lines, respectively. (d) Theoretical concentration dependencies of the average number $\langle m \rangle$ of peptides in single tapes (dotted line) and in ribbons (dash-dot line) and in fibrils made of $p = 4$ ribbons (dashed line). The molecular parameters were chosen to comply with the fit (c) and with the observed lengths of ribbons at $c = 0.2$ mM and with the transition concentration from ribbon to fibrils at $c = 0.6$ mM. Note that the minimum number of peptides in tapes is two, in ribbons is four, and in fibrils is eight (Aggeli et al., 2001b).

These long aggregates then turn out to be double tapes (ribbons) rather than single ones (Figure 5c and d). The CD spectra as a function of concentration have no isodichroic point (Figure 5a), further supporting that more than two states, that is, peptide monomers, β-tapes, and ribbons, are involved in the conformational transition. The best-fit energy values obtained are $\varepsilon_{trans} = (3 \pm 1)\, k_B T$, $\varepsilon_{tape} = (24.5 \pm 1.0)\, k_B T$, and $\varepsilon_{ribbon} = (0.6 \pm 0.3)\, k_B T$.

The estimated ε_{trans} is higher for P$_{11}$-1 than for P$_{11}$-2 (Table 1). Although both peptides have the same length, they may have different propensity to form a random coil and a β-strand, which may account for this difference in ε_{trans}. In particular P$_{11}$-1 has less bulky side-groups, hence it is generally more flexible than P$_{11}$-2, and as a result P$_{11}$-1 should have higher entropy in the coiled conformation. The magnitude of ε_{tape} is also higher for P$_{11}$-1 than for P$_{11}$-2, which indicates that the intermolecular glutamine side-chain interactions between P$_{11}$-1 peptides are more efficient at promoting self-assembly compared to intermolecular aromatic side-chain interactions between P$_{11}$-2 peptides. ε_{ribbon} is at least two orders of magnitude lower for P$_{11}$-1 compared to P$_{11}$-2, as engineered by peptide design. This difference explains the shorter (by one order of magnitude) length of P$_{11}$-1 ribbons compared to P$_{11}$-2 ones (Table 1). It also accounts for the one order of magnitude difference in critical concentrations for ribbon formation between the two peptides. This results in stabilization of single, curly β-tapes in a wide range of P$_{11}$-1 concentrations. In contrast, P$_{11}$-2 tapes are not observed because they convert to ribbons as soon as they are 3–4 peptides long, at very low concentration.

The formation of fibrils (Figure 4b) at higher concentrations of P$_{11}$-2 implies the presence of a weaker attraction between the polar sides of P$_{11}$-2 ribbons ($\varepsilon_{fibril}^{attr}$, Figure 1e'). From the concentration at which fibrils appear, we calculate $\varepsilon_{fibril} = (2.0 \pm 0.3) \cdot 10^{-4}\ k_B T$. Despite this attraction, the fibril dispersions are stable and the fibril diameter is finite (rather than growing indefinitely). Furthermore, the fibril width W_1 corresponds to the expected length of an 11-residue β-strand, whilst W_2 corresponds to roughly the thickness of 4 ribbons (i.e., 8 single tapes, each tape with a thickness of ca 1 nm) per fibril and is concentration independent (at least from 0.6 to 7 mM). The energy required to break such a fibril, scission energy ε_{sc}, is $\varepsilon_{sc} = 8\varepsilon_{tape} \sim 200\ k_B T$ (comparable to covalent bond energies!) and is much higher than that of a single ribbon $\varepsilon_{sc} = 2\varepsilon_{tape} \sim 50\ k_B T$. This results in fibrils of extraordinary predicted equilibrium average length: $L_{fibril} \sim 10^8$ km!, compared to $L_{ribbon} \sim 1$ μm, for $c = 6$ mM.

2.2.3. Effect of chirality

The formation of fibrils can be explained by the theoretical model of stabilization by twist. β-Sheet ribbons have an intrinsic left-handed twist, due to the L-chirality of peptides (Chothia, 1973). The fibrils also exhibit a left-handed twist with a twist pitch h_{fibril} of approximately 120–200 nm (Figure 4c). Based on the observed geometrical characteristics of P$_{11}$-2 ribbons and fibrils in combination with the theoretical model, the following parameters can be estimated: $h_{ribbon} \sim 120$–200 nm, elastic constants k_{bend} and k_{twist}, and twist angle $\gamma_\theta = 1°$ for isolated P$_{11}$-2 ribbons, and $\varepsilon_{fibril}^{attr} \sim 0.015\ k_B T$ for fibrils (Table 1). The magnitude of $\varepsilon_{fibril}^{attr}$ is expected to be similar both for P$_{11}$-1 and P$_{11}$-2, because of the identity of their polar sides. However, P$_{11}$-1 ribbons are three times more twisted than P$_{11}$-2 ones (twist angles γ_θ in Table 1). The higher elastic penalty $\varepsilon_{\text{fibril}}^{\text{elast}}$ associated with untwisting P$_{11}$-1 ribbons compared to P$_{11}$-2 ones seems to result in lower overall magnitude of $\varepsilon_{\text{fibril}}$ for P$_{11}$-1 compared to P$_{11}$-2 and thus prevents the stacking of P$_{11}$-1 ribbons into fibrils. This may explain why P$_{11}$-1 ribbons do not combine into fibrils up to $c = 25$ mM, whilst P$_{11}$-2 ribbons form fibrils at $c < 1$ mM.

2.2.4. Nematic fluids and gels

The fibrils and fibers are rigid and thus they give rise to the formation of a nematic phase at $c \geq 0.9$ mM (0.001 v/v) for P$_{11}$-2. The texture in the optical micrograph (Figure 4f) and its dependence on flow is characteristic of viscoelastic nematic fluids (Figure 6a) of semirigid polymers (Dobb et al., 1977). The isotropic-to-nematic phase separation gap is narrow: 0.8 mM $< c_I < c_N <$ 0.9 mM (relative gap width $w \equiv c_N/c_I - 1 < 0.13$) and is insensitive to temperature variations up to at least 60°C. Polydisperse rigid-rod solutions have much wider phase separation gaps ($w \sim 2$) (Semenov and Khokhlov, 1988). The fibrils behave more like typical semirigid (worm-like) chains with hard-core excluded volume interactions, for which $w \sim 0.09$ (Semenov and Khokhlov, 1988). The isotropic-to-nematic transition of such chains with rectangular cross section $W_1 \times W_2$ is predicted (Semenov and Khokhlov, 1988) to occur at volume fractions $\Phi_{\text{IN}} \approx 5.5W/\tilde{l}_{\text{fibril}}$ [where $W \approx 2W_1W_2/(W_1 + W_2)$, provided that $L \geq \tilde{l}_{\text{fibril}}$]; this yields for P$_{11}$-2, $\Phi_{\text{IN}} \approx 0.0004 - 0.0015$ v/v (corresponding to $c_{\text{IN}} \approx 0.4$–1.5 mM), in agreement with our observations. $\tilde{l}_{\text{ribbon}}$ for P$_{11}$-1 is one to two orders of magnitude shorter than $\tilde{l}_{\text{fibril}}$ of P$_{11}$-2. The isotropic-to-nematic transition of solutions of such semiflexible ribbons of P$_{11}$-1 is predicted to occur at $\Phi_{\text{IN}} \approx 0.015$–0.05 v/v (corresponding to 15–50 mM). Indeed, it is found that P$_{11}$-1 forms nematic phase at $c \sim 13$ mM (Table 1).

At $c \geq 4$ mM, the birefringent solution of P$_{11}$-2 becomes a self-supporting birefringent gel (Figure 4g). Gelation is associated with onset of fiber

Nematic fluid
(a)

Nematic gel
(b)

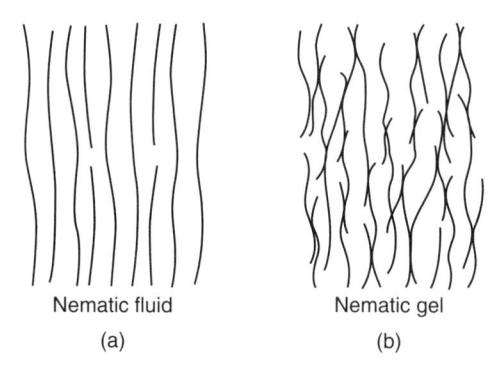

Figure 6 Schematic diagram showing arrangement of fibrils in nematic fluids (a) and the fiber-like junctions in nematic gel states (b) (Aggeli et al., 2003a, b).

formation (Figure 4) which leads to the perception that in the gels, fibrils are linked by fiber-like junctions (Figure 6b). In contrast, tape-based gels are more extendable (Aggeli et al., 1997a, b) and relax slowly with time, behavior indicative of transient gels of semiflexible polymers. It can be concluded that the type of polymer (tape, ribbon, or fibril, each associated with its own characteristic flexibility, contour length and crosslinking mechanism) determines the liquid crystalline and gelation properties of its solution.

3. RESPONSIVENESS TO EXTERNAL TRIGGERS

The magnitudes of energetic parameters that govern and quantify peptide self-assembly and material properties are dependant on a number of variables such as solution conditions (e.g., pH, ionic strength, and relative dielectric constant) and molecular/chemical permutations (e.g., peptide length and amino acid composition). Appropriate choice of solution conditions in combination with rational peptide design allow the control of a number of crucial properties, for example, the magnitudes of critical concentrations for self-assembly of particular types of aggregate, the number of each type of aggregate present and their average length as a function of peptide concentration, the engineering of responsiveness of the self-assembling peptides to a wide range of external triggers, the lifetime and dissolution rate of the aggregates and the material (mechanical, nematic, and gelation) properties.

3.1. pH

A series of systematically varied peptides have been previously designed and studied in order to illustrate that it is possible to engineer responsiveness of self-assembling peptide materials to external chemical triggers (Aggeli et al., 2003a, b). Solutions of P_{11}-2 peptide form stable gels below pH 5, where only arginine is positively charged and the total net charge is +1 per peptide. Above pH 5, the net charge is 0, hence flocculation of the peptide occurs. Thus, it becomes apparent that stabilization of fibrillar dispersions requires a net positive or negative charge per peptide molecule. Incorporation of further charged groups in the primary structure of P_{11}-2 molecule, such as Glu (—CH_2CH_2COOH) or Orn (—$CH_2CH_2CH_2NH_2$), enables self-assembly to be rapidly (seconds) and reversibly controlled by simply changing pH.

This is demonstrated by the behavior of two peptide variants: P_{11}-4 (CH_3CO—QQRFEWEFEQQ—NH_2), which can be switched from its nematic to its isotropic fluid state by increasing pH, and P_{11}-5 (CH_3CO—QQOFOWOFQQQ—NH_2) (O = ornithine), designed to exhibit the opposite

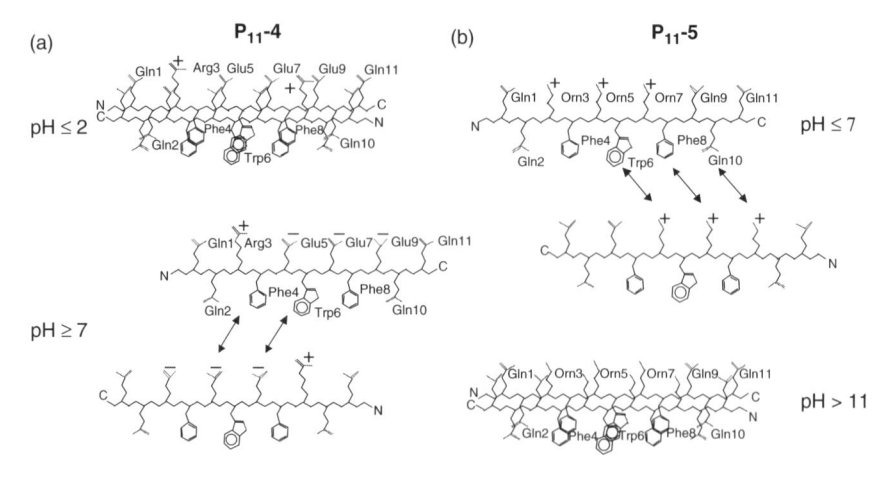

Figure 7 Self-assembly of (a) P$_{11}$-4 and (b) P$_{11}$-5 peptides at low and high pH values showing organization in dimeric tape-like substructures and monomeric states (Aggeli et al., 2003a, b).

pH behavior (Figure 7). In these two peptides, the interpeptide association energies, such as ε_{tape}, are strongly influenced by direct electrostatic forces between γ-COO$^-$ in Glu$^-$ or δ-NH$_3^+$ in Orn$^+$, respectively. This is illustrated by the acid–base titrations of fibrillar dispersions, which reveal that the deprotonation of the γ-COOH of Glu or of the δ-NH$_3^+$ of Orn$^+$ occurs over wide bands of up to 5 pH units, a feature of polyelectrolytes. The values of the energy parameters controlling self-assembly and the values of the critical concentration for self-assembly can therefore be smoothly and continuously varied by changing pH. This enables fast isotropic fluid-to-nematic and fluid-to-gel transitions to be triggered by relatively small additions of acid or base, typically one part in 10^3 by volume of 1 M HCl or NaOH, corresponding to a change of pH by a single unit (Figures 8 and 9).

A number of other pH-responsive β-sheet self-assembling peptides have also been described (Hong et al., 2003; Schneider et al., 2002; Zimenkov et al., 2006). One such class of peptides is EAK16, which includes a repetitive sequence of hydrophobic alanine and charged residues of glutamic acid and lysine (AEAEAKAKAEAEAKAK) (Zhang et al., 1993). The arrangement of the individual amino acids gives rise to two distinct faces: a hydrophobic side comprising solely alanines, and a charged side, of negative glutamic acid residues and positive lysine residues. This motif resulted in the rational design of a class of EKA16 peptides that could respond to a host of external environmental triggers. EAK16 IV (AEAEAEAEAKAKAKAK) is a variant of EAK16 II (AEAEAKAKAEAEAKAK). The subtle change to the peptide primary structure demonstrates vast differences in responsiveness to varying pH when

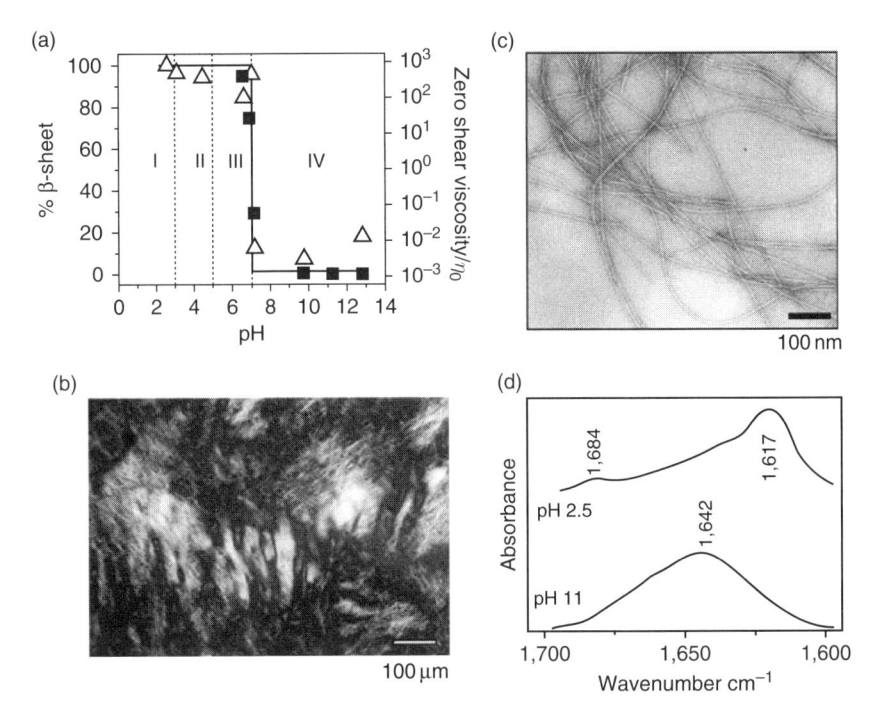

Figure 8 (a) Phase behavior of P_{11}-4 at $c = 6.3$ mM, as a function of pH in water: (I) nematic gel, (II) flocculate, (III) nematic fluid, (IV) isotropic fluid. The continuous line denotes the proportion of peptide in fibrils. (b) Polarizing optical micrograph of a P_{11}-4 gel in water ($c = 6.3$ mM, pH 3) showing a typical thick thread-like viscoelastic nematic texture (path length = 0.2 mm). (c) Transmission electron micrograph of a P_{11}-4 gel in water ($c = 6.3$ mM, pH = 3) showing semirigid fibrils and fibers. (d) FTIR spectrum of amide I' bands showing β-sheet conformation of P_{11}-4 nematic gel ($c = 6.3$ mM) at pH 2.5 and random coil state of P_{11}-4 isotropic fluid ($c = 6.3$ mM) at pH 11 (Aggeli et al., 2003a, b).

compared to EAK II. Observations on the change in morphology were considered by Hong et al. (2003), and it was reported that the original EAK16 II motif showed no change in its morphology between the range of 4.0 < pH < 11. However, the morphology of the EAK16 IV aggregates did depend on its environment. In a pH range of between 6.5 and 7.5, globular assemblies were observed. Below a pH of 6.5 and above a pH of 7.5, the structural morphology had changed to a fibrilar state (Hong et al., 2003). The slight difference in charge distribution is the reason for the observed behavior. In neutral conditions, EAK16 IV does not form nanostructures, this can be attributed to all of the ionizable side groups carrying their respective charge, and thus the electrostatic interactions produce a repulsive effect. When one side chain has become electrostatically neutral, this induced a conformational change.

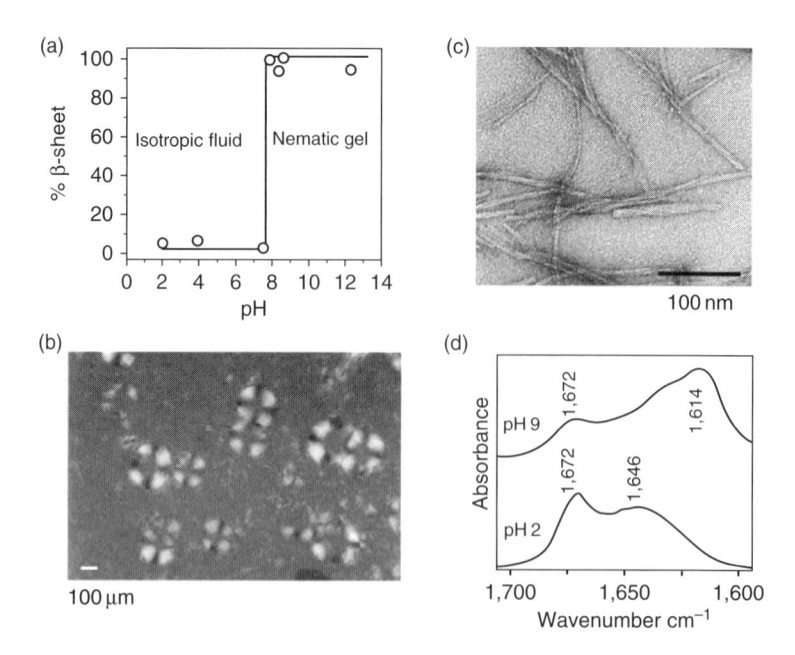

Figure 9 (a) Phase behavior of P_{11}-5 ($c = 13.1$ mM in water) as a function of pH, showing the sharp transition from isotropic fluid to nematic gel states at pH 7.5. (b) Polarizing optical micrograph showing nematic droplets with a radial director distribution (maltese cross) dispersed in an isotropic fluid phase. (c) Transmission electron micrograph of a gel (pH 9, $c = 13.1$ mM) showing semirigid fibrils. (d) FTIR spectrum of amide I′ bands showing β-sheet conformation of P_{11}-5 in the nematic gel state ($c = 13.1$ mM, pH 9), and the random coil conformation in the isotropic fluid state ($c = 13.1$ mM, pH 2) (Aggeli et al., 2003).

3.2. Ionic strength

The control on peptide self-assembly and material properties exerted by electrostatic interactions naturally leads to the question of how the self-assembly would be affected by another related chemical trigger, that is, changes in the ionic strength in solution. An increase in salt concentration in water would be expected to shield the electrostatic forces between side chains, change the magnitudes of energetic parameters, and shift both the critical concentration required for the onset of self-assembly and the observed pH of the transition from gel to fluid. A previous experimental study has revealed the influence of biologically relevant ionic strength on the self-assembly and gelation of pH responsive, systematically varied β-sheet tape-forming peptides, such as P_{11}-4 (Carrick et al., 2007). Another example is peptide P_{11}-9 (CH$_3$CO—SSR$^+$FE$^-$WE$^-$FE$^-$SS—NH$_2$), a variant of P_{11}-4. In order to assess the effect of addition of salt, first the self-assembling properties of P_{11}-9 in water in the absence of salt were established. P_{11}-9 ($c = 7.0$ mM) forms clear self-supporting gels at pD $\leq 3.2 \pm 0.2$ in D$_2$O (Figure 10a). FTIR spectra of the

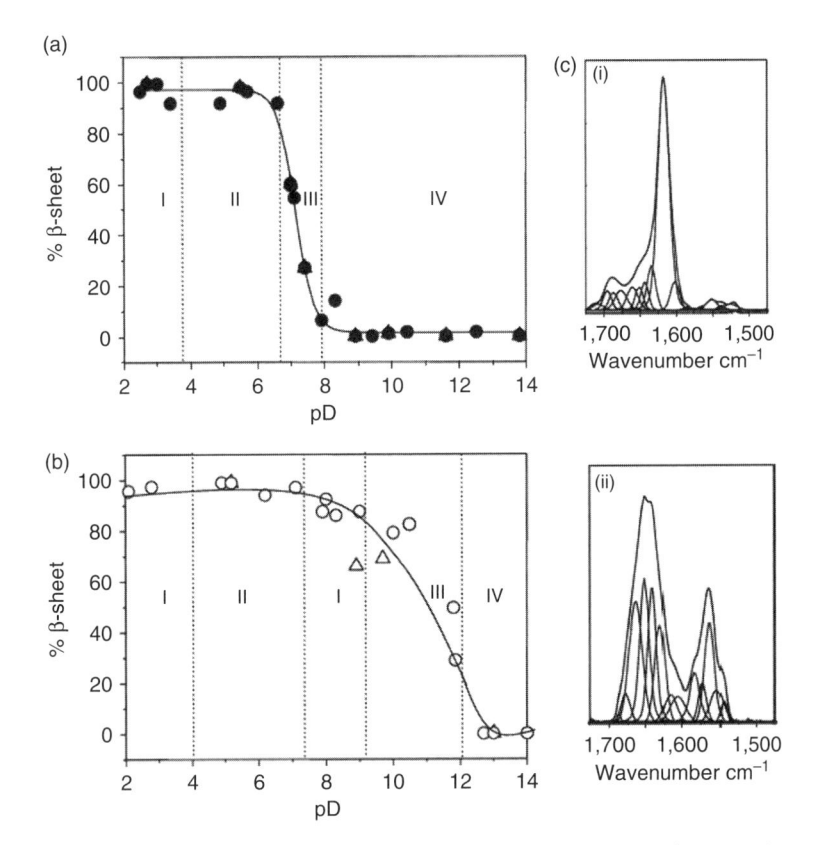

Figure 10 Self-assembly behavior of P_{11}-9 ($c = 7.0$ mM). (a) Percentage β-sheet of P_{11}-9 as a function of pD in D_2O. (b) Percentage β-sheet of P_{11}-9 in 130 mM NaCl in D_2O: (I) nematic gel, (II) flocculate, (III) nematic fluid, (IV) isotropic fluid. (c) FTIR absorption spectra of P_{11}-9: (i) nematic self-supporting gel at pD 2 in D_2O, (ii) monomeric isotropicfluid at pD 10 in D_2O (Carrick et al., 2007).

gels (Figure 10c(i)) have a large absorption band at 1,614 cm^{-1} corresponding to β-sheet aggregates and a weaker band at 1,684 cm^{-1} indicating an antiparallel β-sheet. In addition, a band at 1,710 cm^{-1} corresponding to COOD (protonated glutamic acid side chains) is observed. At $3.2 \leq$ pD ≤ 6.8, flocculates are obtained. The percentage of β-sheet within the flocculate is shown to be comparable to that in the self-supporting gels (Figure 10a), but the fibrillar aggregates are insoluble. At pD ≤ 3.2, the gels have a single positive charge (arginine in position 3). Glutamic acid side chains begin to deprotonate as the pD is increased, and a situation where the peptide molecules have a net charge of 0 (+1 Arg, −1 Glu) will be encountered. Without sufficient repulsion between the fibrils, large insoluble aggregates are formed leading to flocculation. This explanation is evidenced by the disappearance of the very weak band at

1,710 cm^{-1} and the increase in a band at 1,565 cm^{-1} which corresponds to COO$^-$ (i.e., deprotonated glutamic acid). At 6.8 ≤ pD ≤ 7.2, clear viscous fluids are obtained as the peptide molecules begin to have a slight net negative charge and the fibrillar aggregates become soluble and establish gel networks. The FTIR spectra contain similar proportions of random coil (1,645 cm^{-1}) and β-sheet (1,614 cm^{-1}). At pD > 7.2, clear isotropic fluids are observed, accompanied by a characteristic random coil band at 1,645 cm^{-1}. As may be expected from the similarities of the primary structures, the pH-responsive behavior of P$_{11}$-9 is similar to that of P$_{11}$-4 (Figure 8). Addition of 130 mM NaCl shifts the beta-sheet to random coil transition by more than 3 units to higher pH (Figure 10b) and makes the transition broader compared to that in the absence of salt. The gels of P$_{11}$-9 at pD ≤ 3 are shown by electron microscopy to be composed predominantly of micrometer-length fibrils, which are typically 5–7 nm at their widest point and 4 nm at their narrowest point (Figure 11). A twist pitch of

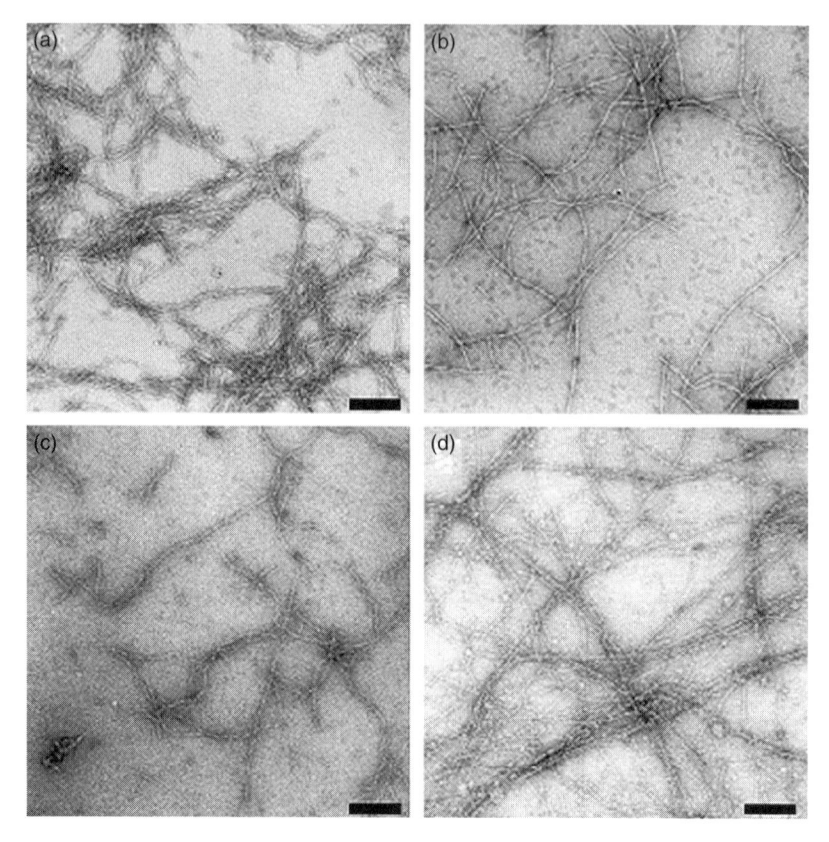

Figure 11 TEM of P$_{11}$-9 ($c = 7.0$ mM): (a) nematic gel at pD 2 in D$_2$O, (b) viscous fluid at pD 7 in D$_2$O, (c) weak gel of P$_{11}$-9 at pD ∼ 3 in 130 mM NaCl in D$_2$O, (d) nematic gel at pD ∼ 8 in 130 mM NaCl in D$_2$O. Scale bars 100 nm (Carrick et al., 2007).

50–80 nm is observed. Most fibrils are composed of two thinner structures, 2–4 nm in width. Flocculates at $3.2 \leq pD \leq 6.8$ and nematic fluids at $6.8 \leq pD \leq 7.2$ also contain fibrils of the 5–10 nm in width and micrometers in length.

In summary, aqueous solutions (~7 mM, equivalent to ~0.7% v/v) of amphiphilic β-sheet tape-forming peptides containing glutamic acid side chains are found to undergo significant changes as a function of pD (2–14) and salt concentration, depending on the deprotonation state of the side chains. Insoluble flocculates of self-assembling fibrils are observed when the net peptide charge is close to zero, nematic gels when there is a small amount of net charge, and monomeric fluid when there is a high net charge per peptide (–2 for solutions with no added salt and –3 for solutions with 130 mM added NaCl). In the absence of added salt, a sharp transition at $6.5 < pD < 8$ is observed from the antiparallel β-sheet at $pD < 6.5$ to the monomeric random coil at $pD > 8$. This is accompanied by a change of the solution properties from a nematic solution at $pD < 8$ to isotropic Newtonian fluid at $pD > 8$. The transition is reversible several times by cycling the pD up and down. In the presence of 130 mM NaCl, the transition becomes broader and shifts to higher pD, by 2–3 pD units. In this way, at physiological-like conditions ($pD \sim 7$–8, 130 mM NaCl), self-supporting nematic gels are now obtained rather than fluids. The shift of the transition to higher pD in the presence of added salt is attributed to screening of electrostatic repulsion between negative charges, as expected from the Derjaguin, Landau, Verwey, Overbeek (DLVO) theory (Carrick et al., 2007), rather than to a change in the pKa of glutamic acid residues in the presence of the added salt.

Peptide P_{11}-12 ($CH_3COSSR^+FO^+WO^+FE^-SSNH_2$) is similar to P_{11}-9 but is designed through the incorporation of ornithine side chains in positions 5 and 7 to have the opposite pH-responsive behavior to P_{11}-9. Solutions of P_{11}-12 at $pD < 9$ have an absorption band at $1,645\,cm^{-1}$ corresponding to peptide in a random coil state. The large absorption band at $1,673\,cm^{-1}$ corresponds to trifluoroacetic acid counterions (Figure 12a). Bands centered at 1,614 and $1,625\,cm^{-1}$ were assigned to peptide in a β-sheet conformation and were seen to increase in magnitude as a function of pD. The weak absorption bands at 1,695 and $1,685\,cm^{-1}$ indicate an antiparallel arrangement. In the range $9.5 < pD < 11.5$, the amount of β-sheet present is nearly 100%. However, at $pD > 11.5$, it starts dropping again. This may be explained by the increased presence of negatively charged carbamates and glutamic acid side chains and the loss of the positive charge on the arginine at $pD > 12.5$. This makes the net peptide charge higher than -1 and partially destabilizes the aggregates at very high pD. Addition of salt causes the conformation transition to shift by 2 pD units to lower pD (Figure 12b). In the absence of added salt at $pD \leq 8$ in D_2O, fibrils of

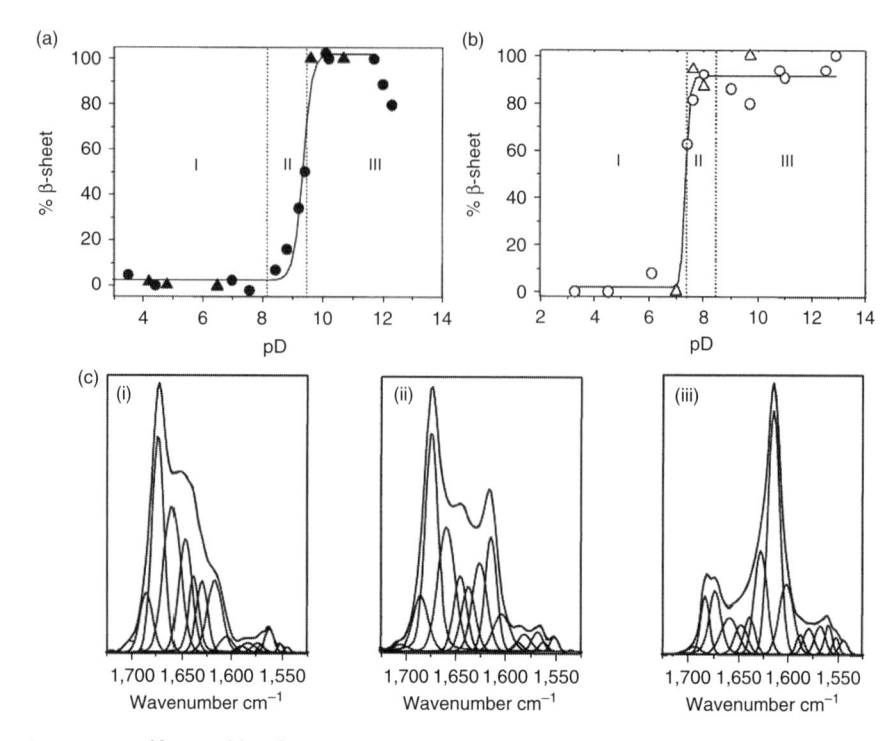

Figure 12 Self-assembly of P_{11}-12 at $c = 7.1$ mM. (a) Percentage β-sheet of P_{11}-12 as a function of pD in D_2O. (b) Percentage β-sheet of P_{11}-12 as a function of pD in 130 mM NaCl in D_2O: (I) isotropic fluid, (II) weakly nematic viscous fluid, (III) weakly nematic gel. (c) FTIR absorption spectra of P_{11}-12 in D_2O showing (i) monomeric peptide at pD 7, (ii) a viscous fluid at pD 9, (iii) a self-supporting gel at pD 10 (Carrick et al., 2007).

P_{11}-12 are not observed. At $8.5 \leq pD \leq 12$, fibrils and bundles composed of 3- to 4-nm wide fibrils are observed (Figure 13), which are similar to those observed for P_{11}-9 (Figure 11).

In summary, amphiphilic tape-forming peptides containing positively charged side chains are found to undergo major charges as a function of pD (2–14) and salt concentration, depending on the deprotonation state of the side chains. In the absence of added salt, a sharp transition at $8 < pD < 10$ is observed from gels with antiparallel β-sheet structure at $pD > 10$ to fluid solutions with monomeric random coil peptides at $pD < 8$. In the presence of 130 mM NaCl, the transition shifts to lower pD by 2–3 pD units. In this way, at physiological-like conditions ($pD \sim 7$–8), viscous nematic fluids are now obtained which convert to self-supporting gels at higher peptide concentration. These data show that the charged peptide solutions are reminiscent of the behavior of polyelectrolyte molecules and complexes (Tsuchida and Abe, 1982).

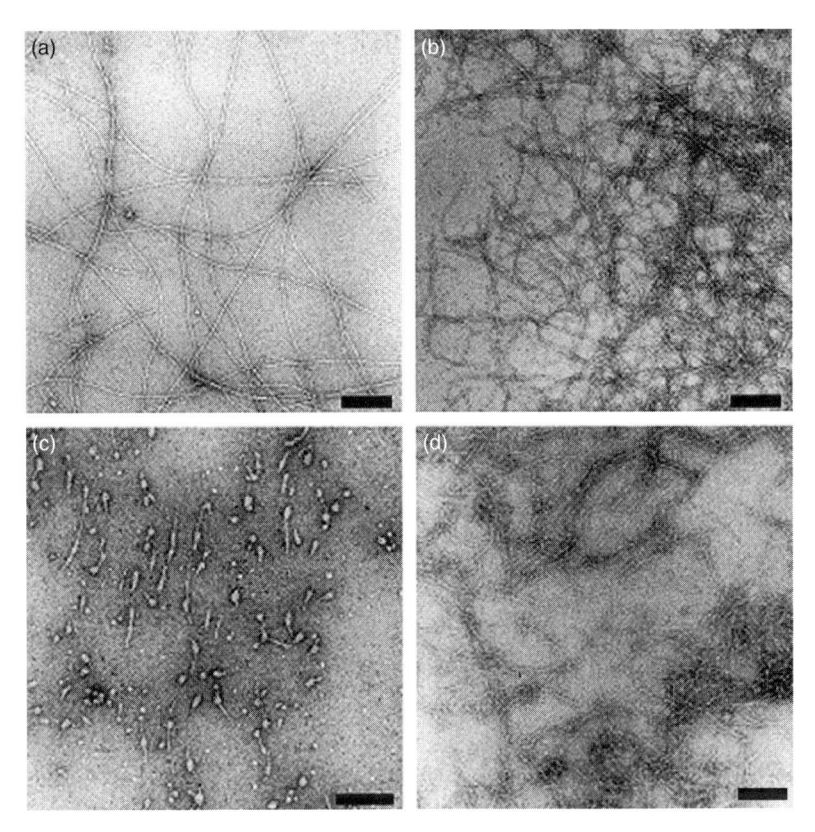

Figure 13 TEM of P$_{11}$-12 ($c = 7.1$ mM): (a) nematic fluid at pD 8.5 in D$_2$O, (b) nematic gel at pD 12 in D$_2$O, (c) isotropic fluid at pD 6 in 130 mM NaCl in D$_2$O, (d) nematic gel at pD 10 in 130 mM NaCl in D$_2$O. Scale bars 100 nm (Carrick et al., 2007).

Systematic studies of the morphology of the self-assembling fibrils by TEM revealed a surprising observation, namely that the most frequently observed fibrillar structures for a given peptide appear to be largely independent of ionic strength, phase (i.e., gel, flocculate, or nematic solution), and pD. In contrast, the characteristics of the fibrils seem to change significantly by modifications of the peptide primary structure and to a lesser extent by the presence of positively or negatively charged side chains.

These observations (Carrick et al., 2007) are also in line with data obtained with other families of gel-forming self-assembling peptides (Collier and Messersmith, 2003; Collier et al., 2001; Mart et al., 2006; Matsumura et al., 2006). Such a class of amphiphilic self-assembling peptides with complementary charges present on each molecule has been previously developed primarily for applications as 3D scaffolds for tissue engineering (Zhang et al., 1993). Aqueous gels formed by one such peptide, KFE12 (FKFEFK-FEFKFE), were stabilized in the presence of salts when the intermolecular electrical double-layer repulsion became less than the van der Waals attraction. This is quantified by the DLVO theory, which predicts that addition of salts would screen charged groups from each other and would thus decrease the Debye length of the solvent (Caplan et al., 2000, 2002).

Another family of peptide amphiphiles consisting of a long hydrocarbon chain segment followed by a peptide segment has also been previously presented. Detailed studies of the rheological properties of aqueous solutions of a typical molecule PA-1 as a function of pH and a wide range of counter ions have been published (Stendahl et al., 2006). The data show that self-assembly of PA-1 and gelation are triggered by counterion screening and are in agreement with the DLVO theory. Another family of peptides designed to adopt a β-hairpin have also demonstrated a pH and ionic strength responsiveness (Schneider et al., 2002). An example of such a peptide is MAX1 (VKVKVKVK^{D}PPTKVKVKVKV—NH$_2$). In a basic environment, the peptide adopts an amphiphilic β-hairpin conformation. One face of the hairpin is arranged with the hydrophobic valine and the opposing face is lined with hydrophilic lysine residues. Self-assembly of the individual monomeric hairpins occurs due to the formation of hydrogen bonds between discrete hairpins. At low pH, interstrand charged–charged repulsions between lysines destabilize individual hairpins and cause unfolding; subsequently this disrupts the conformation of the peptide in solution and results in total disassembly (Ozbas et al., 2004; Schneider et al., 2002). MAX 1 in an aqueous solution of pH ∼ 7.4 required addition of 150 mM KF or 400 mM NaCl for self-assembly to occur.

3.3. Other

This section is by no means an exhaustive account of the various triggers used for self-assembly. It only serves as a quick reminder that apart from pH and ionic strength, which were discussed in the previous sections, a wide range of other factors may be used to control peptide self-assembly and gelation. For example, the ability of an enzyme to act as an external trigger to control self-assembly formation has been demonstrated in recent years (Mart et al., 2006; Ulijn, 2006). Enzyme-responsive self-assembling materials usually contain two main elements. The first is an enzyme-sensitive component which is essentially the trigger, and the second defines the higher order self-assembled structure through its intrinsic molecular characteristics (i.e., controls the weak noncovalent interactions that cause self-assembly) (Ulijn, 2006). A variant of spider dragline silk protein was amongst the first systems to demonstrate enzyme responsiveness. Here enzymatic dephosphorylation controlled the β-sheet formation of the protein (Winkler et al., 2000). The phosphorylation/dephosphorylation has also been used to trigger (Fmoc)-tyrosine into self-assembly. Fluid phosphorylated (Fmoc)-tyrosine was dephosphorylated by a phosphatase causing a phase transition from solution to a gel (Yang et al., 2006).

Another class of "switch" peptides has also been presented (Mutter et al., 2004). This work has developed further into a switchable peptidic material derived from the β-amyloid motif which has an enzymatic triggering system. Controlling the self-assembly of amyloid β-derived peptides by using enzyme-triggered intermolecular acyl migration to promote gel formation has been demonstrated. Modified serine, threonine, and cysteine had enzyme-removable acyl groups attached at the N-terminus which were also modified to have a linked peptide chain. Enzyme deacylation of the *N*-acyl group transferred the linked peptide chain to the amine terminus resulting in a conformation change into a β-sheet (Dos Santos et al., 2005). Consecutive switching of the peptide had been demonstrated and the conformation change was deemed to be spontaneous.

Thermal (Collier et al., 2001; Pochan et al., 2003) and phototriggering (Bosques and Imperiali, 2003; Collier et al., 2001) has also been demonstrated for gel-forming self-assembling peptides. For example, the incorporation of a photoactive compound within the primary structure of a β-hairpin facilitated the use of light to act as an external trigger. The covalent inclusion of a photocage (α-carboxy-2-nitrobenzyl) electrostatically prevented peptide folding, subsequently inhibiting self-assembly. Upon irradiation with ultraviolet light, the photocage was released and self-assembly was triggered (Haines et al., 2005).

3.4. Heteroaggregates formed by complementary peptides

An approach to change the magnitudes of the energetic parameters and greatly affect the tendency for self-assembly and the material properties, without changing the solution conditions, is to mix solutions of complementary peptides together. To demonstrate this principle, two such peptides C (cationic) and A (anionic) have been previously designed (Figure 14) (Aggeli et al., 2003a, b). The two complementary peptides C and A must have a propensity to antiparallel β-sheet formation, appropriate complementarity in the disposition of charged amino acid side chains, and at least one additional charged amino acid per peptide pair to stabilize the peptide fibrillar network against flocculation. Polyelectrolyte β-sheet complexes (PECs) were shown to form on mixing aqueous solutions of such cationic and anionic peptides (Figure 15a and b). This results in the spontaneous self-assembly of fibrillar networks (Figure 15d) and the production of nematic hydrogels (Figure 15c). These complexes have a 1:1 molar stoichiometry, and their networks are robust to variations in pH or peptide concentration. They may be likened to the PECs formed on mixing oppositely charged polymeric polyelectrolytes except that their supramolecular structures are quite different. In the case of peptide complexes, the fibrils have more definitive molecular and mesoscopic structures making it easier to specify the requisite molecular design. Another example of complementary self-assembling peptide amphiphiles was also reported (Niece et al., 2003).

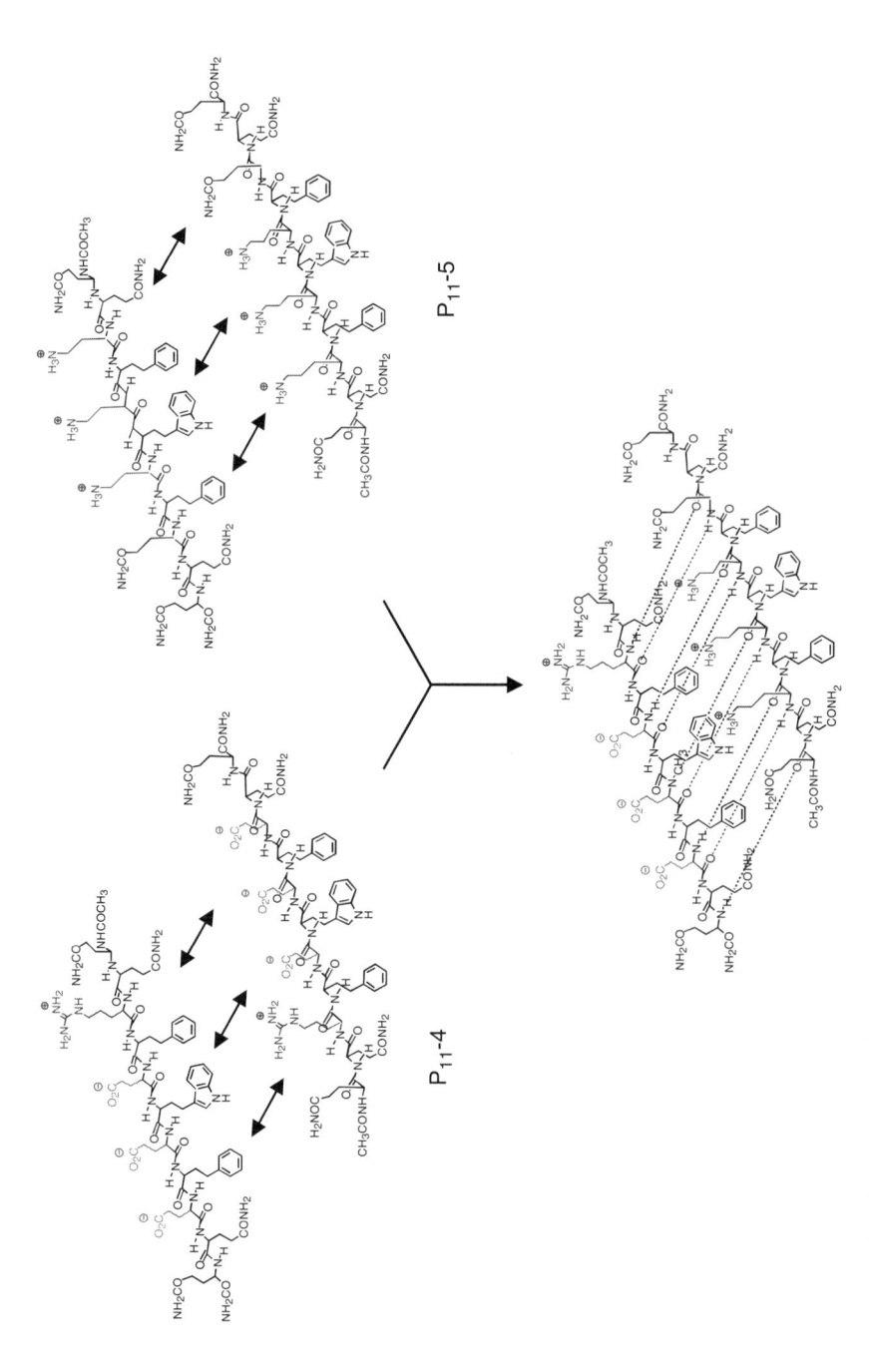

Figure 14 Molecular structures of P_{11}-4, P_{11}-5, and the complex showing the electrostatic charge distributions at pH 7.3 (Aggeli et al., 2003).

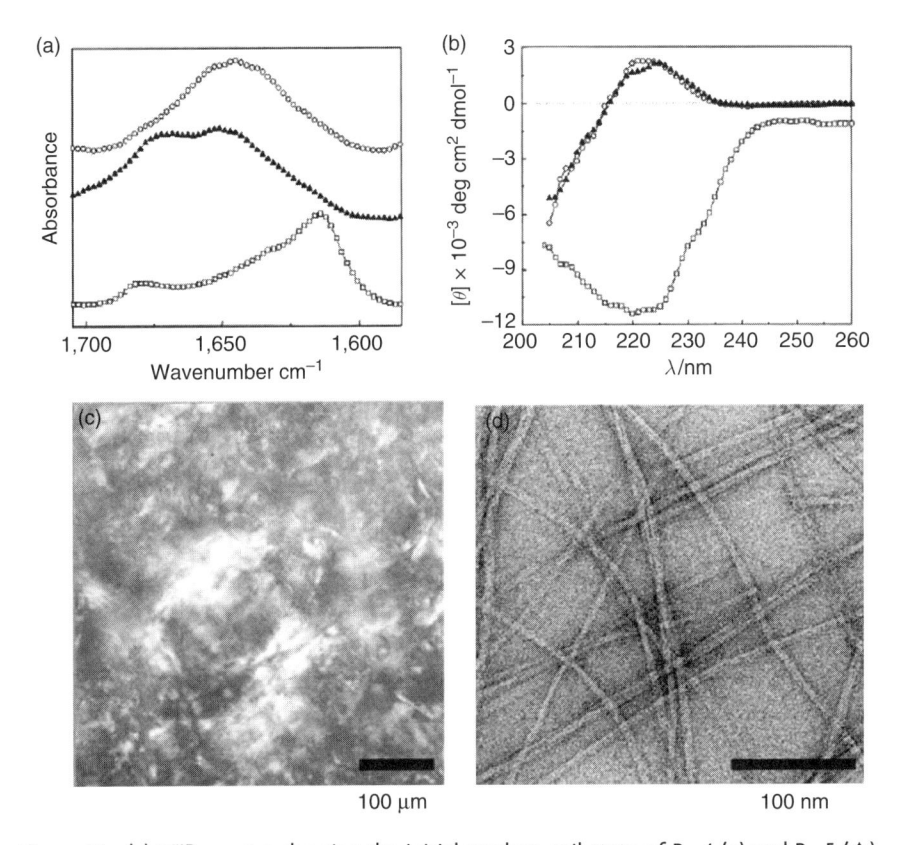

Figure 15 (a) FTIR spectra showing the initial random coil state of P$_{11}$-4 (○) and P$_{11}$-5 (▲) in 6.3 mM aqueous solutions at pH 7.3 prior to mixing, and the β-sheet conformation of the polyelectrolyte complex after mixing (□). (b) Far-UV CD spectra showing the initial random coil states of P$_{11}$-4 (○) and P$_{11}$-5 (▲) in aqueous solutions ($c = 3.1$ mM) prior to mixing, and the β-sheet complex formed after mixing (□). (c) Polarizing optical micrograph of the gel ($c = 6.3$ mM) formed after mixing aqueous solutions of the monomeric peptides at pH 7.3 showing a typical nematic gel texture. (d) Transmission electron micrograph showing mainly fibrils and a few fibers in the nematic gel ($c = 6.3$ mM) (Aggeli et al., 2003a, b).

4. PEPTIDE SELF-ASSEMBLY ON SURFACES

In order to achieve thorough fundamental understanding of biomolecular self-assembly, it is imperative to study 1D tape-like self-assembly not only in bulk solution but also at interfaces. An example of a biologically relevant interface is that of the lipid bilayer. Systematic peptide-lipid studies have begun to offer an insight into the basic principles and mechanisms of interactions of self-assembling peptides with model lipid layers (Protopapa et al., 2006).

Other studies of the tape-forming peptide P_{11}-2 have focused on its interaction with model solid surfaces, such as mica (Whitehouse et al., 2005). It was shown that the presence of the surface has a profound effect on peptide self-assembly. Monomeric peptides in solution below the critical concentration for solution self-assembly were found to interact with the mica surface via the charged amino acid residues Arg and Glu present on the polar side of the peptide and the corresponding negative charges and positively charged counterions on mica surface. Furthermore, adsorbed monomeric peptides went on to self-assemble on the solid surface producing individual tapes, with measured width of 4.9 nm and height of 0.8 nm in agreement with the molecular dimensions (Figure 16). Effectively, the solid surface acts as a template for the induction of peptide self-assembly at concentrations well below their critical concentration for self-assembly in solution. The tapes are stabilized by the intermolecular peptide interactions and also by the interactions of the polar face of the tape with the mica (Figure 17). The tapes on the surface are found to be completely flat, unlike their twisted counterparts in solution. The elastic penalty for untwisting the tape is thought to be compensated for by the gain in enthalpy due to tape–mica interactions. In order for the tapes to maximize their interaction with the surface, they don't overlap with one another, that is, as soon as they grow enough to meet another tape, tape growth stops. In this way, they

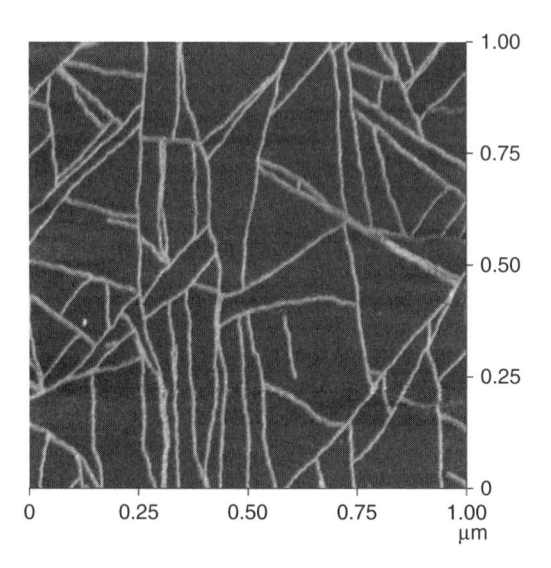

Figure 16 In situ AFM images of single β-sheet tapes grown on mica from 5 μM P_{11}-2 in 10% H_2O in 2-propanol. The tapes are aligned with the hexagonal symmetry of the underlying mica lattice (Whitehouse et al., 2005).

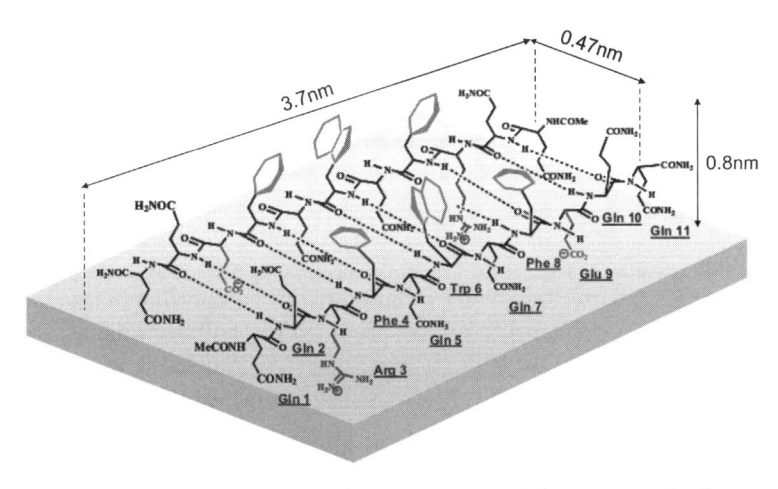

Figure 17 Schematic diagram showing the orientation and dimensions of a dimeric P_{11}-2 peptide tape when adsorbed at the surface of mica (Whitehouse et al., 2005).

form true single-molecule thick surface coatings. The macroscopic arrangement of the tapes on the surface can be controlled: hexagonal arrangement following the underlying mica crystal lattice, macroscopic alignment facilitated by shear field or side by side arrangement into tactoid structures depending on the drying process.

5. CONCLUSIONS AND FUTURE PROSPECTS

A theoretical framework exists that describes well 1D peptide tape self-assembly in solution in terms of a set of five molecular energetic parameters. However, it is currently not known how the magnitudes of the energetic parameters vary with changes of the peptide primary structure and solution conditions. Extensive systematic studies are underway to establish this quantitative information. The combined knowledge of the quantitative energetic parameters as a function of peptide chemistry and external condition, together with the mathematical model, can be a uniquely powerful tool at our disposal for understanding in detail biological self-assembly and for the efficient design of peptides and proteins with well-defined combination of self-assembling properties to fit appropriate applications.

For example, it will be possible to design peptides that aggregate in well-defined morphology, for example, fibrils of specific and controllable width and twist pitch to act as versatile templates for nanoporous inorganic materials (e.g., silica nanotubes) (Meegan et al., 2004). Another opportunity that opens up will be the ability to produce by precise molecular engineering, a range of injectable self-assembling nanostructured networks with tunable activity, mechanical properties, and rates of dissolution to be used, for example, in health care (Bell et al., 2006; Firth et al., 2006; Kirkham et al., 2007; Scanlon et al., 2007) or screening devices. Similar theoretical and quantitative insight needs to be extended to peptide self-assembly on surfaces. This will allow us to produce ultrathin peptide nanocoatings on a variety of substrates, for example, for biomaterial or catalysis applications. The field of protein-like self-assembly is fairly new, but very promising and the opportunities it offers are limitless. The fundamental understanding of mechanisms and principles that drive protein-like self-assembly will become increasingly important as scientists will strive to come up with gradually more functional and complex protein-like building blocks for industrial applications.

REFERENCES

Aggeli, A., Bell, M., Boden, N., Carrick L.M., and Strong A.E. "Self-assembling peptide polyelectrolyte beta-sheet complexes form nematic hydrogels". *Angew. Chem. Int. Ed.* **42**(45), 5603–5606 (2003a).

Aggeli, A., Bell, M., Boden, N., Keen, J.N., Knowles, P.F., McLeish, T.C.B., Pitkeathly, M., and Radford S.E. "Responsive gels formed by the spontaneous self-assembly of peptides into polymeric beta-sheet tapes". *Nature* **386**(6622), 259–262 (1997a).

Aggeli, A., Bell, M., Boden, N., Keen, J.N., McLeish, T.C.B., Nyrkova, I., Radford, S.E., and Semenov, A. "Engineering of peptide beta-sheet nanotapes". *J. Mater. Chem.* **7**(7), 1135–1145 (1997b).

Aggeli, A., Bell, M., Carrick, L.M., Fishwick, C.W.G., Harding, R., Mawer, P.J., Radford, S.E., Strong A.E., and Boden, N. "pH as a trigger of peptide beta-sheet self-assembly and reversible switching between nematic and isotropic phases". *J. Am. Chem. Soc.* **125**(32), 9619–9628 (2003b).

Aggeli, A., Boden N., and Zhang S., "Self-Assembling Peptide Systems in Biology, Medicine and Engineering". Kluwer Academic Publishers, Dordrecht, The Netherlands (2001a).

Aggeli, A., Nyrkova, I.A., Bell, M., Harding, R., Carrick, L., McLeish, T.C.B., Semenov, A.N., and Boden, N. "Hierarchical self-assembly of chiral rod-like molecules as a model for peptide beta-sheet tapes, ribbons, fibrils, and fibers". *Proc. Nat. Acad. Sci. U.S.A.* **98**(21), 11857–11862 (2001b).

Bell, C.J., Carrick, L.M., Katta, J., Jin, Z.M., Ingham, E., Aggeli, A., Boden, N., Waigh T.A., and Fisher, J. "Self-assembling peptides as injectable lubricants for osteoarthritis". *J. Biomed. Mater. Res. Part A* **78A**(2), 236–246 (2006).

Bosques, C.J., and Imperiali B. "Photolytic control of peptide self-assembly". *J. Am. Chem. Soc.* **125**(25), 7530–7531 (2003).

Burkoth, T.S., Benzinger, T.L.S., Jones, D.N.M., Hallenga, K., Meredith, S.C., and Lynn D.G. "C-terminal PEG blocks the irreversible step in ß-amyloid (10–35) fibrillogenesis". *J. Am. Chem. Soc.* **120**(30), 7655–7656 (1998).

Caplan, M.R., Moore, P.N., Zhang, S.G., Kamm, R.D., and Lauffenburger, D.A. "Self-assembly of a beta-sheet protein governed by relief of electrostatic repulsion relative to van der Waals attraction". *Biomacromolecules* **1**(4), 627–631 (2000).

Caplan, M.R., Schwartzfarb, E.M., Zhang, S., Kamm, R.D., and Lauffenburger, D.A. "Control of self-assembling oligopeptide matrix formation through systematic variation of amino acid sequence". *Biomaterials* **23**(1), 219–227 (2002).

Carrick, L.M., Aggeli, A., Boden, N., Fisher, J., Ingham, E., and Waigh, T.A. "Effect of ionic strength on the self-assembly, morphology and gelation of pH responsive beta-sheet tape-forming peptides". *Tetrahedron* **63**(31), 7457–7467 (2007).

Chothia, C. "Conformation of twisted beta-pleated sheets in proteins". *J. Mol. Biol.* **75**(2), 295–302 (1973).

Collier, J.H., Hu, B.H., Ruberti, J.W., Zhang, J., Shum, P., Thompson, D.H., and Messersmith, P.B. "Thermally and photochemically triggered self-assembly of peptide hydrogels". *J. Am. Chem. Soc.* **123**(38), 9463–9464 (2001).

Collier, J.H., and Messersmith, P.B. "Enzymatic modification of self-assembled peptide structures with tissue transglutaminase". *Bioconjugate Chem.* **14**(4), 748–755 (2003).

de Loos, M., van Esch, J., Kellogg, R.M., and Feringa, B.L. "Chiral recognition in bis-urea-based aggregates and organogels through cooperative interactions". *Angew. Chem. Int. Ed.* **40**(3), 613–616 (2001).

Dobb, M.G., Johnson, D.J., and Saville, B.P. "Supramolecular structure of a high-modulus polyaromatic fiber (Kevlar 49)". *J. Polym. Sci. Part B-Polym. Phys.* **15**(12), 2201–2211 (1977).

Dos Santos, S., Chandravarkar, A., Mandal, B., Mimna, R., Murat, K., Saucede, L., Tella, P., Tuchscherer, G., and Mutter, M. "Switch-peptides: Controlling self-assembly of amyloid beta-derived peptides in vitro by consecutive triggering of acyl migrations". *J. Am. Chem. Soc.* **127**(34), 11888–11889 (2005).

Firth, A., Aggeli, A., Burke, J.L., Yang, X.B., and Kirkham, J. "Biomimetic self-assembling peptides as injectable scaffolds for hard tissue engineering". *Nanomedicine* **1**(2), 189–199 (2006).

Gronwald, O., Snip, E., and Shinkai, S. "Gelators for organic liquids based on self-assembly: a new facet of supramolecular and combinatorial chemistry". *Curr. Opin. Colloid Interface Sci.* **7**(1–2), 148–156 (2002).

Haines, L.A., Rajagopal, K., Ozbas, B., Salick, D.A., Pochan, D.J., and Schneider, J.P. "Light-activated hydrogel formation via the triggered folding and self-assembly of a designed peptide". *J. Am. Chem. Soc.* **127**(48), 17025–17029 (2005).

Hanabusa, K., Yamada, M., Kimura, M., and Shirai, H. "Prominent gelation and chiral aggregation of alkylamides derived from trans-1,2-diaminocyclohexane". *Angew. Chem. Int. Ed.* **35**(17), 1949–1951 (1996).

Hong, Y., Legge, R.L., Zhang, S., and Chen, P. "Effect of amino acid sequence and pH on nanofiber formation of self-assembling peptides EAK16-II and EAK16-IV". *Biomacromolecules* **4**(5), 1433–1442 (2003).

Kirkham, J., Firth, A., Vernals, D., Boden, N., Robinson, C., Shore, R.C., Brookes, S.J., and Aggeli, A. "Self-assembling peptide scaffolds promote enamel remineralization". *J. Dental Res.* **86**(5), 426–430 (2007).

Lashuel, H.A., LaBrenz, S.R., Woo, L., Serpell, L.C., and Kelly, J.W. "Protofilaments, filaments, ribbons, and fibrils from peptidomimetic self-assembly: Implications for amyloid fibril formation and materials science". *J. Am. Chem. Soc.* **122**(22), 5262–5277 (2000).

Manning, M.C., Illangasekare, M., and Woody, R.W. "Circular-dichroism studies of distorted alpha-helices, twisted beta-sheets, and beta-turns". *Biophys. Chem.* **31**(1–2), 77–86 (1988).

Marini, D.M., Hwang, W., Lauffenburger, D.A., Zhang, S.G., and Kamm, R.D. "Left-handed helical ribbon intermediates in the self-assembly of a beta-sheet peptide". *Nano Lett.* **2**(4), 295–299 (2002).

Mart, R.J., Osborne, R.D., Stevens, M.M., and Ulijn, R.V. "Peptide-based stimuli-responsive biomaterials". *Soft Matter* **2**(10), 822–835 (2006).

Matsumura, S., Uemura, S., and Mihara, H. "Metal-triggered nanofiber formation of His-containing beta-sheet peptide". *Supramol. Chem.* **18**(5), 397–403 (2006).

Meegan, J.E., Aggeli, A., Boden, N., Brydson, R., Brown, A.P., Carrick, L., Brough, A.R., Hussain, A., and Ansell, R.J. "Designed self-assembled beta-sheet peptide fibrils as templates for silica nanotubes". *Adv. Funct. Mater.* **14**(1), 31–37 (2004).

Mutter, M., Chandravarkar, A., Boyat, C., Lopez, J., Dos Santos, S., Mandal, B., Mimna, R., Murat, K., Patiny, L., and Saucede, L. "Switch peptides in statu nascendi: induction of conformational transitions relevant to degenerative diseases". *Angew. Chem. Int. Ed.* **43**(32), 4172–4178 (2004).

Niece, K.L., Hartgerink, J.D., Donners, J.J.J.M., and Stupp, S.I. "Self-assembly combining two bioactive peptide-amphiphile molecules into nanofibers by electrostatic attraction". *J. Am. Chem. Soc.* **125**(24), 7146–7147 (2003).

Nyrkova, I.A., Semenov, A.N., Aggeli, A., Bell, M., Boden, N., and McLeish, T.C.B. "Self-assembly and structure transformations in living polymers forming fibrils". *Eur. Phys. J. B* **17**(3), 499–513 (2000a).

Nyrkova, I.A., Semenov, A.N., Aggeli, A., and Boden, N. "Fibril stability in solutions of twisted beta-sheet peptides: a new kind of micellization in chiral systems". *Eur. Phys. J. B* **17**(3), 481–497 (2000b).

Ozbas, B., Rajagopal, K., Schneider, J.P., and Pochan, D.J. "Semiflexible chain networks formed via self-assembly of ß-hairpin molecules". *Phys. Rev. Lett.* **93**(26), 268106 (2004).

Perutz, M.F., Johnson, T., Suzuki, M., and Finch, J.T. "Glutamine repeats as polar zippers – their possible role in inherited neurodegenerative diseases". *Proc. Nat. Acad. Sci. U.S.A.* **91**(12), 5355–5358 (1994).

Pochan, D.J., Schneider, J.P., Kretsinger, J., Ozbas, B., Rajagopal, K., and Haines, L. "Thermally reversible hydrogels via intramolecular folding and consequent self-assembly of a de Novo designed peptide". *J. Am. Chem. Soc.* **125**(39), 11802–11803 (2003).

Protopapa, E., Aggeli, A., Boden, N., Knowles, P.F., Salay, L.C., and Nelson, A. "Electrochemical screening of self-assembling beta-sheet peptides using supported phospholipid monolayers". *Med. Eng. Phys.* **28**(10), 944–955 (2006).

Qu, Y., Payne, S.C., Apkarian, R.P., and Conticello, V.P. "Self-assembly of a polypeptide multi-block copolymer modeled on dragline silk proteins". *J. Am. Chem. Soc.* **122**, 5014 (2000).

Scanlon, S., Aggeli, A., Boden, N., Koopmans, R.J., Brydson, R., and Rayner, C.M., "Peptide aerogels comprising self-assembling nanofibrils". *Micro Nano Lett.* **2**, 24–29 (2007).

Schneider, J.P., Pochan, D.J., Ozbas, B., Rajagopal, K., Pakstis, L., and Kretsinger, J. "Responsive hydrogels from the intramolecular folding and self-assembly of a designed peptide". *J.Am. Chem. Soc.* **124**(50), 15030–15037 (2002).

Semenov, A.N., and Khokhlov, A.R. "Statistical physics of liquid-crystalline polymers". *Physics-Uspekhi* **31**(11), 988–1014 (1988).

Stendahl, J.C., Rao, M.S., Guler, M.O., and Stupp, S.I. "Intermolecular forces in the self-assembly of peptide amphiphile nanofibers". *Adv. Funct. Mater.* **16**(4), 499–508 (2006).

Terech, P., and Weiss, R.G. "Low molecular mass gelators of organic liquids and the properties of their gels". *Chem. Rev.* **97**(8), 3133–3159 (1997).

Tsuchida, E., and Abe, K. "Interactions between macromolecules in solution and intermacromolecular complexes". *Adv. Polym. Sci.* **45**, 1–119 (1982).

Ulijn, R.V. "Enzyme-responsive materials: a new class of smart biomaterials". *J. Mater. Chem.* **16**(23), 2217–2225 (2006).

Weiss, R.G., and Terech, P. "Molecular Gels: Materials with Self-Assembled Fibrillar Networks". Springer, Dordrecht, , The Netherlands (2006).

Whitehouse, C., Fang, J.Y., Aggeli, A., Bell, M., Brydson, R., Fishwick, C.W.G., Henderson, J.R., Knobler, C.M., Owens, R.W., Thomson, N.H., Smith, D.A., and Boden, N. "Adsorption and self-assembly of peptides on mica substrates". *Angew. Chem. Int. Ed.* **44**(13), 1965–1968 (2005).

Wilson, M., Kannangara, K., Smith, G., and Simmons, M. "Nanotechnology: Basic Science and Emerging Technologies". CRC Press, Sydney, Australia (2002).

Winkler, S., Wilson, D., and Kaplan, D.L. "Controlling beta-sheet assembly in genetically engineered silk by enzymatic phosphorylation/dephosphorylation". *Biochemistry* **39**(41), 12739–12746 (2000).

Yamada, N., Ariga, K., Naito, M., Matsubara, K., and Koyama, E. "Regulation of beta-sheet structures within amyloid-like beta-sheet assemblage from tripeptide derivatives". *J. Am. Chem. Soc.* **120**(47), 12192–12199 (1998).

Yang, Z., Liang, G., Wang, L., and Xu, B. "Using a kinase/phosphatase switch to regulate a supramolecular hydrogel and forming the supramolecular hydrogel in vivo". *J. Am. Chem. Soc.* **128**(9), 3038–3043 (2006).

Zhang, S.G., Holmes, T., Lockshin, C., and Rich, A. "Spontaneous assembly of a self-complementary oligopeptide to form a stable macroscopic membrane". *Proc. Nat. Acad. Sci. U.S.A.* **90**(8), 3334–3338 (1993).

Zimenkov, Y., Dublin, S.N., Ni, R., Tu, R.S., Breedveld, V., Apkarian, R.P., and Conticello, V.P. "Rational design of a reversible pH-responsive switch for peptide self-assembly". *J. Am. Chem. Soc.* **128**(21), 6770–6771 (2006).

Nucleation and Co-Operativity in Supramolecular Polymers

Paul van der Schoot

Contents		
	1. Introduction	45
	2. Nucleated Assembly	51
	3. Kinetics of Nucleated Assembly	55
	4. Co-Operativity and Nucleation	61
	5. Coarse Graining Reversed	66
	6. Summary and Outlook	70
	Acknowledgement	72
	References	72

1. INTRODUCTION

Supramolecular polymers are polymeric objects that spontaneously arise in solutions or melts from one or more types of molecular building block that can range from the chemically simple to the very complex (Ciferri, 2005). Depending on the shape of these molecules, such self-assembled polymers can have different geometries but invariably are the result of a process usually called *micro phase separation* (Cohen Stuart, 2008; Safran, 1994) although strictly speaking *nano phase separation* would be a more appropriate term. Micro (or nano) phase separation is the spontaneous ordering of molecules on a local (molecular) scale but not on a global (macroscopic) scale; however, the assemblies so formed may themselves self-assemble hierarchically into structures on much larger scales (Cates and Fielding, 2006; Ciferri, 2005). For example, self-assembled helical tapes of specifically designed β-sheet-forming oligopeptides form bundles of various sort, including ribbons, fibrils, and fibers, which at sufficiently high concentrations collectively align and self-assemble into a macroscopic nematic phase (Aggeli, 2001). (see Figure 1).

Group Theoretical and Polymer Physics, Department of Applied Physics, Technische Universiteit Eindhoven, Postbus 513, 5600 MB Eindhoven, The Netherlands

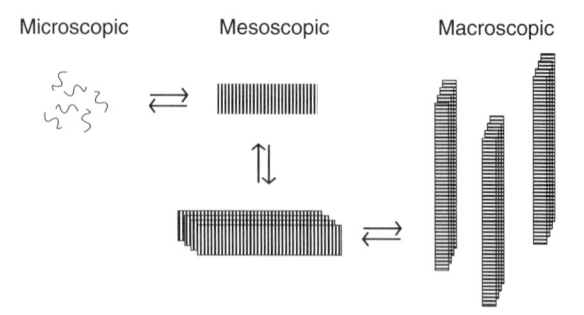

Figure 1 Schematic representation of an example of hierarchical self-assembly at microscopic, mesoscopic, and macroscopic levels. At the microscopic level, molecules assemble into supramolecular polymer-like assemblies. This involves conformational changes to the monomer units that themselves are complex molecules. The polymers assemble into bundles at mesoscopic levels that under appropriate conditions spontaneously align macroscopically along some preferred direction to form a uniaxial nematic liquid–crystalline phase (after Aggeli et al., 2001).

Historically, micro phase separation is understood to give rise to micelles in aqueous surfactant solutions but also to highly regular structures in block copolymer melts (Cohen Stuart, 2008; Gelbart et al., 1994; Safran, 1994). Micro phase separation results from a compromise between antagonistic physical (i.e. noncovalent) interactions between different moieties on the molecular building blocks that as a rule combine parts of different polarity. In surfactant molecules, for instance, highly polar or charged groups are covalently linked to apolar aliphatic moieties. Dissolved in water, the aliphatic groups drive the molecules together in order to shield themselves from contact with the solvent, in this case water, whilst still allowing for the polar or charged parts of the molecules to be solvated by the solvent. The term nano phase separation is quite appropriate as only the solvent-phobic parts of the molecules are guarded from it and the solvent-phylic parts still interact with the solvent.

The molecules that form supramolecular polymers tend to be much more complex than surfactants or block copolymers and are usually designed for the purpose of forming a structure with a predestined geometry, and to perform a certain function (Binder, 2005; Brunsveld et al., 2001; ten Cate and Sijbesma, 2002; Dankers and Meijer, 2007; van Gorp et al., 2002; Vriezema et al., 2005). The ultimate supramolecular polymer formers are, of course, proteins, molecules of which the structure and function are intimately connected. In biology, supramolecular polymerization is a strategy followed when large-scale structures are to be constructed at minimal cost of genetic encoding (Caspar, 1980; Chiu et al., 1997; Kushner, 1969; Lauffer, 1975; Oosawa and Asakura, 1975). A case in point are viruses that package their genome in highly regular (and typically icosahedral or helical) structures, constructed from very many copies of one or a few kinds of the so-called 'coat proteins'. Because infectious viruses can be assembled in vitro by simply mixing the constituents under appropriate physicochemical conditions, virus assembly is considered a thermodynamic process and viruses a kind of supramolecular polymer (Bruinsma et al., 2003; Kegel and van der Schoot, 2004; McPherson, 2005).

Despite large differences in the primary structure, that is, the amino acid sequence, the tertiary or three-dimensional structure of the viral coat proteins of very many viruses, is similar and (usually but not exclusively) characterized by the so-called jelly-roll structure (Chiu et al., 1997). Despite their apparent similarity in structure, these proteins still produce assemblies that can be very different in size and/or shape depending also on the solution conditions. One of the great challenges is to understand the relation between molecular and supramolecular structure, not just in structural and molecular biology but also in the context of supramolecular chemistry. In fact, the field of supramolecular polymers has proven highly successful in constructing complex structures with predestined functionalities utilizing relatively simple molecular building blocks (Bouteiller, 2007; Hagerink et al., 2001; Kato et al., 2006; Lehn, 1990; Moore, 1999; Percec, 1996; Weiss and Terech, 2006; Whitesides and Boncheva, 2002). Not surprisingly, supramolecular chemists are often inspired by examples taken from biology (Binder, 2005; Ciferri, 2002; Moore, 1999).

There are in fact good reasons for the biomimetic design of supramolecular polymers, quite irrespective of the desired topology and/or functionality. Nature has found ways to control the relatively blunt and insensitive instrument provided by the *law of mass action*, a principle that regulates the balance between the assembled and disassembled states of molecules (Ciferri, 2005; Gelbart et al., 1994; Safran, 1994). The mass action of pertinent variable is $X = \phi K$, a measure of the probability that a molecule will attach to an assembly. Here, ϕ denotes a dimensionless concentration (e.g. a mole fraction) and K a dimensionless equilibrium constant. The constant K is temperature dependent and can be associated with the Boltzmann weight of a *free energy of binding*, g, of a single monomer through $K = \exp[-\beta g]$, where $\beta = 1/k_B T$ and $k_B T$ denotes the thermal energy with k_B is Boltzmann's constant and T the absolute temperature.

The mass action variable X is the product of the a priori probability ϕ of a molecule being near the growing assembly and the enhancement K of this probability by the gain of free energy by actually attaching to it (Ciferri, 2005). Unless the supramolecular assembly is highly co-operative, the fraction of bound molecules and their mean aggregation number is a fairly weak function of this variable because it is regulated by a simple Boltzmann distribution between the assembled and free species (Cates and

(a) ○ ⇌^{K} ○○ ⇌^{K} ○○○ ⇌^{K} ○○○○ ⇌^{K} ...

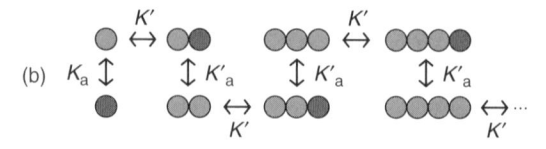

(b) K_a ⇅ ○ ⇌$^{K'}$ ○○   ○○○ ⇌$^{K'}$ ○○○○
    ● ⇅K'_a ●● ⇌$_{K'}$ ●●● ⇅K'_a ●●●● ⇌$_{K'}$ ...

Figure 2 Chemical reaction models for (a) isodesmic and (b) nucleated supramolecular assembly. $K \gg 1$ and $K' \gg 1$ are equilibrium constants for the elongation reactions, and $K_a \ll 1$ and K'_a those for the conversion between assembly active and inactive forms of the monomer units. If $K'_a \gg K_a$, then the nucleated assembly is self-catalyzed ("autosteric") and if $K'_a = K_a$ this is not so.

Candau, 1990; Safran, 1994). For the ideal case of supramolecular assemblies in which each monomer gains a fixed free energy of binding irrespective of the size of the aggregate, schematically depicted in Figure 2a, the polymerized fraction of material, f, obeys the simple relation (Ciferri, 2005)

$$f = 1 - \overline{N}^{-2} \qquad (1)$$

where $\overline{N}$ denotes the mean aggregation number. The latter is a simple function of the mass action variable,

$$\overline{N} = \frac{1}{2} + \frac{1}{2}\sqrt{1 + 4X} \qquad (2)$$

and enters the *mass* or *molecular weight distribution* that within the model can be shown to be exponential (Cates and Candau, 1990; Ciferri, 2005). The *weight distribution* over the molecular weights, however, is peaked around $\overline{N}$.

This model, termed the *isodesmic assembly, multistage open association*, or *ladder model* (Brunsveld et al., 2001; Ciferri, 2002, 2005), is often applied to quasi one-dimensional, *linear* assemblies and found to be quite accurate all the way from the dilute to the very concentrated solutions and even in the melt state, irrespective of whether the chains are rigid or flexible (Ciferri, 2005). Interactions within the chains or between chains modify the predictions only very mildly and can be ignored for most practical applications (Cates and Candau, 1990), except in liquid–crystalline phases where a strong coupling between alignment and growth has been predicted theoretically (Cates and Fielding, 2006; Lü and Kindt, 2006; van der Schoot, 1996; Taylor and Herzfeld, 1993). Also, rings do not form in appreciable quantities if the chains are sufficiently rigid on the scale of the size of the monomer units (Cates and Candau, 1990; Ciferri, 2005; Porte, 1983).

The mass action variable X can be rendered into an experimentally more meaningful form. To do this, let X_* be a reference value for given values of the concentration ϕ_* and temperature T_*. Then, by a linear Taylor expansion of the dimensionless free energy βg around the reference temperature T_*, we have

$$X/X_* \sim (\phi/\phi_*)\exp[h_*(T/T_* - 1)/k_B T_*] \quad (3)$$

with h_* the binding *enthalpy* at the temperature $T = T_*$. For instance, we may choose to set $X_* = (2(1-f_*)^{-1/2} - 1)/4$ by the fraction polymerized material under the arbitrary reference conditions, f_*, often taken to be half-way point for which $f_* = 1/2$. This is done in Figures 3 and 4, representing the fraction polymerized material and the mean aggregation number as a function of X/X_*, showing their relatively weak sensitivity to the value of this parameter.

According to Equation (3), the mass action variable X/X_* depends strongly, that is, exponentially, on the temperature. Depending on whether the assembly is endo or exothermic, that is, whether $h_* \gtrless 0$,

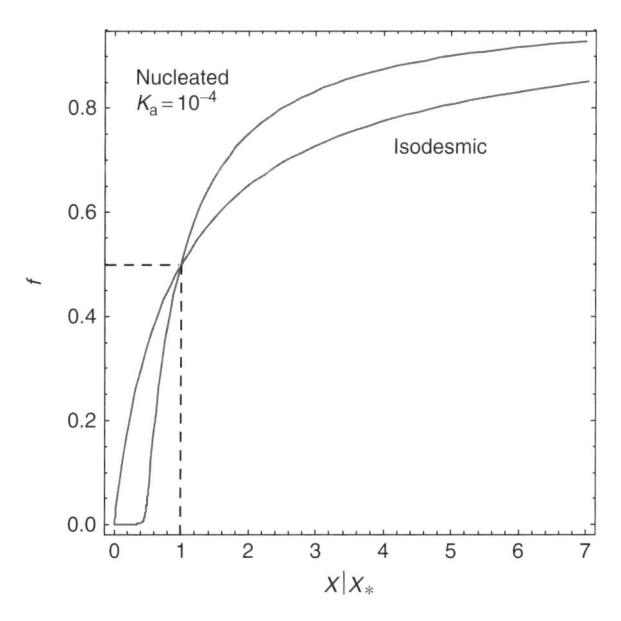

Figure 3 Fraction of material in the polymerized state, f, as a function of the mass action variable X relative to its value X_* at the half-way point $f = 1/2$. Indicated are predictions for the isodesmic and the self-catalyzed nucleated polymerization models. Activation constant of the nucleated polymerization $K_a = 10^{-4}$.

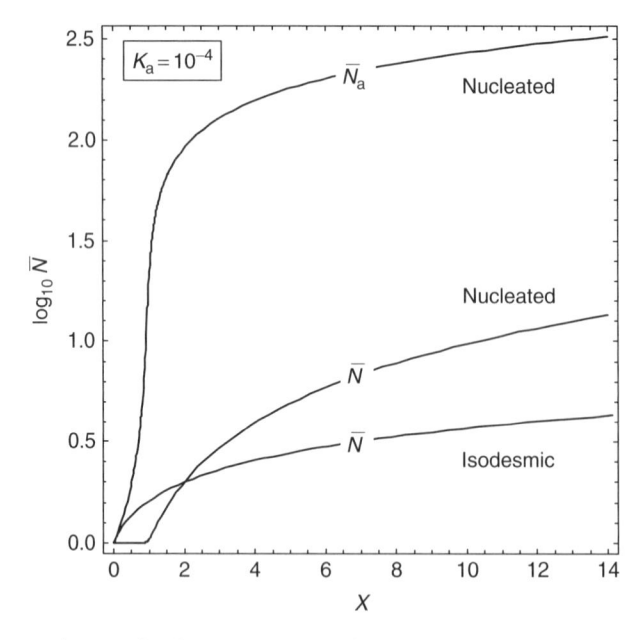

Figure 4 Mean degree of polymerization as a function of the mass action variable X for the isodesmic and the self-catalyzed nucleated polymerization models. $\overline{N}$ is the value averaged over all the monomers in the solution, $\overline{N}_a$ that averaged over the active material only. Activation constant of the nucleated polymerization: $K_a = 10^{-4}$.

the aggregates grow or shrink with increasing temperature. However, unless $|h_*/k_B T_*|$ is very large, the crossover from monomer- to the polymer-dominated temperatures occurs over many tens of degrees on the Kelvin scale. Typical values for $|h_*/k_B T_*|$ range from 25 to $75\,k_B T_*$, in themselves not large enough to produce a sharp crossover to the polymerized state (Würthner et al., 2004).

For the polymerization of the protein *g-actin* into *f-actin* fibers, to name but one example, the transition occurs at a sharp, well-defined temperature and does not seem to follow isodesmic polymerization (Greer, 2002; Hofrichter et al., 1974; Lomakin et al., 1996; Niranjan et al., 2003; Oosawa and Asakura, 1975; Šiber and Podgornik, 2007). In fact, isodesmic assembly cannot explain the huge aggregation numbers nor the highly co-operative and nucleated assembly often seen in supra-molecular polymers from biological origin, including actin, tubulin, and flagellin (Oosawa and Asakura, 1975) but also in biomimetic ones, such as those shown in Figure 6.

Indeed, to get very large aggregation numbers in the tens or hundreds of thousands requires very high free energies g of binding not much below 1 eV, being equivalent to approximately $40\,k_B T$. This, of course, is quite difficult to justify if the interactions between the monomers in an assembly are of physical origin, and typical values seem to be in the range 10–$20\,k_\mathrm{B}T$ (Cates and Candau, 1990; Ciferri, 2005). Higher values are possible, for example, if multiple hydrogen bonds are involved in the binding of each monomer, such as is the case in β-sheet fibril forming oligopeptides (real or synthetic) (Aggeli et al., 2001). We come back to the issue of how to interpret or even predict the quantity g below, a quantity that is of *microscopic* origin, that is, it depends on the chemical structure of the molecules involved as well as that of the solvent, and the way the solvent and the supramolecular polymer formers interact.

In the remainder of this brief overview, several mechanisms are discussed that (i) significantly enhance aggregation numbers and (ii) provide sensitive control over assembled and disassembled states of the supramolecular polymers. As we shall see, co-operativity is required but not sufficient for this purpose. Models at different levels of coarse-grained description shall be presented and a case is made for studying supramolecular assembly both as a function of concentration of molecular building blocks as of the ambient temperature. The latter methodology seems the more practical in distinguishing isodesmic from nucleated assembly. The relation between reversible polymerization in biological and supramolecular chemistry contexts shall be stressed and also their relation with phase transitions in condensed matter physics (Wheeler and Pfeuty, 1981).

2. NUCLEATED ASSEMBLY

Almost 30 years ago, Caspar suggested that switching between conformational states of molecules exerts self-control in the structure and action of protein assemblies in vivo (Caspar, 1980). In practice, this means that a protein molecule can exist in two (or more) conformational states, one of them *assembly inactive* and the other(s) *assembly active*. These conformers are in thermodynamic equilibrium, meaning that they can in principle interconvert. However, if the balance between them is very much in favor of the inactive state, a free energy has to be invested in order to form the assemblies. This free energy is gained by the interactions between the molecules that drive the assembly, part of which is lost to convert these molecules to their active state. If the free-energy cost of producing an active monomer is very high, then their self-assembly into supramolecular aggregates becomes highly co-operative and nucleated, that is, the transition between assembled and disassembled states is sharp and resembles a thermodynamic phase transition (Jahn and Radford, 2008; Scott, 1965; Tobolsky and Eisenberg, 1960; Wheeler and Pfeuty, 1981; Zhao and Moore, 2003).

We can distinguish between two extremes of nucleated assembly, one where the addition of each monomer to the growing assembly is associated with a free-energy cost, $g_a > 0$, of the conversion from the inactive to the active state, and one where only the conversion of the first one costs free energy but the subsequent ones do not (or not as much as the first one) (see also Figure 2b.). The latter kind of binding can be called self-catalyzed or *autosteric* because the conversion of the first molecule catalyzes the conversion of the next one bound to it and so on (Caspar, 1980). [Recent calculations show that allosteric and therefore also autosteric enhancement of binding may result from the coupling of internal fluctuations of the molecules involved in the binding (Hawkins and McLeish, 2006).] It is not so easy to distinguish between the two scenarios of nucleated assembly in practical situations. Both types have been discussed in various kinds of context, most notably the polymerization of a wide variety of proteins (Douglas et al., 2008; Edelstein-Keshet and Ermentrout, 1998; Erickson and Pantaloni, 1981; Goldstein and Stryer, 1986; Greer, 2002; Martin, 1996; Oosawa and Asakura, 1975), so we only outline the predictions of the simplest of coarse-grained models put forward.

Let us start with the nucleated assembly that is *not* self-catalyzed. It turns out useful to distinguish between the mean aggregation number of all the material in the solution, $\overline{N}$, from that in which only the activated species is considered and that we denote by $\overline{N}_a$. If we define the equilibrium constant $K = \exp[-\beta g]$ with g as the binding free energy, and introduce the nucleation constant $K_a = \exp[-\beta g_a]$, then under conditions of thermodynamic equilibrium, mass action gives (Aggeli, 2001; Ciferri, 2005; Nyrkova et al., 2000; Tobolsky and Eisenberg, 1960)

$$f = K_a \overline{N}_a^2 / (1 + K_a \overline{N}_a^2) \tag{4}$$

for the fraction active material in terms of the mean aggregation number, $\overline{N}_a$, of active (polymerized) material that obeys the relation

$$XK_a = 1 - \overline{N}_a^{-1} + K_a(\overline{N}_a - 1)\overline{N}_a \tag{5}$$

We furthermore have the identity

$$\overline{N}^{-1} = 1 - f + f\overline{N}_a^{-1} \tag{6}$$

The smaller the value of the activation constant, K_a, the sharper the transition from the monomer- to the polymer-dominated regime (Douglas et al., 2008; van Jaarsveld and van der Schoot, 2007; Scott, 1965; Tobolsky and

Eisenberg, 1960; Wheeler and Pfeuty, 1981). The distribution of the sizes of the assemblies is bimodal over the inactive and active states of the material, where the latter is an exponential function of the aggregation number. The weight distribution is also bimodal and has peaks centered at the aggregation numbers of unity and $\overline{N}_a$ (Zhao and Moore, 2003).

If we take the formal limit $K_a \to 0$, which in practice implies the less restrictive conditions that $K_a \ll 1$ and that the polymerization is highly co-operative, we find that $f \sim 1 - (K_aX)^{-1}$ and $\overline{N}_a \sim \sqrt{(K_aX - 1)/K_a}$ for $K_aX > 1$, and $f \sim 0$ with $\overline{N}_a \sim 1$ for $K_aX < 1$. We can define a critical polymerization concentration $\phi_p = \phi_p(T) \equiv 1/K_aK$ at fixed temperature T such that $f \sim 1 - \phi_p/\phi$ and $\overline{N}_a \sim \sqrt{(\phi - \phi_p)K}$ for $\phi > \phi_p$. This has to be compared with the predictions for the isodesmic model that gives $\overline{N} \sim \sqrt{\phi K}$ for $X = \phi K \gg 1$. So, for both kinds of equilibrium polymerization, we find the functional dependence of the length of the polymers on the equilibrium constant K and on the concentration ϕ obey the same asymptotic relation deeply in the polymerized regime, that is, at very high concentrations $\phi \gg \phi_p$.

If $K_a \ll 1$, we are able to define a sharp polymerization temperature $T_p = T_p(\phi)$ that is a function of the concentration ϕ, such that

$$f \sim 1 - \exp[-h_p/k_B T_p(T/T_p - 1)] \qquad (7)$$

for $T > T_p$ if the enthalpy $h_p > 0$ or $T < T_p$ if $h_p < 0$, and $f \sim 0$ otherwise. Note that (i) the enthalpy here is defined as the sum of the activation and elongation steps and (ii) T_p is not to be confused with the temperature at the half-way point T_*. A more accurate estimate for the degree of polymerization valid near the polymerization temperature $T = T_p$ is presented at the end of this section. Because the transition can be very sharp indeed, sufficiently many data points have to be taken to be able to tell the difference between isodesmic and nucleated assembly. This is obviously more practical in the temperature than in the concentration domain because the former type of measurement can be automated.

If the nucleated assembly is self-catalyzed, then this modifies only the role of the equilibrium constant K in the non-self-catalyzed model and has to be replaced by the ratio K/K_a. This means that we again obtain for the fraction polymerized material $f = K_a \overline{N}_a^2/(1 + K_a \overline{N}_a^2)$ but that the degree of polymerization averaged over the active material only now obeys (Ciferri, 2005)

$$X = 1 - \overline{N}_a^{-1} + K_a(\overline{N}_a - 1)\overline{N}_a \qquad (8)$$

So, if we again take the limit $K_a \to 0$ to make the polymerization highly co-operative, we find that $f \sim 1 - X^{-1}$ and $\overline{N}_a \sim \sqrt{(X-1)/K_a}$ for $X > 1$, and $f \sim 0$ with $\overline{N}_a \sim 1$ for $X < 1$. In this case, the activation step enhances the growth of the assemblies by a factor $1/\sqrt{K_a}$ for large $X \gg 1$ and reduces the critical concentration by a factor K_a to $\phi_p \equiv 1/K$. We have plotted the fraction active material and the degree of polymerization in Figures 3 and 4 as a function of the value of X relative to the half-way point X_*, showing the strongly enhanced growth of nucleated supramolecular polymerization when compared to that of the isodesmic mode of aggregation.

Now, for both kinds of catalyzed and non-catalyzed nucleated assembly we can write $f \sim 1 - \phi_p/\phi$ and $\overline{N}_a \sim \sqrt{(\phi/\phi_p - 1)/K_a}$ if $\phi > \phi_p$, and $f \sim 0$ and $\overline{N}_a \sim 1$ if $\phi < \phi_p$ in the limit $K_a \to 0$, implying that it is indeed impossible to distinguish between the self-catalyzed and non-self-catalyzed models. Both models predict an actual phase transition in the limit $K_a \to 0$, with a heat capacity that jumps at $\phi = \phi_p$ or, equivalently, at the polymerization temperature $T = T_p$. The jump is typical for mean field theories (Chaikin and Lubensky, 1995). In reality, K_a is not vanishingly small and the polymerization transition not infinitely sharp albeit that it should still be accompanied by a significant peak in the heat capacity (Douglas et al., 2008; Greer, 2002; van Jaarsveld and van der Schoot, 2007).

For small $K_a \ll 1$, the fraction polymerized material can in the vicinity of the polymerization temperature be described by the approximate expression (Jonkheijm et al., 2006)

$$f \sim K_a^{1/3} \exp\left[\frac{2}{3} K_a^{-1/3} h_p (T/T_p - 1)/k_B T_p\right] \quad (9)$$

where $|T/T_p - 1| \ll 1$, allowing for the determination of K_a provided h_p and T_p are fixed by fitting the asymptotic relation $f \sim 1 - \exp[-h_p(T/T_p - 1)/k_B T_p]$ valid more deeply in the polymerized regime (see Figure 5). The same

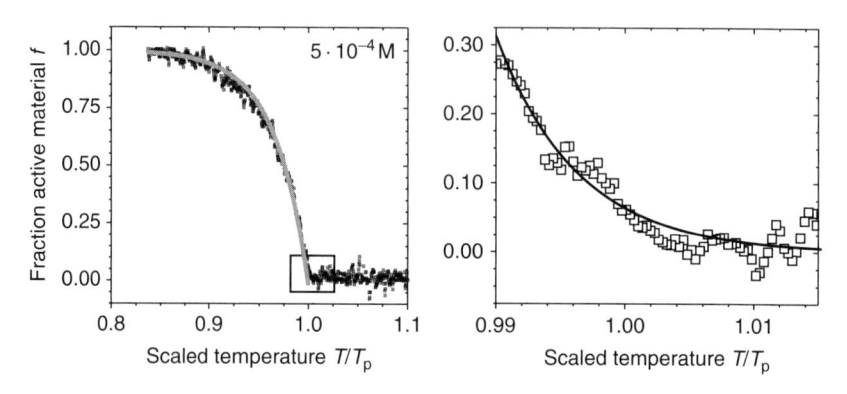

Figure 5 Left: determination of the polymerization temperature T_p by fitting the approximate relation Equation (7) indicated with the drawn line to the experimental data points for $T < T_p$ obtained for an oligo(phenylene vinylene) similar to compound 2 of Figure 6 in dodecane (Jonkheijm et al., 2006). Around T_p, the more accurate expression Equation (9) can be used to obtain the value of $K_a \approx 10^{-4}$ at $T = T_p$.

quantities can be determined isothermally by varying the concentration although this is more cumbersome as already noted above. In that case, one could invoke the expression

$$f \sim K_a^{1/3} \exp\left[\frac{2}{3} K_a^{-1/3}(\phi/\phi_p - 1)\right] \tag{10}$$

for $|\phi/\phi_p - 1| \ll 1$ and fix ϕ_p by fitting the expression $f \sim 1 - \phi_p/\phi$ that applies more deeply in the polymerized regime.

The self-catalyzed model is a simplification of the actin polymerization model of Oosawa and Kasai (1962) [and more recent elaborations of it (Niranjan et al., 2003)], and both equilibrium constants K and K_a, and hence the free energies g and g_a can in principle be obtained by fitting the theory to assembly experiments. Typical values of g for, for example, actin vary between -10 and $-20\,k_BT$ and g_a between $+2$ and $+3\,k_BT$ (Oosawa and Asakura, 1975; Oosawa and Kasai, 1962). For the biomimetic compound oligo(phenylene vinylene) similar to compound 2 shown in Figure 6, dissolved in alkane solvents, similar values were found for g but much larger ones for g_a of $+8$ to $+10\,k_BT$, making the supramolecular polymerization of this compound an extremely highly co-operative process (Jonkheijm et al., 2006).

In fact, studies on oligo(phenylene vinylene)s have exposed an unexpected role of *solvent* in nucleated assembly. Usually, the solvent is thought to be in some sense passive and hence in effect ignored in theoretical studies by absorbing the influence of the solvent in the free energies of binding. However, by (i) using the fact that exactly at the polymerization point, $X = 1$, the average degree of polymerization of the activated species obeys $\overline{N}_a \sim K_a^{-1/3}$ for $K_a \ll 1$, and (ii) comparing values of $\overline{N}_a$ obtained by fitting the activation constant $K_a \approx 10^{-4}$ to assembly experiments for a homologous series of alkanes, Jonkheijm and collaborators found an odd–even effect in the aggregate size at the transition point with increasing length of the solvent molecules (Jonkheijm et al., 2006).

Odd–even effects are usually only encountered in the large scale, that is, *macroscopic* ordering of molecules, for example, in liquid crystals (Chandrasekhar, 1992). This suggests that solvent molecules must not only actively participate in the co-operative formation of the helical assemblies but somehow physically be part of the ordered assembly. The connection between solvent size and the stability of assemblies was also pointed out by Bouteiller and collaborators (2007).

3. KINETICS OF NUCLEATED ASSEMBLY

A tell-tale sign of nucleated assembly is the existence of a critical concentration below which assemblies do not form in measurable quantities. Another is the observation of hysteresis in assembly and disassembly experiments

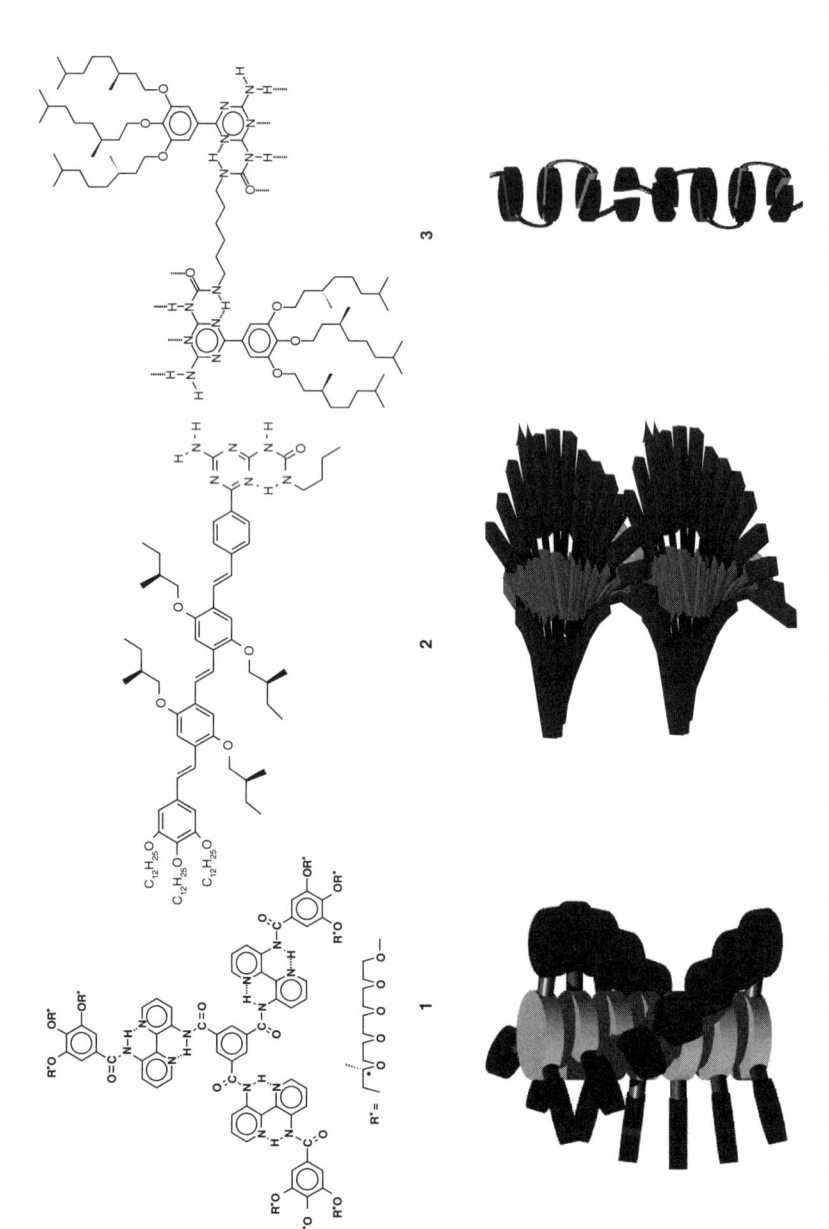

Figure 6 Examples of compounds that in selective solvents produce biomimetic supramolecular polymers that depending on the conditions exhibit a co-operative intramolecular ordering transition from random to highly ordered helical stacks or a nucleated polymerization transition from oligomeric prenuclei to very long, helical polymeric objects (ten Cate and Sijbesma, 2002; Dankers and Meijer, 2007; van Gorp et al., 2002; Brunsveld, 2001; Jonkheijm, 2005; Hirschberg, 2001).

and of *sigmoidal kinetics*, that is, the existence of what can be viewed as a lag time before assembly actually sets in. Kinetic experiments are interesting in their own right because they provide insight in reaction pathways and therefore in the free energy landscape of the system at hand (Auer et al., 2007). Unfortunately, quite often kinetic studies are performed without a clear view of, or reference to, the underlying "phase diagram" of monomer- and polymer-dominated regimes, making the interpretation of experimental observations not so straightforward.

There is a large body of theoretical literature on the kinetics of reversible linear supramolecular polymerization, in particular on the isodesmic assembly of linear surfactant micelles (Cates and Fielding, 2006; O'Shaughnessy and Vauylonis, 2003; O'Shaughnessy and Yu, 1995; Padding and Boek, 2004). These works focus on different kinetic mechanisms known as *reversible scission/recombination, end interchange*, and *end association/evaporation* (Dubbeldam and van der Schoot, 2005; Marques et al., 1993; Marquesa et al., 1994). Nucleated linear assembly has attracted quite a bit of attention too, albeit mostly in the context of the polymerization of proteins (Attri et al., 1991; Edelstein-Keshet and Ermentrout, 1998; Goldstein and Stryer, 1986; Hiragi et al., 1990; Lomakin et al., 1996; Oosawa, 1970; Oosawa and Asakura, 1975; Oosawa and Kasai, 1962; Powers and Powers, 2006; Sept and McCammon, 2001). Almost all theoretical studies available in this field focus on *end association/evaporation* kinetics, where single monomers attach and detach from the chains, but *reversible scission/recombination* kinetics, involving chain breakage and fusion, has also very recently been investigated (Nyrkova and Semenov, 2007).

The starting point of the kinetic models are seemingly simple reaction rate equations that deal with addition and/or shedding of material from polymers, but it so happens that these equations are (i) nonlinear and (ii) there are in principle infinitely many of them. (There is no obvious upper limit to their size other than that set by the system size.) This makes the analysis non-trivial, notwithstanding that in some cases useful analytical results can be obtained, exact or approximate (Dubbeldam and van der Schoot, 2005; Nyrkova and Semenov, 2007; Oosawa and Kasai, 1962; O'Shaughnessy and Yu, 1995; Powers and Powers, 2006; van der Linden and Venema, 2007; Wentzel, 2006). Here, we do not wish to indulge in the often times highly technical kinetic analysis but instead focus on the prevalent physics at hand. As mentioned already, nucleated assembly resembles a thermodynamic phase transition. It makes sense, then, to make use of this fact and apply the phenomenological theory of phase transitions to describe the assembly and disassembly kinetics in this kind of system (Chaikin and Lubensky, 1955; Hohenberg and Halperin, 1977; van der Schoot and Zandi, 2007).

In the phenomenological theory of phase transitions, it is customary to attribute order parameters to the relevant thermodynamic quantities associated with the macroscopic transition of the state of the material. In our case, we are dealing with monomers that self-assembly to form polymer-like objects. There is inactive material and active material, separated by a conversion reaction. It seems sensible, therefore, to define two order parameters, not one, and both are nonconserved. This means that they do not obey any balance equation or sum rule, such as the law of conservation of mass.

Let the order parameter S_1 be associated with the fraction active material, and be defined as $S_1^2 \equiv f$. For the second order parameter we choose $S_2 \equiv \overline{N}_a(\overline{N}_a - 1)K_a/X$ as a measure of the degree of polymerization of the active material and presume that the polymerization is highly co-operative, so $K_a \to 0$. The *macroscopic* Landau free-energy density F that describes the nucleated assembly now reads

$$\beta F = -(1 - X^{-1})S_1^2 + S_1^4 + \frac{1}{2}S_2^2 - S_1^2 S_2 \quad (11)$$

It is easy to verify that in equilibrium, that is, under condition where $\partial f/\partial S_1 = \partial f/\partial S_2 = 0$, $f = 0$ for $X \leq 1$ and $f = 1 - X^{-1}$ for $X \geq 1$. Furthermore, in that case $S_2 = S_1^2$, so under conditions of equilibrium $f = \overline{N}_a(\overline{N}_a - 1)K_a/X$ as expected from the theory of Section 2.

The obvious ("model A") set of relaxational kinetic equations are (Hohenberg and Halperin, 1977)

$$\frac{\partial S_1}{\partial t} = -\Gamma_1 \frac{\partial \beta F}{\partial S_1} \quad (12)$$

and

$$\frac{\partial S_2}{\partial t} = -\Gamma_2 \frac{\partial \beta F}{\partial S_2} \quad (13)$$

where Γ_1 and Γ_2 are phenomenological relaxation rates having no precise physical meaning (van der Schoot and Zandi, 2007). Reasonably, Γ_1 must be related to the speed at which material converts from the inactive to active state and Γ_2 to that at which the activated materials grows into polymers. So, one might call Γ_1 the activation rate and Γ_2 the elongation rate. Even at this level of coarse graining, where we have lost all detailed information about the reaction pathways, the set of coupled nonlinear equations is not easily solved analytically except in two extreme limits.

If the order parameter S_1 relaxes instantaneously, implying fast conversion of the inactive species into the active one and slow elongation, we find for the time dependence of the fraction active material

$$f(t) = f(\infty) - (f(\infty) - f(0))\exp[-\frac{1}{2}\Gamma_2 t] \quad (14)$$

In this case, the relaxation is single exponential and there is no sigmoidal kinetics.

If, on the other hand, S_2 relaxes instantaneously, so the conversion to the active species is now the rate-determining step, then the kinetics can be sigmoidal depending on whether the "quench" (a sudden change of solution conditions) induces a net polymerization or a net depolymerization of material, with

$$f(t) = \frac{f(0)f(\infty)}{(f(\infty) - f(0))exp[-t/t_*] + f(0)} \quad (15)$$

and t_* an inverse rate (a time)

$$t_* = \frac{X}{4(X-1)\Gamma_1} \quad (16)$$

that diverges if $X \to 1$. This makes the polymerization kinetics slow if the system is quenched near the polymerization point. For $X \gg 1$, so for high concentrations, $t_* \sim 1/4\Gamma_1$ levels off in accord with a more elaborate reaction rate theory (Powers and powers, 2006).

In Figure 7, we have plotted the response of the fraction active material to quench promoting assembly and (partial) disassembly. The model seems

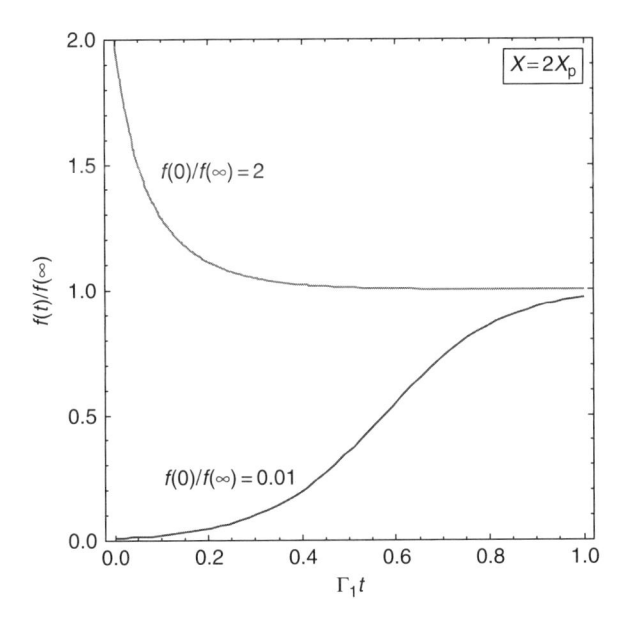

Figure 7 The fraction polymerized material f as a function of the dimensionless time $\Gamma_1 t$ according the kinetic Landau model discussed in the main text, with Γ_1 the nucleation rate. Shown are results valid in the limit where the nucleation reaction is rate limiting, for a quench to $X/X_p = 2$ where in equilibrium $f = 0.5$. Depolymerization is much faster than polymerization.

to capture the fundamentally different assembly and disassembly behavior of nucleation-limited polymerization as seen experimentally (Goldstein and Stryer, 1986; Hiragi et al., 1990; Hofrichter et al., 1974; Lomakin et al., 1996; Wang et al., 1989; Wentzel, 2006). The former is characterized by sigmoidal growth kinetics and the latter by in essence nearly instantaneous fragmentation, that is, slow assembly and fast disassembly. The difference between these diminishes with decreasing asymmetry in the nucleation and elongation rates.

If the assembly is nucleation limited, the obtained relaxation rate $t_*^{-1} > 0$ is positive for $X > 1$, implying net polymer growth if $f(0) < f(\infty)$ and net disassembly if $f(0) > f(\infty)$ (see Figure 8). We find a negative rate $t_*^{-1} < 0$ for $X < 1$, signifying complete polymer disassembly because by construction $f(0) \geq f(\infty) = 0$. The slowing down of the polymerization kinetics near the polymerization point is reminiscent of the phenomenon of *critical slowing down* that occurs in phase transitions near a critical point and in fact near any bound of thermodynamic stability (Chaikin and Lubensky, 1995; Hohenberg and Halperin, 1977). The equivalent of the critical slowing down in the growth of seeds of actin has indeed been observed experimentally (Attri et al., 1991), but todate has not been

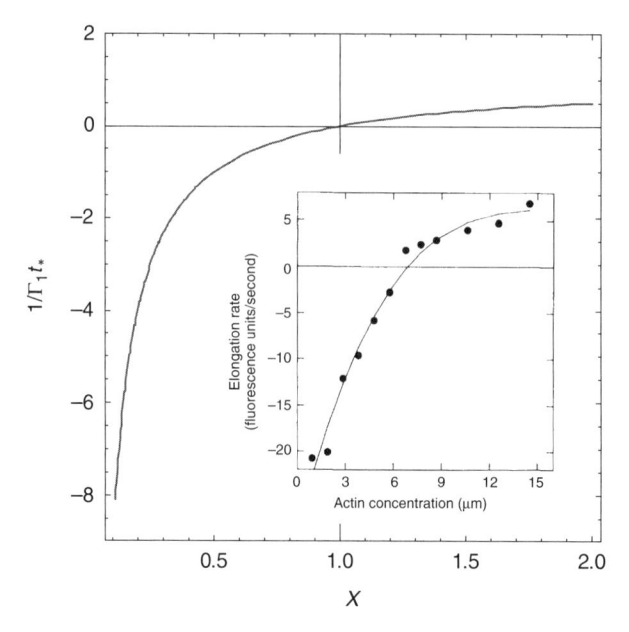

Figure 8 Dimensionless relaxation rate $1/\Gamma_1 t_*$ as a function of the dimensionless mass action variable X according the kinetic Landau model discussed in the main text, with Γ_1 the nucleation rate. Shown is the prediction in the limit where the nucleation reaction is rate limiting. Inset: experimental results from measurements on actin (Attri et al., 1991). Notice the zero growth at $X = 1$, the critical polymerization point.)

studied in supramolecular chemistry where more attention is paid to static properties and less to assembly kinetics.

In our model, this slowing down depends only on the distance from the polymerization point, not on the level of co-operativity. The reason is that the model presumes infinite co-operativity (or activation) for the fraction assembled material, which then behaves as if the polymerization transition were an actual phase transition. More elaborate calculations for reversible scission/recombination kinetics show that if $K_a > 0$: (i) the relaxation time t_* does not actually diverge at $X = 1$ albeit that it can become very large because $K_a \ll 1$ and (ii) that t_* is also sensitive to the precise value of K_a (Nyrkova and Semenov, 2007).

Although not commonly applied in the context of supramolecular assembly, an advantage of the Landau theoretical description is that it can easily be adapted to include a coupling of nucleated polymerizations to *macroscopic* phase transitions (Panizzay et al., 1998), including the isotropic-to-nematic transition. Although explored in the context of isodesmic assembly (van der Schoot, 1996; Taylor and Herzfeld, 1993), no theory has been put forward for nucleated assembly even though that the nematic transition is observed in, for example, actin solutions (Janmey et al., 1999) and in solutions of tape-forming oligopeptides (Aggeli et al., 2001). Perhaps even more interesting, the Landau description allows for straightforward application to the description of nonequilibrium phenomena such as shear banding under flow (Olmsted, 2008).

4. CO-OPERATIVITY AND NUCLEATION

Nucleated assembly emerges if intermediates stages between single monomers and long supramolecular polymers involve high-energy species. A high-energy intermediate can be a single assembly-active conformer but also a conglomerate of molecules that undergoes a conformational and bonding transition. In the cylindrical tobacco mosaic virus, for instance, conformational switching between a disk assembly consisting of 34 proteins and a protohelical lockwasher structure seems crucial in the co-operative encapsulation of the viral RNA (Butler, 1999). The switching involves not only a conformational change of the individual monomer units but also the way that they bond in the assembly. Calculations suggest that the free energy difference between a disk and a lock washer at neutral pH amounts to $1 k_B T$ per protein subunit, corresponding to a activation energy g_a of $34 k_B T$ for the complete protohelix (Kegel and van der Schoot, 2006). For actin, a similar scenario seems to be followed albeit that the intermediates involved in the nucleation event involve low-aggregation number species, that is, dimers and trimers (Janmey et al., 1999; Oosawa and Kasai, 1962; Oosawa and Asakura, 1975).

Biomimetic molecules such as those shown in Figure 6 behave very much like tobacco mosaic virus coat proteins in that the building blocks can convert between free monomers in solution, highly ordered helical aggregates and disordered linear assemblies, as evidenced, for example, by UV-vis, CD, NMR, and fluorescence decay spectroscopy (Brunsveld, 2001; Jonkheijm, 2005). For oligo(phenylene vinylene) in alkane solvents, the polymerization transition is sharp and seems to involve a helical transformation of disordered linear assemblies consisting of about 30 monomers (Jonkheijm and van der Schoot, 2006). For the discotic compound 1 shown in Figure 6, the polymerization, which can also switch between helical and nonhelical polymerized states, on the other hand, seems isodesmic yet the transition from disordered to helical state is very co-operative. Clearly, a less coarse-grained model than the nucleated assembly model is required to explain this.

A model that can describe conformational ordering transitions observed in supramolecular polymers is the so-called *self-assembled Ising chain*. It marries a simple mass action theory to the one-dimensional Ising model of ferromagnetism (Chaikin and Lubensky, 1995), which itself can be mapped onto the well-known Zimm–Bragg theory for the helix-coil transition in biopolymers (Ciferri, 2005; van Gestel, 2004a; van Gestel et al., 2003a; van der Schoot et al., 2000; Weiss and Terech, 2006). Basic ingredients are three free energies of bonding associated with two bonded states, so one more than the simple nucleated assembly model. To describe the bonded state of two neighboring monomers in an assembly, we assign an order parameter $S_i = \pm 1$ to each bond $i = 1, \ldots, N-1$ of an assembly consisting of N monomers. For instance, if $S_i = +1$ then the bond is a helical one and if $S_i = -1$ then it is nonhelical.

For the nonhelical bonded state, we assign a free energy g and for the helical one a free energy $h + g$. If $h < 0$, then the helical bond is more favorable than the nonhelical bond, and if $h > 0$ the opposite is true. It is reasonable to presume that in order to accommodate a particular kind of bond with a neighbor, the conformational state of the monomer must be affected. This implies that if a monomer is involved in two types of bond (helical and nonhelical) with its two neighbors, this should lead to *conformational frustration*. To account for this, a free energy penalty $j > 0$ is introduced every time a molecule has to accommodate two unequal types of bonding. Within the model, the internal energy of a single aggregate then obeys

$$E(N) = -\frac{1}{2} j \sum_{i=1}^{N-2} [S_i S_{i+1} - 1] + \frac{1}{2} h \sum_{i=1}^{N-1} [S_i + 1] + g(N-1) \qquad (17)$$

which depends on the bound state of the monomers $\{S_i\}$. Note that j takes over the role of the coupling constant in the Ising model, and h that of the magnetic field (Chaikin and Lubensky, 1995).

From statistical mechanics, we deduce that the (dimensionless) number density $\rho(N)$ of aggregates of degree of polymerization N must obey

$\rho(N) = Z(N)\lambda^N$ with $Z(N) = \Sigma_{\{S_i = \pm 1\}} \exp[-\beta E(N)]$ the partition function of a single aggregate and λ the fugacity of the monomers (Ciferri, 2005). The latter is the Boltzmann weight of the chemical potential of the monomers and fixed by the conservation of mass, that is, by the condition that the overall concentration of material is fixed, $\phi = \Sigma_{N=1}^{\infty} N\rho(N)$. The partition function can be calculated exactly and allows one to calculate the mean aggregate size, $\overline{N}$, as well as the conformational state of the assemblies, say, the fraction helical bonds $f \equiv (\langle S \rangle + 1)/2$. Here, the ensemble averaging $\langle \ldots \rangle$ is implied over all bonds of all assemblies in the solution.

In the model, co-operativity enters in two ways: (i) through the conformational frustration that dictates the value of the free-energy penalty j and (ii) through the boundary conditions that we impose on the preferred state of the first and last bond, $S_i = \pm 1$ for $i = 1, N - 1$. The coupling constant j produces correlations between bonded states, implying on average larger consecutive sequences of helical or nonhelical conformers the larger the value of j, and a sharper (more co-operative) transition from the nonhelical to the helical state of the assemblies. The boundary conditions dictate the extent to which the polymerization transition becomes nucleated. Because the first and last monomers are less strongly bound to the assemblies than monomers in the middle part of the chains, it seems plausible that either one or both should be nonhelical. Fitting of the theory to experiments confirms this scenario for a number of compounds, including compounds 1 and 2 of Figure 6 (Ciferri, 2005; van Gestel et al., 2003a; Weiss and Terech, 2006).

If the first bond can for whatever reason only be of the nonhelical (disordered) kind, then a helical polymer has to be nucleated if this nonhelical bond represents an excited state, so $g > 0$. The size of the critical nucleus depends in that case on how favorable the helical bond is relative to the nonhelical bond, in other words, on how strongly negative h is, and on the frustration free energy $j > 0$ because there is at least one monomer involved in two types of bonding. Within the model, the assemblies of size $2 < N < 2 - (j+g)/(g+h)$ with $h < -g < 0$ are high-energy structures and are therefore statistically suppressed.

Three regimes emerge from the model, where monomers, nonhelical polymers, and helical polymers dominate. These regimes that have indeed been observed experimentally in a variety of systems, such as for those molecules depicted in Figure 6 and tobacco mosaic virus (Kegel and van der Schoot, 2006).

1. If $X = \phi \exp[-\beta g] \gg 1$, we have long self-assembled polymers that obey simple *isodesmic* polymerization. For $h > 0$, the assemblies are disordered whilst for $h < 0$ they are ordered. The conformational transition at $h = 0$ is sharper and more *co-operative* the larger the frustration free energy j. The fraction helical bonds f that can be obtained from circular dichroism

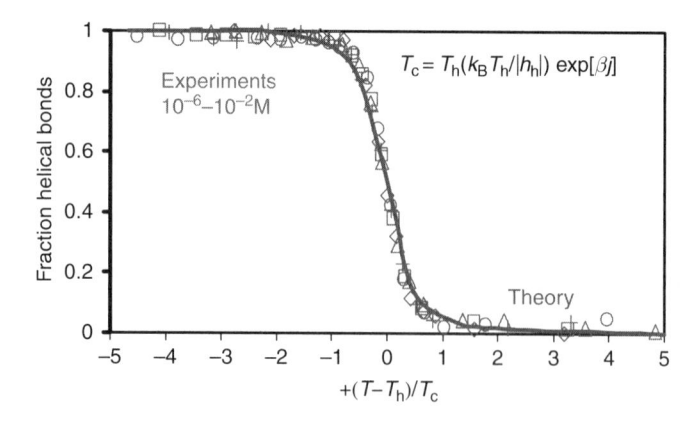

Figure 9 Fraction helical bonds as a function of the temperature T of the discotic compound 1 of Figure 6 in the solvent n-butanol. Symbols indicate experiments done in the range of concentrations of 10^{-6}–10^{-2} M (van der Schoot et al., 2000). The temperature has been scaled to the helical transition temperature T_h and a temperature $T_c = T_h(k_B T_h/|h_h|)\exp(j/k_B T_h)$ that depends on the frustration energy j and the excess enthalpy of the helical bond h_h. Remarkably, data taken over four orders in concentration collapse onto the theoretical curve, indicated by the drawn line.

spectroscopy (van der Schoot et al., 2000) follows a universal law as a function of the helical transition temperature T_h and a temperature scale T_c that depends on the coupling constant j and the excess enthalpy of the helical bond h_h. See Figure 9.

2. If $h > 0$, then helical assemblies do not form. There is an *isodesmic* polymerization from monomers to "disordered" supramolecular polymers that takes place around $X \approx 1$. The theory of Section 1 applies and the equivalent equilibrium constant obeys $K = \exp[-\beta g]$. Its value can be probed by means of radiation scattering, fluorescence decay, and UV absorption spectroscopy (Brunsveld et al., 2001; Jonkheijm et al., 2006).

3. If $X \ll 1$, disordered (nonhelical) assemblies do not form in any appreciable quantities. For $h < 0$, there is a polymerization transition from monomers to helical assemblies that is of the self-catalyzed nucleation type provided $\beta j \gg 1$. In the language of the coarse-grained self-catalyzed nucleated assembly model, the transition takes place near $X_p \approx \exp[\beta h]$ and we are able to assign an activation constant $K_a \approx \exp[-\beta j + \beta h]$. The theory of Section 2 approximately applies.

In practice, the relevant control parameters are not the concentration ϕ, and free energies g, h, and j but, for example, the concentration ϕ, the temperature T, and the type of solvent. In more complex systems additional variables can become important. By varying ϕ and T, we influence all these parameters via the various enthalpies that are therefore amenable to experimental determination. This, however, is not a simple task because there are six free parameters associated with the three binding free energies: each free energy can be separated into contributions from the enthalpy and entropy. A very large number of very different kind of experiments have to be done as a function of ϕ and T to independently fix all the parameters (van Gestel et al., 2003a; van der Schoot et al., 2000).

To date, there are only a few molecules for which this has been done systematically, in particular the discotic and oligo(phenylene vinylene) compounds shown in Figure 6 (van Gestel et al., 2003; Jonkheijm et al., 2006), albeit that only for the former the diversity of the experiments allowed for independent cross-checking of the fitting parameters. The phase diagram of the discotic compound dissolved in n-butanol is depicted in Figure 10, showing good agreement between theory and experiment. The conformational switching exhibited by the assemblies at the helical transition temperature is highly co-operative and also resembles a phase transition. That this is indeed the case is illustrated in Figure 11, showing the tell-tale peak in the measured and calculated heat capacities (Brunsveld et al., 2000).

The model outlined in this section seems specific but in actual fact is very versatile. It applies not only to helical transitions in supramolecular polymers but can easily be adapted to model other types of structural reorganization (Bouteiller et al., 2005), chirality amplification (Palmans and Meijer, 2007) of both the sergeants-and-soldiers (van Gestel et al., 2003b), and the majority-rules type (van Gestel, 2004b) and the so-called *templated* assembly. We come back to the phenomenon of templated assembly below.

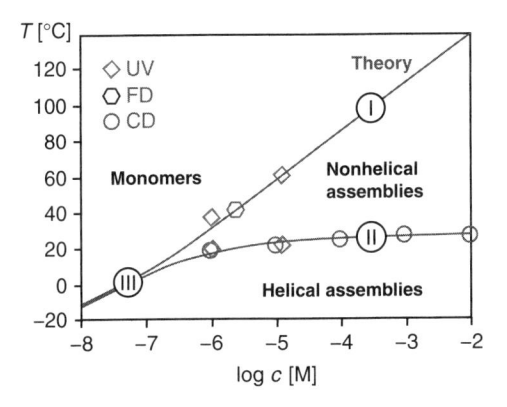

Figure 10 Assembly diagram of compound 1 of Figure 6 in the solvent n-butanol (adapted from van Gestel, 2004a; Weiss and Terech, 2005). Symbols represent results from UV-vis absorption, UV, fluorescence decay, FD, and circular dichroim spectroscopy, CD, and the drawn line the theoretical fits to the data. There are three types of transition: I, isodesmic polymerization; II, helical transition of long supramolecular polymers; and III, nucleated helical assembly of the monomer units.

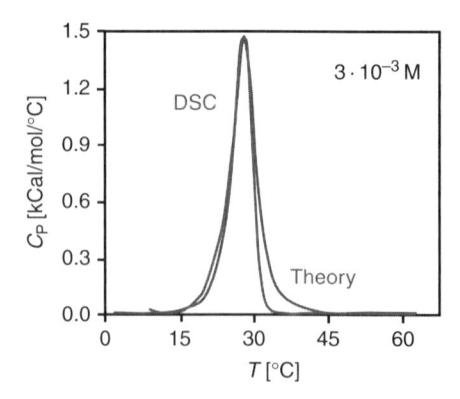

Figure 11 Comparison of prediction of the self-assembled Ising model with the experimentally determined heat capacity of compound 1 of Figure 5 around the co-operative helical transition point (Brunsveld, 2001; Brunsveld et al., 2000).

5. COARSE GRAINING REVERSED

The assembly models discussed in the preceding sections are presumably the simplest ones that one can set up for co-operative supramolecular polymerization. Their advantage is the relatively small number of adjustable parameters and conceptual simplicity. Disadvantage is the lack of a detailed description of the processes that actually led to the assembly becoming nucleated and that are system specific, that is, that depend on the details of the molecules involved and how precisely they interact.

The nucleated assembly model has the highest level of abstraction in which processes that potentially involve a hierarchy of high-energy intermediates are represented by as few as two free energies that give rise to four thermodynamic parameters: two binding enthalpies and two binding entropies. To interpret values of the model parameters as obtained by fitting the model to experimental data, one would need to go beyond the bounds of the coarse-grained model and explicitly consider the molecular details of the system and, in a way, "fine grain" the model post hoc.

Not surprisingly, post hoc fine graining is system specific and no *general* recipe can be provided how to do this. By way of illustration we discuss in more detail two relevant examples. The simplest situation arises if conformational switching upon binding occurs at the level of the individual monomeric building blocks. A molecular interpretation of the model in the light of experimental data is in that case straightforward. This turns out so for the assembly of short oligopeptide chains into β-sheet fibrils, a supramolecular polymerization that belongs to the class of nucleated assemblies we focus attention on (Nyrkova et al., 2000) (see Figure 1).

For this type of monomer unit, the polymerization requires the stretching of the chains in free solution for the hydrogen bonds to be able to form upon attachment of these chains to the growing β-sheet tape. The stretching of the coil-like chains reduces their conformational freedom and hence their entropy. Clearly, this loss of conformational entropy must be compensated for by the gain in free energy through the formation of the hydrogen bonds.

Within this kind of "reconstruction" of a molecular picture from the abstract model, an estimate based on ideal-chain statistics would suggest that the free energy cost associated with the bonding of the monomeric units would amount to $+(n-1)\varepsilon_a$, where n is the number of bonds involved in the backbone of the chain and $\varepsilon_a > 0$ a model-dependent constant of order unity. The parameter $\varepsilon_a > 0$ may itself again be quantified using, for example, a rotational isomeric state model commonly applied in polymer physics (Mattice and Suter, 1994). As for the driving force for the tape polymerization, let $m \propto n$ be the number of hydrogen bonds that each oligopeptide chain can form. One would surmise that the free energy of bonding must then be equal to $-m\varepsilon$, where $\varepsilon > 0$ is the free energy gained when a hydrogen bond is formed.

In conclusion, if an assembly consists of N bound oligopeptide chains, the overall free energy of binding would be equal to $N(n-1)\varepsilon_a - (N-1)m\varepsilon$. This in turn implies that in order to describe the equilibrium between the free oligopeptides and the very polydisperse tapes, we have in terms of the parameters of the nucleated assembly model of Section 2 $g_a = (n-1)\varepsilon_a > 0$ and $g = (n-1)\varepsilon_a - m\varepsilon < 0$ as a function of the length of the chains, n, and the number of hydrogen bonds, m, formed per chain. Note that the tapes self-assemble into a hierarchy of higher order structures including ribbons, fibrils, and fibers that require separate theoretical description outside of the scope of this review (Aggeli et al., 2001; Nyrkova et al., 2000).

In this particular example, estimates for both free energy parameters g and g_a of the coarse-grained model are obtainable from more detailed molecular models albeit that strictly speaking they depend not only on the chemical composition and structure of the molecular building blocks but in principle also on the solvent properties. It is important to stress again for it is often ignored, the solvent molecules not only drive the assembly but have been shown to play an active role in structural reorganizations of supramolecular assemblies (Bouteiller et al., 2005; Jonkheijm et al., 2006). Ideally, their influence should not be absorbed in adjustable parameters as is almost always done.

Much more sophisticated models are needed to explicitly deal with the role of the solvent and presumably the only sensible way to make headway here is by means of detailed, that is, atomistic computer simulations. Unfortunately, detailed computer simulations of the self-assembly of large polymeric objects are often not very practical because they require excessive computer processing times, in particular if the solvent molecules are explicitly included or if timescales are to be probed comparable to those of the actual assembly kinetics in experiments.

Nonetheless, atomistic or quasi-atomistic simulations remain highly valuable indeed because they (i) provide insight into the three-dimensional structure of the assemblies, (ii) produce estimates for the binding energies of the monomer units that can be used in coarse-grained models, (iii) aid guide model building at higher levels of coarse graining, and (iv) give insight in predominant assembly pathways (Auer et al., 2007; Chennubhotla et al., 2005; Cellmer, et al., 2007; Foster, 2002; Jahn and Radford, 2008; Lehn, 2004; Phelps et al., 2000; Sept and McCammon, 2001).

One might presume that the larger the molecular building blocks are, the more difficult a sensible theoretical description becomes. Whilst this is true at the atomistic level, the opposite is actually true for coarse-grained descriptions because these become *more* accurate the larger the molecules involved in the assembly. The reason is that with increasing size of the molecules that make up a supramolecular assembly, the number of chemical groups on them that are engaged in the intermolecular interactions also increases. This (in a sense) averages out the contributions of the individual chemical moieties to the overall binding free energy. It explains why interactions between protein molecules, which belong to the chemically most complex of molecules, can often quite successfully be described in terms of relatively simple, effective potentials in which chemical detail enters only statistically, for example, in some average of the surface properties of the molecules (Kegel and van der Schoot, 2004, 2006; Prinsen and Odijk, 2004; Sear, 2006).

For example, the main driving force for the self-assembly in aqueous solution of protein molecules to produce larger scale structures such as fibers or virus coats results from hydrophobic interactions between apolar patches or functional groups on the surfaces of these molecules (Lauffer, 1975). The hydrophobic interaction is of an entropic nature and becomes stronger with increasing temperature. It has its origin in the statistics of the short-range structure of fluid water. A plausible estimate for the bare binding free energy of two proteins would then be $-2\gamma a_h$ where $\gamma > 0$ is the macroscopic surface tension of the hydrophobic patch and a_h its area. This is a reasonable estimate: if two water-loathing surfaces of area a_h are removed from contact with water, this must liberate two times the surface energy of each patch (Kegel and van der Schoot, 2004, 2006).

However, proteins are almost always charged (Dello'Orco et al., 2005), for otherwise they would drop out of the solution, that is, phase separate macroscopically. Indeed, this is what usually happens near the isoelectric point of the proteins, i.e. their point of zero net charge. If two proteins bind to become part of an assembly, charged patches on them get on average closer together and hence repel each other when of the same sign. In other words, Coulomb repulsion between the proteins should actually oppose the formation of an assembly (Kegel and van der Schoot, 2004). The strength of the Coulomb repulsion depends on the net surface charge density, σ_c, that itself depends on the pKs of all water-exposed acidic and basic groups and on the solution pH (Kegel and van der Schoot, 2006).

An estimate of the free energy of the Coulomb repulsion between two bound proteins is $+k\sigma_c^2 a_c \lambda_B \lambda_D$, where k is a geometrical factor, σ_c the *effective* surface charge density, a_c the area of the water-exposed part of the proteins, λ_B the so-called Bjerrum length, and $\lambda_D = 1/\sqrt{8\pi\lambda_B I}$ the Debye screening length with I the ionic strength of the solution (Kegel and van der Schoot, 2004). The Bjerrum length is the distance at which two point charges interact with a Coulomb energy equal to the thermal energy $k_B T$, in water about 0.7 nm. For salts of monovalent cat- and anions, the Debye length obeys $\lambda_D \approx 0.3/\sqrt{c_s}$ nm with c_s the concentration of salt in moles per liter. It measures how far Coulomb interactions reach before they are "screened" (made ineffective) by the presence of small, mobile ions in the solution.

This gives the following estimate for the protein–protein binding energy in an assembly,

$$g \approx -2\gamma a_h + k\sigma_c^2 a_c \lambda_B \lambda_D \tag{18}$$

provided the salt concentration is not very low. Although derived for weak surface charge densities, it seems to be reasonably accurate too at surface charge densities high enough for nonlinear screening effects to take over (Šiber and Podgornik, 2007), and describes the pH, salt, and temperature dependence of the in vitro assembly of the coat proteins of a number of viruses (Kegel and van der Schoot, 2004, 2006). Here, the details of the chemical composition of the coat proteins come in via the parameters γ, a_h, σ_c, and a_c. The last three of these can be estimated from information present in protein databases, where the pH dependence of σ_c follows from the Henderson–Hasselbalch relation,

$$\sigma_c a_c = \sum_{i_b} (1 + 10^{\text{pH}-pK_{a,i_b}})^{-1} - \sum_{i_a} (1 + 10^{pK_{a,i_a}-\text{pH}})^{-1}, \tag{19}$$

where i_a and i_b refer to the acidic and basic groups on the surface of the protein with $pK_{a,j}$ the (negative logarithm of the) acid dissociation constants for $j = i_a, i_b$ (Kegel and van der Schoot, 2006). Additional dependence on the solvent conditions might come in via the activation step of the polymerization, that is, via g_a (Dell'Orco et al., 2005; Niranjan et al., 2003). Application of this kind of seemingly simplistic modeling to the in vitro

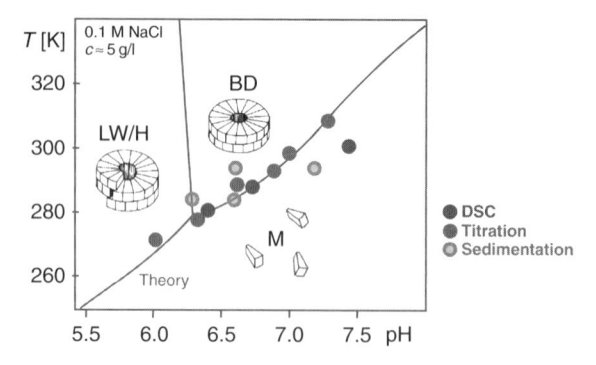

Figure 12 Diagram of assembled states of the coat protein of tobacco mosaic virus: M free monomers, BD cylindrical disks, LW/H protohelices and helices. Symbols: results from differential scanning microscopy DSC, titration, and sedimentation experiments, lines: theory. The theory is based on binding energy Equation (9) and presumes competing repulsive Coulomb and attractive hydrophobic interactions (Kegel and van der Schoot, 2006).

assembly of the coat protein of tobacco mosaic virus proves quite successful (see Figure 12).

The solution conditions enter the description through the pH, λ_B, and λ_D. Expression (19) provides an explanation for why self-assembled protein structures become more stable with increasing concentration of salt, which seems to be true for a wide variety of systems, including spherical and cylindrical viruses (Kegel and van der Schoot, 2004, 2006).

6. SUMMARY AND OUTLOOK

Reversible supramolecular polymerization takes place in solutions of melts or molecules that combine moieties of quite different polarity. This process can be driven by specific interactions, such as hydrogen bonds and/or generic types of (solvophobic) interaction. Very often, supramolecular polymerization is isodesmic. This points to the existence of a single predominant energy scale associated with the binding of a monomer unit to a growing polymer chain. Isodesmic self-assembly is a relatively weak function of the external conditions such as temperature or concentration and large aggregation numbers imply large binding energies.

Nucleated supramolecular polymerization, on the other hand, is a much more sensitive function of the external conditions. Indeed, a sharp polymerization point can be identified below which almost no material is in the polymerized state and above which the self-assembled polymers exhibit a strong variation of their mean size with varying concentration, temperature, and so on. Nucleated equilibrium polymerization requires the existence of at least two competing energy scales. A competing energy scale emerges naturally with conformational switching.

Three kinds of conformational switching may be identified:

1. Individual monomeric building blocks interconvert between assembly-inactive and assembly-active conformational states;
2. Assembled states interconvert between different conformers characterized by differently bound states of monomers that themselves remain inert;
3. Switching between different aggregated states resulting from a switching between different conformational states of the monomer units.

If the conformational switching is that between high-molecular weight polymeric species, then the structural transition between them can be highly co-operative but the assembly remains by and large isodesmic. So, co-operativity is a required but not a sufficient condition for creating nucleated supramolecular polymerizations.

Simple, coarse-grained models of nucleated assembly have a number of advantages over more detailed ones. First, coarse-grained models have only a limited number of free energy parameters. Second, analytical predictions can be obtained, if not in general then usually in practical limits. Third, they apply to a wide range of materials that may differ in detail but not in principle. This allows for a relatively straightforward fitting to experiments and comparison between materials, where the analysis can be made system specific by post hoc fine graining.

Both isodesmic and nucleated assembly produce polymeric species of a wide range of aggregation numbers and hence molecular weights. This makes self-assembled polymers very polydisperse and in fact is a result of the law of mass action. Self-assembled polymers cannot be made monodisperse, unless forced to form on a template of fixed and uniform length (Hannah and Armitage, 2004). The binding of the molecular building blocks to a "template" or "tape measure" molecule, itself a polymer, sets a maximum to the self-assembled polymer length. If all the binding sites on the monodisperse tape measure molecule are occupied, then the self-assembled polymer is by construction also monodisperse.

Complete coverage of the template molecule, for example, by hydrogen bonding is not so easily achieved, however, at least in theory. The reason is that even if the polymer is a three-dimensional object, templated assembly is in essence a one-dimensional adsorption process. One-dimensional ordering phenomena are dominated by spontaneous fluctuations that kill the fully ordered state (Chaikin and Lubensky, 1995), here the fully covered template. A simple model equivalent to the self-assembled Ising chain discussed in Section 4 shows that complete coverage should indeed be nearly impossible (McGhee and von Hippel, 1974), in particular if the binding free energy is not high and/or the monomeric building blocks do not vastly outnumber the binding sites on all of the template molecules.

Yet, nature has somehow worked itself around the entropy problem, because the coat proteins of linear viruses do spontaneously assemble around single-stranded viral RNA (Butler, 1999), and so do the so-called movement proteins involved in the shuttling of viral RNA from one cell to another (Citousky et al., 1990; Kiselyova et al., 2001). Interestingly, the length of the tails of tailed bacteriophages seems also to be set by co-assembly of the tail proteins with a tape measure molecule (Chiu et al., 1997; Kushner, 1969). The question arises how nature circumvents the entropy problem? In other words, what mechanisms are available that can make template-based assembly productive? The answer may plausibly again be found in high-energy intermediates (Butler, 1999). In biology, many molecular self-assembly processes are controlled by helper molecules such as scaffold molecules and molecular chaperones, usually but not exclusively proteins (Kentsis and Borden, 2004). Hence, the problem merits attention and a better understanding of it may boost the application of templates within supramolecular science.

Of course, with the eye on potential technological applications (Lehn, 2004) of supramolecular polymers in, for example, responsive gels (Kato et al., 2006), smart coatings (van der Gucht et al., 2004; Zweistra and Besseling, 2006) and opto-electronic devices (Flynn et al., 2003; Meegan et al., 2004; Würthner et al., 2004) and tissue engineering (Dankers and Meijer, 2007), more insight is also needed in how these materials respond to external pertubation, phase transitions, and so on (Auvray, 1981; Blankschtein et al., 1985; Feng and Fredrickson, 2006; Odijk, 1996; Oosawa, 1970). Whilst a large body of literature exists on how isodesmic supramolecular polymers behave, in particular in the form of linear surfactant micelles, considerably less is known about the macroscopic properties of nucleated self-assembled polymers. Systematic physicochemical investigation of the physical properties of the latter is much needed, and simple, coarse-grained models can prove valuable in analyzing experimental observations and further our understanding of this fascinating class of materials.

ACKNOWLEDGEMENT

I express my gratitude to Luc Brunsveld, Alberto Ciferri, Pim Janssen, Pascal Jonkheijm, Jeroen van Gestel, Willem Kegel, Tom McLeish, Bert Meijer, Albert Schenning, Rint Sijbesma, Maarten Smulders, Joachim Wittmer, and Roya Zandi for illuminating discussions on the subject of supramolecular polymers.

REFERENCES

Aggeli, A., Nyrkova, I.A., Bell, M., Harding, R., Carrick, L., McLeish, T.C.M., Semenov, A.N., and Boden, N. "Hierarchical self-assembly of chiral rod-like molecules as a model for peptide-sheet tapes, ribbons, bris and bers". *Proc. Nat. Acad. Sci. U.S.A.* **98**, 11857–11862 (2001).

Attri, A.K., Lewis, M.S., and Korn, E.D. "The formation of actin oligomers studied by analytical ultracentrifugation". *J. Biol. Chem.* **266**, 6815–6824 (1991).

Auer, S., Dobson, C.M., and Venndruscolo, M. "Characterisation of the nucleation barriers for protein aggregation and amyloid formation". *Hum. Front. Sci. Program J.* **1**, 137146 (2007).
Auvray, L. "Solutions de macromoles rigided: eets de paroi, de connement et d'orientation par un ecoulement". *Journal de Physique* **42**, 79–95 (1981).
Binder, W.H. "Polymeric Ordering by H-bonds. Mimicking Nature by Smart Building Blocks". *Monatsh. Chem.* **136**, 119 (2005).
Blankschtein, D., Thurston, G.M., and Benedek, G. "Theory of phase separation in micellar solutions". *Phys. Rev. Lett.* **54**, 955–958 (1985).
Bouteiller, L. "Assembly via hydrogen bonds of low molar mass compounds into supramolecular polymers". *Adv. Polym Sci.* **207**, 79112 (2007).
Bouteiller, L., Colombani, O., Lortie, F., and Terech, P. "Thickness transition of a rigid supramolecular polymer". *J. Am. Chem. Soc.* **127**, 8893–8898 (2005).
Bruinsma, R.F., Gelbart, W.M., Reguera, D., Rudnick, J., and Zandi, R. "Viral self-Assembly as a thermodynamic process". *Phys. Rev. Lett.* **90**, 248101, 1–4 (2003).
Brunsveld, L.Supramolecular chirality: from molecules to helical assemblies in polar media, Ph.D. Thesis, Eindhoven University of Technology (2001).
Brunsveld, L., Folmer, B.J.B., Meijer, E.W., and Sijbesma, R.P. "Supramolecular polymers". *Chem. Rev.* **101**, 4071–4097 (2001).
Brunsveld, L., Zhang, H., Glasbeek, M., Vekemans, J.A.J.M., and Meijer, E.W. "Hierarchical growth of chiral self-assembled structures in protic media". *J. Am. Chem. Soc.* **122**, 6175–6182 (2000).
Butler, P.J.G. "Self-assembly of tobacco mosaic virus: the role of an intermediate aggregatein generating both specicity and speed". *Philos. Trans. R. Soc. Lond. B Biol. Sci.* **354**, 537–550 (1999).
Caspar, D. "Movement and self-control in protein assemblies. Quasi equivalence revisited". *Biophys. J.* **32**, 103–138 (1980).
ten Cate, A.T., and Sijbesma, R.P. "Coils, rods and rings in hydrogen-bonded supramolecular polymers". *Macromol. Rapid Commun.* **23**, 10941112 (2002).
Cates, M.E., and Candau, S.J. "Statics and dynamics of worm-like surfactant micelles". *J. Phys. Condens. Matter* **2**, 6869–6892 (1990).
Cates, M.E., and Fielding, S.M. "Rheology of giant micelles". *Adv. Phys.* **55**, 799879 (2006).
Cellmer, T., Bratko, D., Prausnitz, J.M., and Blanch, H.W. "Protein aggregation in silico". *Trends Biotechnol.* **25**, 254–261 (2007).
Chaikin, P.M., and Lubensky, T.C. "Principles of Condensed Matter Physics". Cambridge University Press, Cambridge (1995).
Chandrasekhar, S. "Liquid Crystals", 2nd edition. Cambridge University Press, Cambridge (1992).
Chennubhotla, C., Reader, A.J., Yang, L.-W., and Bahar, I. "Elastic network model for understanding biomolecular machinery: from enzymes to supramolecular assemblies". *Phys. Biol.* **2**, S173–S180 (2005).
Chiu, W., Burnett, R.M., and Garcea, R.L. "Structural Biology of Viruses". Oxford University Press, Oxford (1997).
Ciferri, A. "Supramolecular polymerizations". *Macromol. Rapid Commun.* **23**, 511 (2002).
Ciferri, A. (Ed.), "Supramolecular Polymers". CRC Press, Boca Raton, FL (2005).
Citovsky, V., Knorr, D., Schuster, G., and Zambryski, P. "The P30 movement protein of tobacco mosaic virus is a single-strand nucleic acid binding protein". *Cell* **60**, 637–647 (1990).
Cohen Stuart, M.A. "Supramolecular perspectives in colloid science". *Colloid Polym. Sci.* **286**, 855864 (2008).
Dankers, P.Y.W., and Meijer, E.W. "Supramolecular biomaterials. A modular approach towards tissue engineering". *Bull. Chem. Soc. Jpn.* **80**, 20472073 (2007).
Dell'Orco, D., Xue, W.-F., Thulin, E., and Linse, S. "Electrostatic contributions to the kinetics and thermodynamics of protein assembly". *Biophys. J.* **88**, 1991–2002 (2005).

Douglas, J.F., Dudowicz, J., and Freed, K.F. "Lattice model of equilibrium polymerization. Understanding the role of cooperativity in self-assembly". *J. Chem. Phys.* **128**, 224901, 1–17 (2008).
Dubbeldam, J.L.A., and van der Schoot, P. "End-evaporation dynamics revisited". *J. Chem. Phys.* **123**, 144912, 1–14 (2005).
Edelstein-Keshet, L., and Ermentrout, G.B. "Models for the length distribution of actin laments". *Bull. Math. Biol.* **60**, 449–475 (1998).
Erickson, H.P., and Pantaloni, D. "The role of subunit entropy in co-operative assembly. Nucleation of microtubules and other two-dimensional polymers". *Biophys. J.* **34**, 293–309 (1981).
Feng, E.H., and Fredrickson, G.H. "Confinement of equilibrium polymers: a field-theoretic model and mean-field solution". *Macromolecules* **39**, 2364–2372 (2006).
Flynn, C.E., Lee, S.-W., Peele, B.R., and Belcher, A.M. "Viruses as vesicles for growth, organisation and assembly of materials". *Acta Mater.* **51**, 5867–5880 (2003).
Foster, M.J. "Molecular modelling in structural biology". *Micron* **33**, 365–384 (2002).
Gelbart, W.M., Ben-Shaul, A., and Roux, D. "Micelles, Membranes, Microemulsions and Monolayers". Springer, Berlin (1994).
van Gestel, J. Theory of helical supramolecular polymers, Ph.D. Thesis, Eindhoven University of Technology (2004a).
van Gestel, J. "Amplication of chirality in helical supramolecular polymers: the majority rules principle". *Macromolecules* **37**, 3894–3898 (2004b).
van Gestel, J., van der Schoot, P., and Michels, M.A.J. "Role of end effects in helical aggregation". *Langmuir* **19**, 1375–1383 (2003a).
van Gestel, J., van der Schoot, P., and Michels, M.A.J. "Amplication of chirality in helical supramolecular polymers". *Macromolecules* **36**, 6668–6673 (2003b).
Goldstein, R.F., and Stryer, L. "Cooperative polymerisation reactions". *Biophys. J.* **50**, 583–599 (1986).
van Gorp, J.J., Vekemans, J.A.J.M., and Meijer, E.W. "C3-Symmetrical supramolecular architectures: fibers and organic gels from siscotic trisamides and trisureas". *J. Am. Chem. Soc.* **124**, 14759–14769 (2002).
Greer, S.A. "Reversible polymerizations and aggregations". *Annu. Rev. Phys. Chem.* **53**, 173–200 (2002).
van der Gucht, J., Besseling, N.A.M., and Fleer, G.J. "Equilibrium polymers at interfaces: analytical self-consistent-field theory". *Macromolecules* **37**, 3026–3036 (2004).
Hagerink, J.D., Zubarev, E.R., and Stupp, S.I. "Supramolecular one-dimensional objects". *Curr. Opin. Solid State Mater. Sci.* **5**, 355–361 (2001).
Hannah, K.C., and Armitage, B.A. "DNA-templated assembly of helical cyanine dye aggregates: a supramolecular chain polymerization". *Acc. Chem. Res.* **37**, 845–853 (2004).
Hawkins, R.J., and McLeish, T.C.M. "Coupling of global and local vibrational modes in dynamic allostery proteins". *Biophys. J.* **91**, 2055–2062 (2006).
Hiragi, Y., Inoue, H., Sano, Y., Kajiwara, K., Ueki, T., and Nakatani, H. "Dynamic mechanism of the self-assembly process of tobacco mosaic virus studied by rapid temperature-jump small-angle x-ray scattering using synchrotron radiation". *J. Mol. Biol.* **213**, 495–502 (1990).
Hirschberg, K. Supramolecular polymers, Ph.D. Thesis, Eindhoven University of Technology (2001).
Hofrichter, J., Ross, P.D., and Eaton, W.A. "Kinetics and mechanism of deoxyhemoglobin S gelation: a new approach to understanding sickle cell disease". *Proc. Natl. Acad. Sci. U.S.A.* **71**, 4864–4868 (1974).
Hohenberg, P.C., and Halperin, B.I. "Theory of dynamic critical phenomena". *Rev. Mod. Phys.* **49**, 435–479 (1977).
van Jaarsveld, J., and van der Schoot, P. "Scaling theory of interacting thermally activated supramolecular polymers". *Macromolecules* **40**, 2177–2185 (2007).

Jahn, T.R., and Radford, S.E. "Folding versus aggregation: polypeptide conformation on competing pathways". *Arch. Biochem. Biophys.* **469**, 100–117 (2008).

Janmey, P.A., Tang, J.X., and Schmidt, C.F. Actin Filaments, In: Biophysics Textbook, On-Line (V. Bloomeld, Ed.), Sponsored by the Biophysical Society (1999), http://www.biophysics.org/education/janmey.pdf.

Jonkheijm, P.Nanoscopic supramolecular architectures based on Pi-conjugated oligomers, Ph.D. Thesis, Eindhoven University of Technology (2005).

Jonkheijm, P., van der Schoot, P., Schenning, A.P.H.J., and Meijer, E.W. "Probing the solvent-assisted nucleation pathway in chemical self-assembly". *Science* **313**, 80–83 (2006).

Kato, T., Mizoshita, N., and Kishimoto, K. "Functional liquid-crystalline assemblies: self-organized soft materials". *Angew. Chem. Int. Ed.* **45**, 38–68 (2006).

Kegel, W.K., and van der Schoot, P. "Competing hydrophobic and screened Coulomb interactions in hepatitis b virus capsid assembly". *Biophys. J.* **86**, 3905–3913 (2004).

Kegel, W.K., and van der Schoot, P. "Physical regulation of the self-assembly of tobacco mosaic virus coat protein". *Biophys. J.* **91**, 1501–1512 (2006).

Kentsis, A., and Borden, K.L.B. "Physical mechanisms and biological significance of supramolecular protein self-assembly". *Curr. Protein Pept. Sci.* **5**, 125–134 (2004).

Kiselyova, O.I., Yaminsky, I.V., Karger, E.M., Frolova, O.Yu., Dorokhov, Y.L., and Atabekov, J.G. "Visualization by atomic force microscopy of tobacco mosaic virus movement proteinRNA complexes formed in vitro". *J. Gen. Virol.* **82**, 1503–1508 (2001).

Kushner, D.J. "Self-assembly of biological structures". *Bacteriological Rev.* **33**, 302–345 (1969).

Lauer, M.A. "Entropy-Driven Processes in Biology". Chapman and Hall, London, (1975).

Lehn, J.M. "Perspective in supramolecular chemistry—from molecular recognition towards molecular information-processing and self-organization". *Angew. Chem. Int. Ed.* **29**, 1304–1319 (1990).

Lehn, J.M. "Supramolecular chemistry: form molecular information towards self-organization and complex matter". *Rep. Prog. Phys.* **67**, 249–265 (2004).

Lomakin, A., Chung, D.S., Benedek, G.B., Krischner, D.A., and Teplow, D.B. "On the nucleation and growth of amyloid-protein brils: detection of nuclei and quantitation of rate constants". *Proc. Nat. Acad. Sci. U.S.A.* **93**, 1125–1129 (1996).

Lu, X., and Kindt, J.T. "Theoretical analysis of polydispersity in the nematic phase of self-assembled semiexible chains". *J. Chem. Phys.* **125**, 054909, 1–3 (2006).

Marques, C.M., Turner, M.S., and Cates, M.E. "End-evaporation kinetics in living-polymer systems". *J. Chem. Phys.* **99**, 7260–7266 (1993).

Marquesa, C.M., Turner, M.S., and Cates, M.E. "Relaxation mechanisms in worm-like micelles". *J. Non-Cryst. Solids* **172–174**, 1168–1172 (1994).

Martin, R.B. "Comparisons of indefinite self-association models". *Chem. Rev.* **96**, 3043–3064 (1996).

Mattice, W., and Suter, U. "Conformational Theory of Large Molecules: The Rotational Isomeric State Model in Macromolecular Systems". Wiley-Interscience, New York (1994).

McGhee, J.D., and von Hippel, P.H. "Theoretical aspects of DNA-protein interactions: cooperative and non-cooperative binding of large ligands to a one-dimensional homogeneous lattice". *J. Mol. Biol.* **86**, 469–489 (1974).

McPherson, A. "Micelle formation and crystallization as paradigms for virus assembly". *Bioessays* **27**, 447–458 (2005).

Meegan, J.E., Aggeli, A., Boden, N., Brydson, R., Brown, A.P., Carrick, L., Brough, A.R., Hussain, A., and Ansell, R.J. "Designed self-assembled beta-sheet peptide brils as templates for silica nanotubes". *Adv. Funct. Mater.* **14**, 31–37 (2004).

Moore, J.S. "Supramolecular polymers". *Curr. Opin. Colloid Interface Sci.* **4**, 108–116 (1999).

Niranjan, P.S., Yim, P.B., Forbes, J.G., Greer, S.C., Dudowicz, J., Freed, K.F., and Douglas, J. "The polymerisation of actin: thermodynamics near the polymerisation line". *J. Chem. Phys.* **119**, 4070–4084 (2003).

Nyrkova, I.A., and Semenov, A.N. "Non-linear scission-recombination kinetics of living polymerization". *Eur. Phys. J.* **24**, 167–183 (2007).
Nyrkova, I.A., Semenov, A.N., Aggeli, A., Bell, M., Boden, N., and McLeish, T.C.M. "Self-assembly and structural transformationsin living polymers forming brils". *Eur. Phys. J. B* **17**, 499–513 (2000).
Odijk, T. "Ordered phases of elongated micelles". *Curr. Opin. Colloid Interface Sci.* **1**, 337–340 (1996).
Olmsted, P.D."Perspectives on shear banding in complex uids". *Reologica Acta* **47**, 283–300 (2008).
Oosawa, F. "Size distribution of protein polymers". *J. Theor. Biol.* **27**, 69–86 (1970).
Oosawa, F., and Asakura, S. "Thermodynamics of the Polymerization of Protein". Academic Press, New York (1975).
Oosawa, F., and Kasai, M. "A theory of linear and helical aggregations of macromolecules". *J. Mol. Biol.* **4**, 10–21 (1962).
O'Shaughnessy, B., and Yu, J. "Rheology of wormlike micelles: two universality classes". *Phys. Rev. Lett.* **74**, 4329–4332 (1995).
O'Shaughnessy, B., and Vavylonis, D. "Dynamics of living polymers". *Eur. Phys. J. E* **12**, 481496 (2003).
Padding, J.T., and Boek, E.S. "Evidence for diffusion-controlled recombination kinetics in model wormlike micelles". *Europhys. Lett.* **66**, 756762 (2004).
Palmans, A., and Meijer, E.W. "Amplication of chirality in dynamic supramolecular aggregates". *Angew. Chem. Int. Ed.* **46**, 8948–8968 (2007).
Panizzay, P., Cristobal, G., and Curely, J. "Phase separation in solution of worm-like micelles: a dilute n $\to$ 0 spin-vector model". *J. Phys. Condens. Matter* **10**, 11659–11678 (1998).
Percec, V. "Self-assembly of viruses as models for the design of new macro molecular and supramolecular architectures". *J. Macromol. Science Pure Appl. Chem.* **A33**, 1479–1496 (1996).
Phelps, D.K., Speelman, B., and Post, C.B. "Theoretical studies of viral capsid proteins". *Curr. Opin. Struct. Biol.* **10**, 170–173 (2000).
Porte, G. "Giant micelles in ideal solutions. Either rods or vesicles". *J. Phys. Chem.* **87**, 3541–3550 (1983).
Powers, E.T., and Powers, D.L. "The kinetics of nucleated polymerisations at high concentrations: amyloid bril formation near and above the supercritical concentration". *Biophys. J.* **91**, 122–131 (2006).
Prinsen, P., and Odijk, T. "Optimized Baxter model of protein solutions: electrostatics versus adhesion". *J. Chem. Phys.* **121**, 6525–6537 (2004).
Safran, S. "Statistical Thermodynamics of Surfaces, Interfaces, and Membranes". Addision Wesley, Reading, MA (1994).
van der Schoot, P. "The hexagonal phase of wormlike micelles". *J. Chem. Phys.* **104**, 1130–1139 (1996).
van der Schoot, P., Michels, M.A.J., Brunsveld, L., Sijbesma, R.P., and Ramzi, A. "Helical transition and growth of supramolecular assemblies of chiral discotic molecules". *Langmuir* **16**, 10076–10083 (2000).
van der Schoot, P., and Zandi, R. "Kinetic theory of virus capsid assembly". *Phys. Biol.* **4**, 296–304 (2007).
Scott, R.L. "Phase equilibria in solutions of liquid sulfur. I Theory". *J. Phys. Chem.* **69**, 261–270 (1965).
Sear, R. "Interactions in protein solutions". *Curr. Opin. Colloid Interface Sci.* **11**, 35–39 (2006).
Sept, D., and McCammon, J.A. "Thermodynamics and kinetics of actin filament nucleation". *Biophys. J.* **81**, 667–674 (2001).
Siber, A., and Podgornik, R. "Role of electrostatic interactions in the assembly of empty spherical viral capsids". *Phys. Rev. E* **76**, 061906, 1–10 (2007).
Taylor, M.P., and Herzfeld, J. "Liquid crystal phases of self-assembled molecular aggregates". *J. Phys. Condens. Matter* **5**, 2651–2678 (1993).

Weiss, R.G., and Terech, P. (Eds.), "Molecular Gels, Materials with Self-Assembled Fibrillar Networks". Springer, Berlin (2005).
Tobolsky, A.V., and Eisenberg, A. "A general treatment of equilibrium polymerization". *J. Am. Chem. Soc.* **82**, 289–293 (1960).
van der Linden, E., and P. Venema, E. "Self-assembly and aggregation of proteins". *Curr. Opin. Colloid Interface Sci.* **12**, 158–165 (2007).
Vriezema, D.M., Cornellas Aragnes, M., Elemans, J.A.A.W., Cornelissen, J.J.L.M., Rowan, A.E., and Nolte, R.J.M. "Self-assembled nanoreactors". *Chem. Rev.* **105**, 1445–1489 (2005).
Wang, F., Sampogna, R.V., and Ware, B.R. "pH Dependence of actin self-assembly". *Biophys. J.* **55**, 293–298 (1989).
Weiss, R.G., and Terech, P. (Eds.), "Molecular Gels: Materials with Self-Assembled Brillar Networks". Springer, Dordrecht (2006).
Wentzel, R. "Kinetics and thermodynamics of amyloid bril assembly". *Acc. Chem. Res.* **39**, 671–679 (2006).
Wheeler, J.C., and Pfeuty, P. "The $n \to 0$ model and equilibrium polymerization". *Phys. Rev. A* **24**, 1050–1061 (1981).
Whitesides, G.M., and Boncheva, M. "Beyond molecules: self-assembly of mesoscopic and macroscopic components". *Proc. Nat. Acad. Sci. U.S.A.* **99**, 4769–4774 (2002).
Würthner, F., Chen, Z., Hoeben, F.J.M., Osswald, P., You, C.-C., Jonkheijm, P., van Herrikhuyzen, J., Schenning, A.P.H.J., van der Schoot, P., Meijer, E.W., Beckers, E.H.A., Meskers, S.C.J., and Janssen, R.A.J. "Supramolecular p-n-heterojunctions by co-self-organisation of oligo(p-phenylene vinylene) and perlene bisimide dyes". *J. Am. Chem. Soc.* **126**, 10611–10618 (2004).
Zhao, D., and Moore, J.S. "Nucleation-elongation: a mechanism for cooperative supramolecular polymerization". *Org. Biomol. Chem.* **1**, 3471–3491 (2003).
Zweistra, H.J.A., and Besseling, N.A.M. "Adsorption and desorption of reversible supramolecular polymers". *Phys. Rev. E* **74**, 021806, 1–10 (2006).

CHAPTER 4

Recombinant Production of Self-Assembling Peptides

Michael J. McPherson[1,2,*], **Kier James**[1], **Stuart Kyle**[1], **Stephen Parsons**[2], and **Jessica Riley**[1]

Contents			
	1. Introduction		80
	2. Recombinant Protein Production		80
		2.1 Overview	80
		2.2 Expression and protein recovery	81
	3. Host Organisms for Recombinant Expression		87
		3.1 Bacteria	87
		3.2 Yeasts	90
		3.3 Filamentous fungi	91
		3.4 Transgenic plants	92
	4. Repetitive and Self-Assembling Proteins and Peptides in Nature		97
		4.1 Overview	97
		4.2 Recombinant production of self-assembling proteins	97
	5. Recombinant Peptide Production		100
		5.1 Overview	100
		5.2 Tandem repeat strategy	101
		5.3 Recombinant bioactive peptide examples	102
	6. Recombinant Expression of Self-Assembling Peptides		108
		6.1 Overview	108
		6.2 Recombinant self-assembling peptide examples	109
	7. Perspective		113
	Acknowledgments		113
	References		113

[1] Astbury Centre for Structural Molecular Biology, University of Leeds, Leeds LS2 9JT, United Kingdom
[2] Centre for Plant Sciences, University of Leeds, Leeds LS2 9JT, United Kingdom

[*] Corresponding author.
E-mail address: m.j.mcpherson@leeds.ac.uk

1. INTRODUCTION

Self-assembly is ubiquitous in Nature; it describes the spontaneous association of molecular building blocks into coherent and well-defined structures without external instruction (Zhang, 2002). One aspect of bionanotechnology encompasses the synthesis of novel biomaterials based on an understanding of natural self-assembly. Proteins and peptides are of particular interest due to the number of their constituent amino acids, allowing the generation of a highly diverse family of biomolecules. Nature has taken full advantage of this diversity to produce a vast number of different structural proteins with well known examples including silk and collagen (Mitraki and van Raaij, 2005). Examples of structures that involve biomolecules complexed with other organic or inorganic components also exist, including bones (Giraud-Guille, 1988) and teeth (Paine and Snead, 1997). In all these cases Nature uses a "bottom-up" approach by building molecular assemblies atom by atom, molecule by molecule, and macromolecule by macromolecule. The concepts of this self-assembling procedure are well documented but are not easy to replicate (Zhang, 2003).

A fundamental principle of engineering molecular self-assembly is to manufacture the appropriate molecular building blocks that are capable of spontaneous stepwise interactions to generate larger assemblies. A key requirement for industrial scale implementation is our capacity to scale-up systems for the production of these molecular building blocks. Much of the work on self-assembling peptides has been based on chemically synthesized peptides. While useful in the discovery phase of identifying self-assembling sequences, this approach is costly, requires hazardous chemicals, and the size of peptide that can be synthesized is limited (Vandermeulen and Klok, 2004). To produce peptides on a larger, and ultimately an industrial scale, biological systems are being explored using recombinant DNA technology. Potential host organisms, include bacteria, yeasts, fungi, animal and plant cells, transgenic animals and plants. In this review we focus upon microbial and plant-based expression systems. First we consider general features of systems used to express recombinant proteins including the requirements for effective transcription and translation leading to production of the recombinant protein. We then consider aspects of protein recovery and the uses of various fusion and purification tags as well as mechanisms for their removal from the final product. These general issues are of importance for any expression system, although some specific details will vary depending upon the host organism.

2. RECOMBINANT PROTEIN PRODUCTION

2.1. Overview

Recombinant DNA technology allows a DNA sequence encoding the protein/peptide of interest to be cloned into an organism which acts as a "biofactory" to produce substantial quantities of the recombinant product. The DNA coding sequence for the protein/peptide, which is often custom designed and manufactured synthetically, is cloned into an expression vector which provides the necessary regulatory signals for efficient expression of the protein by the host organism. The key components of an expression vector are (a) a promoter and terminator to direct transcription and termination of the mRNA, (b) a selectable marker, often for an antibiotic resistance trait to ensure retention of the vector by host cells, and (c) a multiple cloning site allowing the easy directional insertion of the coding region to be expressed. In addition, the coding region must also have a suitable ribosome binding region and an initiating codon (ATG). The precise origin of these components will depend upon the host organism as regulatory signals and selectable markers vary from host to host.

The resulting expression construct must then be introduced into the host cells by some form of chemical treatment, electroporation, or in plants biolistics, agrobacterium-mediated transfer or the use of viral infection. Following introduction of the expression vector DNA, transformed cells are grown in the presence of an appropriate antibiotic, which selects cells expressing the selectable marker protein. In bacteria and some yeasts as well as during transient expression in plant cells, the expression vector remains as an autonomous plasmid within the cell. In other cases, including some yeasts, fungi and stable plant lines the vector becomes integrated into the host genome. An outline of the processes for generating cells that express a recombinant protein and for subsequent recovery of the target recombinant product is shown in Scheme 1.

In addition to the basic features of an expression vector outlined above, other considerations include methods for inducing expression of the target gene, the subcellular location to which the protein/peptide is targeted, and fusion proteins or tags that can either enhance expression or solubility and/or facilitate detection, and purification of the recombinant product.

2.2. Expression and protein recovery

2.2.1. Transcription

Generally a key objective is to maximize the level of expression of the transgene to maximize the level of recombinant protein recovered. To achieve this, a promoter that can direct efficient transcription of the target DNA to produce the corresponding mRNA should be selected. Since recombinant products can sometimes be toxic to the host cells, it is common to employ an inducible promoter. This is normally a promoter taken from a gene that is highly expressed in the host organism, but which, under normal conditions, is either repressed by a repressor protein or is activated by an activator protein. The addition of a chemical inducer then either relieves repression or results in activation of the transcription process. During normal cell growth, the promoter is therefore inactive allowing cells to grow

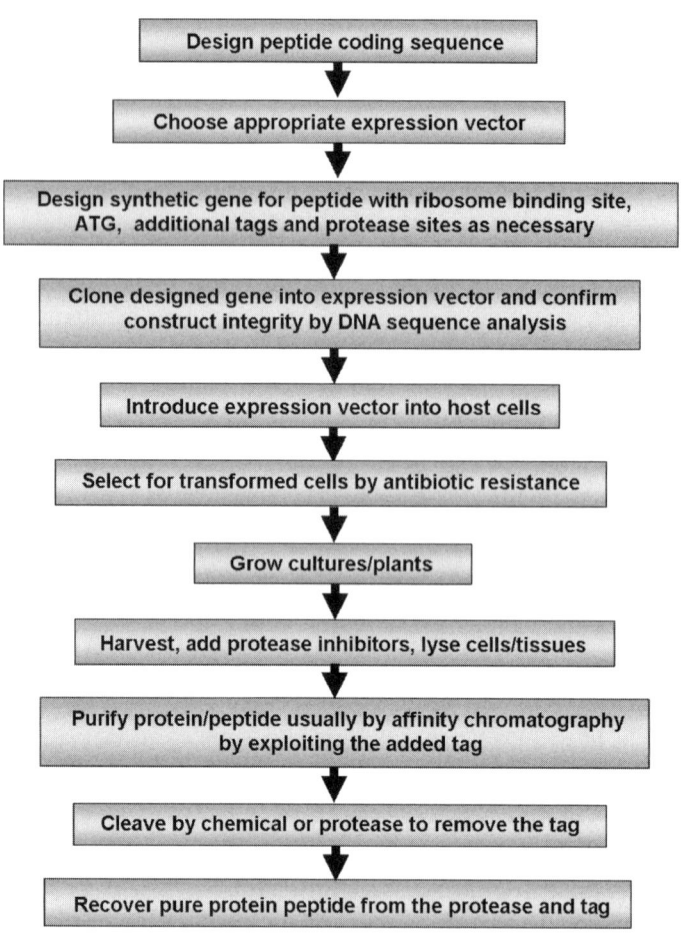

```
                    ┌─────────────────────────────────────┐
                    │   Design peptide coding sequence    │
                    └─────────────────────────────────────┘
                                      ▼
                    ┌─────────────────────────────────────┐
                    │   Choose appropriate expression vector │
                    └─────────────────────────────────────┘
                                      ▼
        ┌──────────────────────────────────────────────────────────┐
        │ Design synthetic gene for peptide with ribosome binding site, │
        │   ATG, additional tags and protease sites as necessary    │
        └──────────────────────────────────────────────────────────┘
                                      ▼
        ┌──────────────────────────────────────────────────────────┐
        │  Clone designed gene into expression vector and confirm   │
        │       construct integrity by DNA sequence analysis        │
        └──────────────────────────────────────────────────────────┘
                                      ▼
                ┌──────────────────────────────────────────┐
                │  Introduce expression vector into host cells │
                └──────────────────────────────────────────┘
                                      ▼
              ┌────────────────────────────────────────────────┐
              │ Select for transformed cells by antibiotic resistance │
              └────────────────────────────────────────────────┘
                                      ▼
                         ┌───────────────────────┐
                         │   Grow cultures/plants │
                         └───────────────────────┘
                                      ▼
           ┌──────────────────────────────────────────────────┐
           │ Harvest, add protease inhibitors, lyse cells/tissues │
           └──────────────────────────────────────────────────┘
                                      ▼
          ┌────────────────────────────────────────────────────┐
          │ Purify protein/peptide usually by affinity chromatography │
          │             by exploiting the added tag              │
          └────────────────────────────────────────────────────┘
                                      ▼
              ┌────────────────────────────────────────────┐
              │  Cleave by chemical or protease to remove the tag │
              └────────────────────────────────────────────┘
                                      ▼
            ┌──────────────────────────────────────────────────┐
            │ Recover pure protein peptide from the protease and tag │
            └──────────────────────────────────────────────────┘
```

Scheme 1 Outline of steps necessary for cloning a target-coding region, introducing the expression construct into the host, expression and product purification including removal of any fusion partner.

efficiently. At an appropriate time the inducing chemical is added to the cells so that they then begin to express the target protein/peptide.

2.2.2. Translation

The context of the mRNA sequence around the AUG codon translational start site, where the ribosome initiates translation of protein synthesis, is an important consideration. An appropriate consensus ribosome-binding region for the host organism should be used and potentially inhibitory RNA secondary structure, that may affect the ability of the ribosome to access this translation start site, avoided. Various RNA structure prediction programs such as

RNADraw (Matzura and Wennborg, 1996) can be useful in this regard. A method to alleviate this problem is to engineer silent mutations into the first few codons of the mRNA to reduce the stability of such RNA structures (Ivanovski et al., 2002). Translation of the protein by the ribosome relies upon the decoding of the mRNA in nucleotide triplets called codons. The frequency of specific codons in an organism reflects the availability of aminoacyl-charged tRNAs. With the exceptions of the amino acids Met (M) and Trp (W), all amino acids can be encoded by multiple codons, with Ser (S), Leu (L), and Arg (R) each having six possible codons. Different organisms, therefore, preferentially use different subsets of codons and so a coding region taken from one organism may not be appropriate for efficient translation by another host organism due to their differences in codon usage. For *Escherichia coli*, certain strains are available that contain additional copies of normally poorly represented tRNAs, and so these strains (such as CodenPlus* strains, Stratagene) can be useful for expressing heterologous sequences. However, the simplest approach, especially for short coding sequences such as those for peptides, is to design a DNA sequence that encodes the necessary amino acid sequence but which is optimized for the codon preference of the host organism, and to have this synthesized as an artifical gene. Codon optimization thus ensures that the availability of tRNAs does not reduce the efficiency of translation of a target protein. Coding sequences can be optimized for many organisms based on known genome sequences and useful web-based programs can be used to design optimized coding regions, for example, JAVA Codon Adaptation Tool (http://www.jcat.de/). Alternatively the commercial companies who provide synthetic genes can optimize the codon usage for a specificed host.

2.2.3. Protein recovery and purification

If a protein/peptide is secreted into the culture medium then it must be recovered from a dilute solution. This could involve precipitating the protein/peptide under nondenaturing conditions, such as using ammonium sulfate and resolubilizing in a smaller volume. Alternatively, a more convenient method is through the use of affinity materials (see below) with an appropriate matrix added to the supernatant to allow batch binding of the target protein.

Commonly, it is necessary to recover the protein from within host cells. This involves harvesting, usually involving some centrifugation or filtration step. Subsequently, the cells must be lysed, which can be done by physical approaches such as sonication or pressure systems including French press or cell disrupters. Simpler methods involve the use of detergent mixtures which can allow recovery of both soluble proteins in the supernatant and of inclusion bodies (Lee et al., 2004). Inclusion body proteins can also be recovered by other whole cell dissolution methods (Choe and Middelberg, 2001; Falconer et al., 1999). It is common to add protease inhibitor mixtures at the lysis stage to prevent damage to the target protein/peptide by proteases released from the host cells during the lysis process.

Subsequent protein purification usually relies upon the presence of an appropriate fusion tag incorporated at the stage of expression vector construction. Thus, the recombinant protein/peptide has an integral purification handle, in theory facilitating the rapid isolation of substantial amounts of recombinant product. Some purification tags require the use of specialist and expensive affinity reagents, and even though a substantial level of purification may be achieved, the protein/peptide may still be contaminated with other proteins and biomolecules. Alternatively, two tags could be used allowing an independent purification step, but this obviously results in increased processing times and cost. A common purification tag is a 6–8 histidine or His (H)-tag added at either the N- or the C-terminus of the protein. The cell extract containing the target protein/peptide can then be subjected to metal chelation chromatography, often using nickel or cobalt-charged affinity matrix. However, some endogenous host proteins often show a degree of affinity for such matrices, which in any case are rather expensive. An obvious approach is to use cheaper protein-binding materials such as cellulose which has been used, but not exploited to the extent one would expect (Reed et al., 2006; Rodriguez et al., 2004). Table 1 provides a summary of various protein fusion components and the method of purification.

It is usually desirable to design the expression construct so that the protein/peptide can be separated from the tag that may otherwise interfere with biological or physical function; this can be particularly important in the production of therapeutic reagents. The cleavage mechanism may be chemical or enzymatic. Unfortunately in many cases some additional amino acids remain associated with the target protein/peptide. For example, cyanogen bromide (CNBr) is a chemical that cleaves at peptide bonds C-terminal to methionine residues. This leaves a homoserine lactone residue at the C-terminus (Kaiser and Metzka, 1999). This is not ideal if trying to produce a therapeutic peptide sequence, although it could be useful for processes such as surface immobilization. However, CNBr is highly toxic and therefore may be undesirable for commercial use (Plunkett, 1976).

An alternative to chemicals for cleaving protein/peptides from fusion partners is enzymatic cleavage. Proteases can have high sequence specificity or substrate structure specificity although it can be difficult avoiding the incorporation of unwanted residues in the final product. Proteases tend to cleave towards the C-terminal end of their cleavage site, and therefore, it is advantageous to design constructs in which the protease site precedes the target protein/peptide. In this way often only one or two residues are added to the N-terminus of the product. Depending upon the protease and recognition sequence it may be even possible to produce natural N-terminii. By contrast, when the cleavage site is located at the C-terminal end of a protein, 5–6 additional residues from the recognition site can remain associated with the target protein. A widely used protease which can itself be produced recombinantly in high yield is tobacco etch virus (TEV) protease (Blommel

Table 1 Recombinant proteins/peptides used as fusion partners for recombinant protein/peptide expression

A. Fusion partners for recombinant protein/peptide purification

Fusion partner	Purification method	Size (kDa) or aa (sequence)
Glutathione-S-transferase (GST)	Glutathione-Sepharose	26 kDa
Maltose-binding protein (MBP)	Amylose resin	40 kDa
Thioredoxin	ThioBond resin	109 aa; 11.7 kDa
Chitin-binding domain	Chitin	~110 aa
Cellulose-binding domain	Cellulose	~110–160 aa
Protein A	IgG	14 kDa
FLAG™ peptide	Monoclonal antibody	7 aa (DYKDDDDK)
Poly-arginine	Cation exchange	6 aa (RRRRRR)
S.Tag	S.protein (RNAse)	15 aa (KETAAAKFERQHMDS)
His-tag	Divalent metals (Ni^{2+} or Co^{2+})	6 or 8 aa (HHHHHHHH)
StrepTag II	StrepTactin	8 aa (WSHPQFEK)
MAT-tag	Divalent metals (Ni^{2+} or Co^{2+})	7 aa (HNHRHKH)
AviD-tag	Avidin or Streptavidin	6 aa (DRATPY)
T7-tag	Monoclonal antibody	MASMTGGQQMG

Table 1 (Continued)

Fusion partner	B. Fusion partners for other purposes	
	Property	Size (kDa) or aa (sequence)
Ketosteroid isomerase (KSI)	Produces insoluble fusion protein	125 aa
T7gene10	Produces insoluble fusion protein	260 aa
GFP	Fluorescent reporter protein	220 aa
NusA	Enhanced solubility	495 aa (54.8 kDa)
Ubiquitin	Enhanced solubility	76 aa
Sumo	Target for SUMO protease	117 aa
Ruby tag	Red reporter protein	53 aa (6.1 kDa)

Table 2 Approaches for specific cleavage of recombinant proteins

Cleavage agent	Cleavage specificity
Enzymatic	
Thrombin	-LVPR▼GS-
Factor Xa	-IDGR▼X-
Enterokinase	-DDDDK▼X-
Endoproteinase GluC	-E▼X-
Endoproteinase Lys-C	-K▼X-
TEV protease	-ENLYFQ▼(S,G)-
HRV 3C protease	-LEVLFQ▼GP-
Sumo protease	SUMO-GG▼XXX-
IGase	-PP▼YP-
Furin	-RX(R/K)R▼X-
Chemical	
Cyanogen bromide	-M▼X-
Formic acid	-D▼X-
Hydroxylamine	-N▼G-
2-nitro-5-thiocyanobenzoate (NTCB)	-C▼X-
1-cyano-4-dimethylaminopyridium tetrafluoroborate (CDAP)	-C▼X-
3-bromo-3-methyl-2-(2-nitrophenylthio)-3H-indole	-W▼X-
2-iodosobenzoic acid	-W▼X-

Amino acids are shown in single letter code; X is any amino acid; ▼ denotes the cleavage site; - indicates the remainder of the protein/peptide chain.

and Fox, 2007; Dougherty and Parks, 1989; Dougherty et al., 1989; Fang et al., 2007). Table 2 provides examples of chemical and protease cleavage approaches.

3. HOST ORGANISMS FOR RECOMBINANT EXPRESSION

3.1. Bacteria

A range of bacteria have been used for heterologous protein expression (Terpe, 2006). *E. coli* is a popular bacterial expression host due to (a) its rapid growth rate on relatively simple growth media (doubling time ~20 min), (b) its well-characterized genome and genetics resulting in a wide range of expression host strains, and (c) the availability of a range of expression vectors. Yields of heterologous proteins can be potentially very high as in the case of a repeat peptide system reported at 65% total cellular protein (Metlitskaia et al., 2004), although this is exceptional. *E. coli* can be grown in a fermenter, but a major advance for laboratory study is the recent introduction of the autoinduction process which can provide high cell-density cultures in shake flasks (Studier, 2005).

3.1.1. Autoinduction provides a new tool for high-density cultures of bacteria

Many inducible promoters in *E. coli*-based expression vectors are regulated by the lac operator/repressor system. Traditionally, the lac system has been induced by the addition of the expensive lactose analogue isopropyl-β-D-thiogalactopyranoside (IPTG) to the cell culture during mid-log phase when the cell density has reached a typical value of around 0.6, measured as optical density at 600 nm (OD_{600}). Under such condition the final OD_{600} of the culture would perhaps reach a value of 2–3. Autoinduction relies upon the natural induction of the lac system by conversion of lactose in the growth medium to allolactose. However, the growth medium also contains glucose, the preferred carbon source of the cells which allows initial growth under noninducing conditions. Following glucose depletion, a metabolic switch occurs to allow the cells to utilize lactose simultaneously inducing recombinant protein expression. The benefits of this approach are (a) no monitoring of cell growth to assess when to add inducer, (b) no addition of expensive IPTG is required, and (c) cultures reach very high cell densities, typically OD_{600} 20–30 resulting in higher potential yield of recombinant protein. The levels of cell density achievable by autoinduction in shake flasks are comparable with those that can be achieved by some fermentation processes. Further enhancements to the autoinduction system resulting in enhanced yields have recently been reported (Blommel et al., 2007).

3.1.2. Targeting protein production in bacteria

E. coli, a Gram-negative bacterium has both an inner and an outer cell membrane. Typically, proteins are expressed as cystoplasmic protein. The addition of a signal sequence to the N-terminus of a protein may allow it to be secreted into the periplasm: the compartment between the two membranes. Secretion into the periplasm represents a purification step as the cells separate the target protein from the cytoplasmic components. An example of a protein successfully secreted from *E. coli* is hirudin, a 65-amino acid anticoagulant peptide/protein produced in the salivary glands of medicinal leeches. Using an L-asparaginase II signal sequence around 60 mg/L were obtained (Tan et al., 2002).

Furthermore, other bacteria have been shown to have superior innate abilities to secrete proteins including *Bacillus subtilis*, a Gram-positive bacterium, which lacks membrane proteases. This indicates that prokaryotes could be viable expression hosts for larger scale production of self-assembling peptides (Olmos-Soto and Contreras-Flores, 2003). Further evidence has been provided by Schmidt (2004) who has presented data on high levels of expression of proteins indicating achievable yields in the order of 1–2 g/L.

Even within the cytoplasm there are, in effect, two compartments. Soluble proteins accumulate in the cytoplasm, but for a number of heterologous proteins the lack of appropriate intracellular chaperones to prevent misfolding can often result in such proteins aggregating as insoluble inclusion bodies. Recovery of protein from inclusion bodies requires solubilization in a strong chaotropic agent such as 8M urea or 6M guanidinium hydrochloride. Usually some renaturation step is required for restoring normal structural and functional properties but this recovery process is not straightforward and can result in poor yields of functional protein (Marston and Hartley, 1990; Rai and Padh, 2001). However, in the case of peptides which do not form 3D structures, targeting to inclusion bodies can serve as a useful purification system. A number of fusion partners have been specifically used to direct recombinant peptides to inclusion bodies (discussed in Sections 5 and 6). Indeed, inclusion body localization can be important for protecting the peptides from degradation, and to prevent them exerting any toxic effects upon the host cell, for example, in the production of antimicrobial peptides.

3.1.3. Ralstonia eutropha

Ralstonia eutropha is a Gram-negative bacterium that has substantial capacity for bioremediation and is found in soil and water. This bacterium is being increasingly used as a host for recombinant protein production. For example, a soluble active organophosphohydrolase (OPH) was produced at levels greater than 10 g/L (Barnard et al., 2004). Previously, this enzyme formed inclusion bodies when heterologously expressed in *E. coli*. This high level of expression was achieved by randomly intergrating three copies of the gene into the bacterial chromosome under the control of a strong inducible promoter. It is likely to have good potential for recombinant protein production, but currently suffers from a lack of genetic tools particularly for development of autonomously replicating plasmids, which could be used as expression vectors.

3.1.4. Disadvantages of bacteria

Although bacterial hosts can be useful for protein production, *E. coli* and other prokaryotic expression systems have some limitations. For example, they are incapable of posttranslational modifications such as glycosylation and as a consequence many eukaryotic proteins produced in prokaryotic expression systems are nonfunctional (Rai and Padh, 2001). However, in the case of most peptides, including self-assembling peptides, such modifications are currently not required. Of course, this limitation may have implications for future attempts to functionalize, or generate composite materials based on recombinant self-assembling peptides. Furthermore, a major issue with respect to downstream processing of E. *coli*-derived products is the potential release of endotoxins, derived from the cell wall lipopolysaccharides. Removal of these from the protein product is particularly important where the peptides or proteins are to be used for medical applications.

3.2. Yeasts

3.2.1. Advantages of yeasts

The yeasts *Saccharomyces cerevisiae* and *Pichia pastoris* are useful expression hosts and others such as *Kluyveromyces lactis* are being developed (van Ooyen et al., 2006). Yeasts are popular for expression of recombinant eukaryotic proteins as levels often exceed those of other eukaryotic expression hosts such as mammalian and insect cells. As yeasts are eukaryotes they are capable of posttranslational modification of proteins such as *N*- and *O*-linked glycosylation, and can ensure correct folding and disulfide bond formation (Cereghino and Cregg, 2000; Cregg et al., 2000; Higgins and Cregg, 1998). Yeasts exhibit some of the advantages of prokaryotic expression systems such as a rapid doubling time and easy culturing to high cell densities on simple media (Rai and Padh, 2001). They are capable of secreting proteins to high levels given a suitable signal sequence. In addition, the genetics of commonly used species is relatively well understood. It has also been reported that yeasts are capable of coping with repetitive gene sequences, an important consideration for many peptides and structural proteins (Strausberg and Link, 1990).

On an industrial scale, safety is also an important consideration. Some yeast species have already been used extensively in the brewing and food industries, and as a result, these have been classified as generally recognized as safe (GRAS) by the US Food and Drugs Administration. Consequently, the guidelines for their use are less stringent, potentially reducing start-up costs. In addition, yeasts are good candidates for fermentation with appropriate fed-batch strategies in a bioreactor having the potential to greatly enhance the levels of target protein production. An example illustrating this is the production of 600 mg/L of human interferon α-2b in high cell-density cultures (Ayed et al., 2008). In the same study, exchanging and supplementing the medium also allowed protease degradation issues to be overcome. Evidence for large-scale protein production by a yeast has also been provided by Yamawaki et al. (2007) who concluded that continuous culture of *P. pastoris* with methanol feeding by a concentration control method held good industrial scale potential for recombinant protein production. Using this method yields of 810 mg/L of a single chain antibody (scFv) were achievable compared with only 198 mg/L by a dissolved oxygen control method, despite the methanol consumption rates being the same under both conditions.

The highest reported heterologous production of a protein in yeast comes from the work by Hasslacher et al. (1997). The enzyme hydroxynitrile lyase (Hnl) from the tropical rubber tree *Hevea brasiliensis* was reported to produce levels of 22 g/L intracellularly in *P. pastoris*. In the same study, *S. cerevisiae* and *E. coli* were tested in parallel experiments but were not competitive. Levels of proteins produced by yeasts have more typically been in the range of 1–15 g/L (Schmidt, 2004). A list of reported yields for expression of proteins in *P. pastoris* is provided by Cregg at http://faculty.kgi.edu/cregg/index.htm

3.2.2. Disadvantages of yeasts

There are also some disadvantages of yeasts. For instance, the procedures required for DNA transfection and lysis of yeasts are more difficult than for bacteria. They can also produce substantial quantities of proteases (Schmidt, 2004), requiring strategies to reduce the impact of proteolytic degradation on cellular, vacuolar, and secretory proteins (Zhang et al., 2007). There is considerable variation in the patterns of glycosylation between eukaryote species which can result in *O*- and *N*-linked glycosylation patterns that are considerably different to those of the native protein (Kukuruzinska et al., 1987). This glycosylation may not only have an impact on the activity of the protein but could also affect folding and secretion particularly if the protein is hyperglycosylated, a problem particularly associated with *S. cerevisiae* (Schmidt, 2004).

3.3. Filamentous fungi

3.3.1. Advantages of fungi

Fungi are potentially useful expression hosts. In contrast to yeasts fungi are capable of more complex posttranslational modifications with some proteins being produced with their natural glycosylation patterns, a good example being tissue plasminogen activator (t-PA) produced by *Aspergillus nidulans* (Schmidt, 2004). Their extensive use in the food and beverage industries has resulted in many species of the Aspergillus and Penicillium genera being granted GRAS status (Iwashita, 2002). Fungi offer excellent protein secretion potential with many species having the innate ability to secrete homologous proteins such as cellulases and amylases at levels of 30–40 g/L. As with yeast and bacteria, fungi also offer the potential of inducible gene expression systems. For example, the presence of pH regulatory systems such as that in *A. nidulans* where the *pacC* gene encodes a transcription factor and the *palA*, *palB*, *palC*, *palF*, *palH*, and *palI* genes encode components of a pH signal transduction pathway (Denison, 2000). This is a particularly appealing mechanism for controlling the expression of genes in fungi due to their ability to grow and thrive over a wide pH range (Arst and Penalva, 2003).

3.3.2. Disadvantages of fungi

Despite the potential for very high levels of secretion of homologous proteins, replicating these levels with heterologous proteins from other sources has yet to be realized. Further problems include the difficulty in transfecting fungi due to the degradation of foreign DNA and random genomic integration that can affect endogenous genes. Expression rates are also restricted by RNA instability, incorrect folding of mRNA, and incorrect folding and secretion of the resulting protein, which is influenced by glycosylation. If the resulting proteins are incorrectly glycosylated and/or secreted, then degradation by the host is often rapid (Punt et al., 2002). For fungal expression systems to be fully exploited, a better understanding of fungal genetics and physiology is required.

3.4. Transgenic plants

Transgenic plants are potent protein bio-factories. The ability of plants to produce complex heterologous proteins was demonstrated when a monoclonal IgM antibody was expressed and fully assembled in transgenic tobacco (During et al., 1990). The first peptide to be produced in a transgenic plant was cecropin B, a small antibacterial peptide from the giant silkmoth *Hyalophora ceropia* (Florack et al., 1995). It proved to be extremely sensitive to the plant endogenous proteases and was degraded within seconds in various plant cell extracts. Later, an analog carrying a valine (V) to methionine substitution was compared for protease degradation and was found to show a sixfold lower degradation rate indicating that the peptide sequence will significantly influence levels of degradation.

There are a number of examples of the development of plants for pharmaceutical protein production including antibodies, vaccines, and other bioactive proteins (Daniell et al., 2001; Daniell, 2006; Ma et al., 2005a, b, c; Twyman et al., 2003) as well as consideration of the issues surrounding regulatory issues of the use of transgenic plants for pharmaceutical protein applications (Ma et al., 2005b; Sparrow et al., 2007; Spok, 2007).

A few well-established crop species are preferentially used for transformation experiments, including tobacco, rice, corn, and soybean. From a therapeutic perspective producing pharmaceutical protein in crops that can be eaten raw, such as banana, has advantages particularly in developing countries (Ma et al., 2005a, c). Protein yield is an important factor, and so, plants that produce high biomass and high quantities of total protein are of particular interest. Soybean has a total protein content of 40%, which is higher than that of other crops meaning fewer plants are needed for the required protein quantity reducing handling costs and downstream processing (Kusnadi et al., 1997). A problem with soybean and many other crop plants is that they are food crops. In some regions, such as the EU, media attention has focussed on the possible contamination of traditional food lines with transgenic material in the U.S.A.

(Fox, 2003). Although there are arguments in favour of the use of crop plants, such as corn, for molecular farming purposes (Ramessar et al., 2008). By contrast, tobacco is a nonfood and nonfeed crop and so is not consumed as food by either humans or farm animals, thus transgenic materials could not inadvertently enter the food chain. Tobacco produces over 40 tons of leaf fresh weight per acre and also produces high seed quantities of up to one million seeds per plant (Daniell, 2006). Therefore, a single plant line can be very rapidly scaled-up which makes tobacco a good candidate for commercial scale recombinant protein production. As a result of these factors, tobacco is the most popular species for plant transformation with more transgenes being introduced into tobacco than into all other plant species combined (Daniell, 2006).

Four basic strategies for recombinant protein expression in plants are transient expression, stable nuclear transformed plants, chloroplast transformed plants, and suspension cultures derived from stable transgenic lines.

3.4.1. Transient expression
This can be achieved using infiltration of whole plants or plant parts (e.g., leaves) with *Agrobacterium tumefaciens* (Boehm, 2007; Sparkes et al., 2006; Vaquero et al., 1999). The transfer of T-DNA into plant cells occurs but in the absence of selective pressure there is no chromosomal integration of the T-DNA. After 2–3 days transient expression levels of the target protein can be examined by extraction of leaf protein and detection of the protein by Western blot analysis. This can be useful for assessing the most appropriate constructs to use for the development of stable transgenic lines, and as such is an analytical rather than production level system. A second transient method involves fusing the target gene to the coat protein gene of tobacco mosaic virus together with an endoplasmic reticulum retention signal, and using this construct for plant infection and propagation of the target protein (Spiegel et al., 1999). Due to the efficiency of viral infection, it is possible to generate substantial levels of recombinant proteins. For example a vector, based on a deconstructed TMV called a MagnICON viral vector, has been used to express hepatitis B surface antigen for use as a vaccine (Gleba et al., 2007). Using this system yields of nearly 300 μg/g leaf tissue in a tobacco species were obtained. This process could also be scaled up by the use of a leaf vacuum infiltration approach (Huang et al., 2008).

3.4.2. Stable nuclear transformed plants
The generation of stable transgenic plants normally involves Agrobacterium-mediated transfer of T-DNA into the plant cell where it undergoes random integration into the nuclear genome. Selective chemical pressure is applied, often hygromycin or G418, to select transformants expressing the selectable

marker protein. Due to its random nature, integration may occur in essential functional genes, whose disruption can lead to abnormal phenotypes, or alternatively into transcriptionally silent regions reducing transgene expression. It is therefore necessary to generate a large number of independent transgenic lines, to self-pollinate these and to select for stable integrants, and eventually homozygous lines. Screening of the lines for expression of the transgene, over several generations, to assess stability of the expression profile is then required. As a consequence, a significant time (6–12+ months) is necessary for the generation and characterization of suitable lines. However, once established these provide an almost unlimited and sustainable production capacity by standard farming practice.

Proteins produced in plants can be targeted to various subcellular organelles or cellular locations through the addition of suitable leader or retention sequences. For example, an N-terminal signal sequence plus C-terminal KDEL or HDEL signals direct and retain the protein in the endoplasmic reticulum (Gomord et al., 1997). Other locations are the apoplast, the vacuole, although this environment may be unsuitable due to its low pH and high proteolytic activity, the mitochondria, peroxisomes, and chloroplasts. Proteins can be expressed constitutively (using, for example, the CaMV 35S promoter or by use of appropriate spatially restricted promoters, can be expressed in specific plant organs).

Transgenic plant systems have the potential to produce recombinant proteins on a commodity scale (Kusnadi et al., 1997) due to the low cost of growing plants and because scale-up of production simply requires sewing seeds over a greater field area. As such they offer almost unlimited scalability (Giddings, 2001). It is estimated by Kusnadi et al. (1997) that transgenic plants can produce pharmaceutical proteins at between 10 and 50-fold lower cost than microbial fermentation systems, and 1,000 times lower than mammalian cell culture systems (Hood et al., 2002).

The process of protein production must be optimized for maximum yield while allowing the plant to function correctly without growth being adversely affected. When considering construct design one must consider carefully a number of factors as outlined below.

(a) *Transgene integration.* A common problem with nuclear transformed transgenic plants is gene silencing. This may be due to the methylation of the gene due to repeated homologous sequences (Meyer, 1996) and can drastically reduce the level of protein production by affecting mRNA transcript levels. Constructs must be designed to avoid silencing triggers such as prokaryotic DNA sequences, sequence repeats, and secondary RNA structure formation (Kohli et al., 2003). In addition, systems have been devised that inhibit the inherent plant-silencing mechanism (Johansen and Carrington, 2001; Voinnet et al., 2003).

(b) *Transcription.* In dicotyledonous plants such as tobacco, the strong constitutive cauliflower mosaic virus 35S (CaMV 35S) promoter (Odell et al., 1985) is commonly used to drive transcription. This may be appropriate if the plant is able to cope with continuous accumulation of protein in a range of its tissues, but for many proteins this could be detrimental to growth (Neumann et al., 2005). Use of a promoter that provides spatial and temporal control over transgene expression, and/or inducibility may be preferable. For example, inducible promoter systems respond to a chemical stimulus (Moore et al., 2006) such as ethanol (Caddick et al., 1998; Salter et al., 1998) or hormones (Padidam, 2003; Padidam et al., 2003; Zuo and Chua, 2000; Zuo et al., 2000). Ideally, the inducer should be a cheap and safe chemical that could be applied to the plants with minimal effort and which is environmentally benign. The mechanical gene activation (MeGA) system, developed by Cramer et al. (1999) for CropTech Corp., uses a promoter inducible by mechanical stress. The shearing of leaves when tobacco is harvested causes a rapid induction of protein expression. This means that any detrimental effect the protein may have had on the plant is negligible, and so one expects good biomass yield, although it is not clear whether appropriate levels of protein will accumulate within the damaged tissue postharvest. Therefore, it seems more appropriate to select promoters that will direct expression to appropriate locations where the product is stored safely and stably before harvest.

(c) *Targeting to subcellular locations.* Targeting a protein to specific compartments can increase yield and protein stability. Given the correct targeting signal, the reporter protein, green fluorescent protein (GFP), is accumulated to 3% in secreted fluid (Komarnytsky et al., 2000). Dual targeting of the same protein to both the chloroplast and the peroxisome, simultaneously, increased accumulation approximately twofold over targeting to either organelle alone (Hyunjong et al., 2006). In terms of posttranslational processing, proteins have different requirements. Complex proteins, such as antibodies, are best suited to the secretory pathway (Schillberg et al., 1999) where the addition of an ER retention H/KDEL C-terminal tag has been shown to enhance accumulation, and enable longer storage in harvested leaf material (Schillberg et al., 1999). ER retention also means the protein will not enter the Golgi apparatus where unwanted modifications such as glycosylations can occur.

3.4.3. Chloroplast transformation

Plastids provide a useful location for the production of human therapeutic proteins (Nugent and Joyce, 2005). It is possible to target integration within the plastid genome minimizing adverse effects that can occur through random integration into the host genomic DNA. Transformation of the chloroplast genome can result in 7–8,000 copies of the gene per cell and has resulted in target proteins accounting for up to 47% of total soluble protein in leaf tissue (De Cosa et al., 2001). Proteins tend to fold properly within chloroplasts. These organelles also lack the ability to posttranslationally glycosylate proteins. Another major advantage of plastid transgenes is that in plants such as tobacco, choroplasts are maternally inherited meaning that there are no transgenes present in pollen thus enhancing the biological containment and reducing environmental risks (Daniell, 2006; Ruf et al., 2007).

3.4.4. Plant cell culture

Plant cell cultures are established from cells derived from stable transgenic plants. The level of recombinant protein that can potentially be generated by whole plant systems, whether by nuclear or by chloroplast transformation, cannot be rivaled by plant cell culture systems. However, cell culture systems offer some advantages including the ability to secrete protein into the culture supernatant, good control over cell growth conditions, reproducibility between batches, the ability to comply with GMP standards, and excellent containment of transgenes offering good commercial opportunities for certain types of recombinant protein production (Hellwig et al., 2004). Adventitious root and hairy root cultures can also be exploited for recombinant protein production in suitable bioreactors (Sivakumar, 2006).

Numerous examples of protein production from plant cell cultures exist with a good example being that of human-secreted alkaline phosphatase, which accumulated to 27 mg/L in the culture medium of transformed tobacco NT1 cells. Interestingly, the activity of the enzyme decreased during stationary phase not through protein degradation, but due to protein denaturation caused by uncharacterized factors generated by cell growth. Bacitracin, a protein stabilizer was shown to be effective at stabilizing activity over a 17-day culture period (Becerra-Arteaga et al., 2006). Tobacco was also used to produce a human monoclonal antibody against the rabies virus. The culture produced relatively low levels of functionally assembled antibody of 0.5 mg/L, but this was threefold greater than was produced by the original transgenic plant, and levels of production were maintained over a 3-month culture period (Girard et al., 2006). Sharma et al. (2006) also used tobacco to produce human interleukin-18 in the culture medium at a level of 166 µg/L and which proved to be bioactive. A range of contructs were tested for the expression of hepatitis B surface antigen (HBsAg) from tobacco cell suspension culture with maximum yields of 31 µg/L being achieved (Kumar et al., 2006).

Although tobacco is probably the most common host, a range of plant hosts have been shown to function in cell suspension cultures. For example, rice cell cultures produced human growth hormone in the medium at 57 mg/L and showed similar biological activity to human growth factor produced by *E. coli* (Kim et al., 2008b). Lee et al. (2007) produced biologically active human cytotoxic T-lymphocyte antigen 4-immunoglobulin in a rice suspension culture at a level of 31.4 mg/L in the culture medium. One issue with secretion of recombinant proteins into the culture medium is the low concentration of the protein. McDonald et al. (2005) used a two-compartment membrane bioreactor in which secreted recombinant α-1-antitrypsin inhibitor was retained in the small volume cell compartment and was recovered at a concentration of up to 247 mg/L representing some 10% of total extracellular protein after 6 days. Multiple cycles of induction could be performed. Moss also shows potential for cell culture applications. The genome sequence has been determined (http://www.cosmoss.org/), and it has a haploid genome and a homologous recombination mechanism allowing targeted gene replacement or specific integration. To reduce immunogenicity issues in humans due to the differences in glycosylation patterns between plants and animals, two enzymes responsible for incorpotaion of xylose and fucose have been knocked out in moss leading to a "humanized" glycosylation pattern. With the ability for production of antibodies and the availability of photoreactors, moss promises to provide a useful system for future recombaint protein production (Deckor and Reski, 2007).

4. REPETITIVE AND SELF-ASSEMBLING PROTEINS AND PEPTIDES IN NATURE

4.1. Overview

In Nature, there are many examples of protein and peptide molecular self-assembly. Of the genetically engineered fibrous proteins, collagen, spider silks, and elastin have received attention due to their mechanical and biological properties which can be used for biomaterials and tissue engineering.

4.2. Recombinant production of self-assembling proteins

4.2.1. Silks

Silk proteins, with comparable mechanical and biological properties to native silk, are being produced recombinantly to improve cost-effective production. Expression of synthetic silk has been explored in many eukaryotic hosts including tobacco and potato plants (Scheller et al., 2001), as well as in mammalian epithelial cells (Lazaris et al., 2002) and transgenic goats (Williams, 2003). More recently, expression of spider dragline silk protein was demonstrated in the milk of transgenic mice to an estimated level of 11.7 mg/L (Xu et al., 2007a).

Cappello et al. (1990) pioneered the genetic engineering of repetitive building blocks. Working with the silkworm fibroin, they found that by starting with short oligonucleotide repeats, the size of the protein generated was easily controlled. Synthetic genes encoding the dragline silk from *Nephila clavipes* have also been designed based upon native gene sequences. Highly repetitive regions exist containing consensus sequences, entitled NCMAG1 and NCMAG2 (formerly referred to as spidroin 1 and spidroin 2) (Foo and Kaplan, 2002). These consensus sequences are

NCMAG1: n-GQGGYGGLGGQGAGRGGLGGQGAGA(A)$_n$GGA-c
NCMAG2: n-GPGGYGPGQQGPGGYGPGQQGPSGPGS(A)$_n$-c

Prince et al. (1995) used the expression vector pQE-9 in which synthetic genes, generated from the consensus sequences, were placed under the control of the bacteriophage T5 promoter for expression in *E. coli*. A His-tag was added to the N-terminus of the recombinant protein to allow purification by nickel-affinity chromatography. However, the resulting yields of purified protein were low at 15 mg/L. By contrast, Fahnestock and Bedzyk (1997) achieved greater success using the methylotrophic yeast *P. pastoris* obtaining yields of 1 g/L prior to purification.

Yeast and bacterial systems often give low levels of expression of silks, and this has led to the development of production systems in tobacco and potato. Scheller et al. (2001) have shown that spider silk proteins can be produced in transgenic plants. They inserted synthetic spider silk protein (spidroin) genes into transgenic plants under the control of the CaMV35S promoter. Using this system they were able to demonstrate the accumulation of recombinant silk proteins to a level of at least 2% of total soluble protein in the endoplasmic reticulum of tobacco leaves, and potato tubers.

There are several reports of the use of silk in biomaterials. Kluge et al. (2008) provide a good overview of application of spider silks including recombinant versions. In addition to spider silks there are other types of silks that provide distinct and useful properties, such as those derived from mussels which will presumably become targets for recombinant protein production (Carrington, 2008).

4.2.2. Elastins

Elastin is an extracellular matrix protein found in connective tissue (Keeley et al., 2002), and as the name suggests provides elasticity and resilience to tissues which require extensibility and recoil. Elastin is a highly insoluble, cross-linked polymer synthesized as a soluble monomer, tropoelastin, which is cross-linked in the extracellular matrix by the enzyme lysyl oxidase to generate a polymeric structure. It is composed of hydrophobic domains rich in glycine (G), valine (V), and proline (P) residues, and cross-linking domains rich in alanine and lysine residues (Bellingham et al., 2001). Once polymerized elastin shows little degeneration with aging, and elastin formed in early development remains the same in the later stages of life (Urry, 1993).

Urry and colleagues pioneered the design and production of elastin-like proteins, which were (VPGVG)-based, with cell attachment sequences (GRGDSP) and showed their biocompatibility using immunogenicity studies and cytotoxicity studies with bovine aortic endothelial cells and fibroblasts (Nicol et al., 1991, 1992). Elastin-like materials of a polypeptide with the sequence G-(VPGVG)$_{19}$-VPGV fused to glutathione S-transferase has been recombinantly produced in *E. coli*. From a fermentation culture, the fusion protein was affinity purified and cleaved with protease Xa with a final yield of 1.15 mg/L of G-(VPGVG)$_{19}$-VPGV (McPherson et al., 1992). Guda et al. (1995) further reported on the hyper-expression of poly(GVGVP) in *E. coli* that exhibited inverse temperature transitions of hydrophobic folding and assembly in which the transitions ranged from hydrogel, elastic, and plastic conformations.

Bellingham et al. (2001) have investigated the self-assembling behaviour of recombinant human elastin polypeptides expressed in *E. coli*. They designed elastin polypeptide constructs as glutathione S-transferase (GST) fusion proteins, and introduced specific hydrophobic and cross-linking domains proposed to be important in elastin self-assembly. This elastin polypeptide was separated from the GST-fusion protein by CNBr cleavage or by thrombin cleavage. They found the hydrophobic domains, which contained a similar consensus sequence to human elastin (PGVGVA repeated 7 times), to be essential for elastin polypeptide self-assembly. It is very promising from a biomaterials perspective that recombinant elastin can undergo physiochemically induced transitions from monomer to fibrillar polymer, a property which is of great benefit at the biomaterial–tissue interface.

Girotti et al. (2004) have designed and bio-produced elastin-like protein polymers (ELP) which contain biofunctional motifs with cell adhesion sequences required for tissue-engineering applications. The protein polymer contained periodically spaced fibronectin CS5 domains enclosing the cell attachment sequence REDV. The overall sequence,

(VPGIG)$_2$VPGKG(VPGIG)$_2$<u>EEIQIGHIP**REDV**DYHLYP</u>
(VPGIG)$_2$VPGKG(VPGIG)$_2$(VGVAPG)$_3$

contains elastin-like units of the type (VPGIG)$_n$, with an adjacent VPGKG in which a lysine (K) was introduced in place of an isoleucine (I) so that cross-linking could occur. The CS5 human fibronectin domain was subsequently introduced (underscored) with a REDV recognition sequence (bold) with further VPGXG units completing the bioactive domain. These authors have also begun to exploit the properties of elastin-like polymers and their "smart" behavior in response to temperature. Below the transition temperature the noncross-linked polymer chains remained soluble in water, yet when the temperature was increased above the transition temperature they formed an aggregated self-assembled complex. The cross-linking of lysine residues enabled the production of scaffold-like biomaterials for use in regenerative medicine. Recombinant approaches offer a more sustainable method for producing large quantities of ELPs with the incorporation of large block polymers and cell recognition sites, compared with production through solid-state synthesis.

5. RECOMBINANT PEPTIDE PRODUCTION

5.1. Overview

Despite burgeoning interest in the self-assembling peptide field, there are few published examples of attempts at recombinant production. This is probably because those working on self-assembling peptides are focussed more on the nature and properties of the peptides than on large-scale production. The production of short peptides in biological hosts commonly provokes intracellular degradation by proteases or if the peptide is antimicrobial or particularly hydrophobic they may be toxic to the host cell, resulting in low yields. Normally a fusion protein (Table 1) is used to both protect the peptide and/or the cell and allow recovery of the target peptide before being cleaved (Table 2) and removed in downstream processing steps to acquire pure active/functional peptide. These downstream processes can lead to a significant loss of target peptide. Recombinant production of peptides, therefore, is not trivial. Due to their potential for medical applications, therapeutic peptides such as antimicrobial peptides have been the subject of sustained attempts at recombinat production. Consideration of some recent examples provides some indication of useful aspects and potential difficulties for the design of self-assembling peptide systems.

Since the fusion protein partner is substantially larger than the size of the required peptide, overall peptide yield represents only a fraction of the purified fusion protein, even before losses due to subsequent purification. For example, for a fusion protein of 150 amino acids and a peptide of 15 amino acids, the relative levels of products are 91 and 9%, respectively. So, it is useful to increase the proportion of the final fusion protein that comprises peptide sequences. This can be achieved either by using a smaller fusion protein or by increasing the number of peptide sequences cloned in tandem with the fusion partner. As shown in the following examples both approaches have been adopted, but with variable results. Finally, the separation of the tandem repeats of peptides into monomers can be achieved by either chemical cleavage or enzymatic cleavage (Section 1.1.1.3).

5.2. Tandem repeat strategy

The concept of expression of tandem protein/peptide repeats interspersed by a single cleavable residue was first introduced by Shen (1984). Monomer, dimer, and trimer repeats of proinsulin were generated with Met residues between each repeat to allow CNBr cleavage. The trimer gave the greatest yield of proinsulin. Of more direct relevance for production of short peptides was the work of Kempe et al. (1985) who expressed tandem repeats of the neuropeptide Substance P (APKQQFFGLM-NH$^+$) joined to a β-galactosidase fusion partner. β-galactosidase was separated from the peptides by a Met residue, and each peptide contained a C-terminal Met, so upon CNBr cleavage the peptide monomers were released with a C-terminal homoserine lactone, which was then chemically converted to a homoserine amide. The fusion partner in this case is large at 1,024 amino acids and fusions of 4 (SP4), 16 (SP16), or 64 (SP64) repeats of the Substance P peptide were expressed. Due to the relative sizes of the β-galactosidase and the peptides, the theoretical levels of recovery would be expected to be 3.8%, 13.2% and 33.5% for the 4, 16 and 64 repeat constructs respectively. Following CNBr cleavage and high-performance liquid chromatography (HPLC) purification 95%, 79% and 70% recovery were reported. To put this into perspective, from a 10 mg sample of the SP64 a final yield of peptide of 2.36 mg was achieved compared with the theoretical yield of 3.35 mg.

A further example of effective expression of tandem repeats of peptide sequences associated with a fusion partner was provided by Kuliopulos and Walsh (1994). Their construct design was based on use of the 125 amino acid, hydrophobic protein fusion protein, ketosteroid isomerase (KSI). The vector pET31b is available from Novagen® (brand of EMD Chemicals Inc., and affiliate of Merck GmbH) and uses the T7 promoter to express a KSI-peptide repeat His-tag protein. Both the vector components of KSI and His-tag are separated from the peptide repeats by methionine residues. The peptide repeats are cloned into the vector by using the restriction enzyme *Alw*NI, which recognizes the sequence CAGNNN/CTG. The unspecified NNN sequence permits the use of ATG, which encodes the amino acid methionine, allowing subsequent CNBr cleavage. Figure 1 shows a schematic representation of the cloning of three tandem repeat sequences using this approach and subsequent protein/peptide products. Kuliopulos and Walsh, 1994 expressed a KSI-(peptide)$_5$-His tag protein in *E. coli* which was then purified by nickel-affinity chromatography and cleaved with CNBr. Cleavage resulted in the release of insoluble KSI, His-tag, and peptide monomers, with yields of peptide reported to be 50–55 mg/L following HPLC purification.

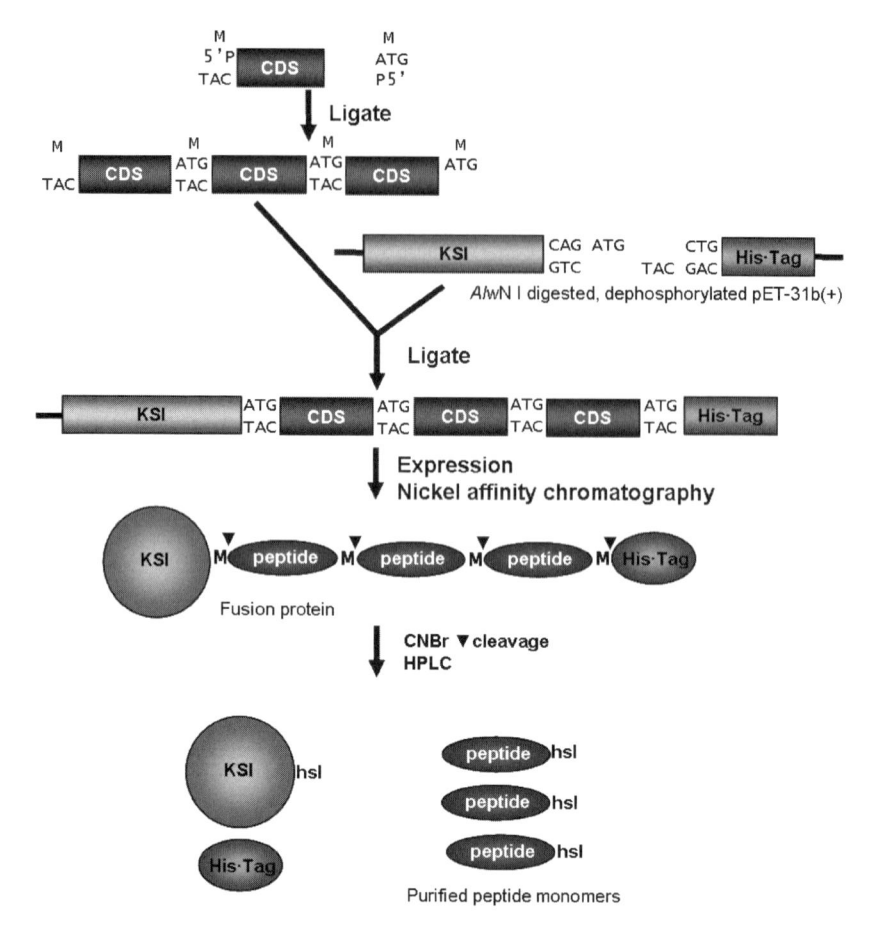

Figure 1 Strategy for cloning a peptide-coding sequence (CDS) as tandem repeats in the vector pET31b. The resulting fusion protein, comprising the ketosteroid isomerase (KSI), peptide repeats, and His-tag, is targeted to inclusion bodies. The fusion protein can be recovered and cleaved, in this case, with cyanogen bromide (CNBr) which acts at the methionine (M) residues allowing further separation of pure peptide from the other fusion components. The cleavage by CNBr results in a C-terminal homoserine lactone (hsl) on each peptide monomer.

5.3. Recombinant bioactive peptide examples

Recent examples of the production of bioactive peptides, often antimicrobials, in bacterial and yeast host systems highlight examples of various fusion and cleavage strategies.

5.3.1. Multiple antimicrobial peptide expression

Niu et al. (2008) used the methylotrophic yeast *P. pastoris* vector pPICZα-A in which transcription of the target protein is controlled by the very powerful methanol-inducible alcohol oxidase 1 (AOX1) promoter. They expressed a hybrid protein comprising four different antimicrobial peptides. The peptides were Protegrin-1, 4 kDa Scorpion Defensin, Metalnikowin-2A, and Sheep Myeloid Antibacterial Peptide SMAP-29. The coding sequences were synthesized based on *P. pastoris* codon bias to enhance translational efficiency, and each peptide was preceded by a KEX2 protease cleavage site. The artificial protein had the sequence

EKR*RGGRLCYCRRRFCVCVGR**EKR***RGLRRLGRKIAHGVKKYG
PTVLRILRIAG**EKR***GFGCPLNQGACHRHCRSIRRRGGYCAGFF
KQTCTCYRN**EKR***VDKPDYRPRPWPRPNSR

with the recognition sequence (EKR) and cleavage site (*) for Kex-2 protease highlighted in bold. Upon expression and secretion into the growth medium the endogenous KEX2 activity catalyzed cleavage at the recognition sites to give the four peptides each with its native N-terminus. Results from expression trials indicated that KEX2 cleavage was not complete, with products larger than the size range (1.9–4.3 kDa) of the intended final cleaved peptide products. It was estimated that some 271 mg/L of the fusion protein was produced. The peptide material was not purified, but growth assays using diluted growth medium indicated inhibition of growth of both Gram-positive and Gram-negative bacteria, but no hemolytic activity on eukaryotic cells which was taken to indicate no toxic effect. This preliminary study is encouraging and demonstrates the potential for exploiting an endogenous proteolytic cleavage mechanism to generate a correct N-terminal sequence, although the C-terminus contains the additional cleavage site residues. Obviously, further refinement of the system and detailed characterization of purified products and their respective yields are required.

5.3.2. Cecropins

Cecropins are antimicrobial peptides first identified in insects; they are small cationic peptides produced in fat bodies and hemocytes in response to bacterial infections or injury. The peptide ABP-CM4 from the silkworm *Bombyx mori* is a 35-amino acid probable amphiphilic alpha-helical peptide that has been shown to kill bacteria, tumors, and fungi by permeabilization of cell membranes but does not display toxicity toward normal mammalian cells. Zhang et al. (2006) produced the synthetic coding region for the peptide optimized for *P. pastoris* codon usage, which was cloned directly into pPICZαA under the control of the AOX1 promoter. Under the most appropriate conditions, 20 °C, 2% Casamino acids, 0.5% methanol induced for 72 h some 40 mg/L of ABP-CM4 was secreted into the growth medium. Following a 30-min boiling step the peptide was further purified by Sephadex G-50 size-exclusion chromatography, resulting in a yield of

15 mg/L peptide. This purified peptide showed antimicrobial activity against both *E. coli* K12 D31 and the pathogenic fungi *Aspergillus niger*, *Trichoderma viride*, *Gibberella saubinetii*, and *Penicillium chrysogenum*.

In another study by Li et al. (2007a), the peptide ABP-CM4-coding sequence was synthesized and cloned into the *E. coli* vector pET32a to generate a fusion protein comprising Thioredoxin (Trx), a His-tag, and CM4. The fusion protein had the format **Trx-His-tag-Asp/Pro-CM4**. The CM4 being preceded by an Asp-Pro dipeptide which is the target for formic acid digestion (Table 2) resulting in release of a CM4 peptide with an N-terminal proline residue. The gene was not codon optimized for expression in *E. coli* and was expressed in *E. coli* BL21 (DE3) as a soluble protein at a level of 25 mg/L. The resulting protein was purified by nickel-affinity chromatography and cleaved by 1% formic acid (40 °C, 72 h). After lyophilization to remove the formic acid, a further nickel-affinity step to remove the fusion partner and any uncleaved protein was performed. Finally, the peptide sample was subjected to reverse-phase HPLC (RP-HPLC) yielding 1.2 mg/L of peptide which was shown to posess antimicrobial activities against strains of *E. coli* $K_{12}D_{31}$, *P. chrysogenum*, *A. niger*, and *G. saubinetii*.

Chen et al. (2008) have expressed a codon-optimized form of the CM4 peptide in *E. coli*. In this case, two alternative expression/purification systems were examined namely the widely used glutathione-*S*-transferase (GST) and a chitin-binding domain (CBD) system with associated intein splicing (New England Biolabs Inc.). The GST system failed to allow recovery of expressed fusion protein. In contrast, the CBD/intein system allowed the recovery of 110 mg/L fusion protein with a final RP-HPLC purification step yielding 2.1 mg/L of pure peptide which displayed antimicrobial activity against *E. coli* $K_{12}D_{31}$ and *Salmonella*.

Another study on expression of an insect cecropin in *E. coli* used a fusion to the thioredoxin protein (Xu et al., 2007b). In this case the *Musca domestica* (house fly) cecropin termed Mdmcec was cloned downstream of the thioredoxin with an intervening His-tag for purification and enterokinase (Table 2) cleavage site (bold) for release of peptide

```
TRX-HisTag-DDDDK/MKKIGKKIERVGQHTRDATIQTIGVAQ
QAANVAATLKG
```

The fusion protein was expressed in BL21 (DE3) cells by IPTG induction and subsequently purified by nickel-affinity chromatography, resulting in fusion protein yields of 48 mg/L. Following enterokinase cleavage and HPLC, 11.2 mg/L of pure recombinant Mdmcec was recovered. The recombinant peptide was shown to have the expected molecular mass by electrospray ionization-mass spectrometry and a predominantly helical conformation by circular dichroism. It also displayed antimicrobial activity against a range of Gram-negative and Gram-positive bacteria and various fungi. It is proposed that this expression system will provide a reliable and simple method for production of different cationic peptides (Xu et al., 2007b).

5.3.3. Dermcidin

Most antimicrobial peptides are cationic but dermcidin is an anionic antimicrobial peptide secreted in human sweat (Cipakova et al., 2006). It comprises 110 amino acids and posesses a 19-amino acid N-terminal signal peptide which directs secretion. The 110-amino acid peptide is proteolytically processed to form C-terminal peptides of predominantly 48 or 47 amino acids in length which are responsible for antimicrobial activity. In this study, the expression of the 48-amino acid form was achieved by fusing the peptide-coding region to the C-terminus of the KSI sequence present in the expression vector pET31 (Novagen®) outlined previously (Figure 1).

```
KSI-MSSLLEKGLDGAKKAVGGLGKLGKDAVEDLE
SVGKGAVHDVKDVLDSVM-Histag
```

The peptide coding region was synthesized with *Alw*NI restriction enzyme sites encoding the amino acid methionine to allow subsequent CNBr cleavage. Following IPTG induction of the *E. coli* BL21(DE3)pLysS cells carrying the plasmid, the inclusion bodies were recovered by centrifugation, the pellet was dissolved in 6M guanidinum hydrochloride and the soluble fraction loaded onto a nickel-affinity column. The eluted protein was dialysed and the protein that precipitated was subjected to CNBr cleavage. Following lyophilization, the peptide was recovered by repeated extraction with distilled water. The recovered material showed antimicrobial activity against both bacterial and fungal samples. In this example, the dermicin extract would also be expected to contain the His-tag peptide as this would also be soluble.

5.3.4. Cathelicidins

Cathelicidins are host defense peptides found in mammals including LL-37, the only cathelicidin found in humans. Bacterial infection leads to release of the LL-37 precursor hCAP-18 which contains a conserved cathelin domain and C-terminal antimicrobial peptides. These peptides are referred to by their two N-terminal residues plus the length in amino acids. Thus, LL-37 starts with Leu-Leu and is 37 amino acids in length. Posttranslational processing of LL-37 in human sweat and skin generates further antimicrobial peptides, and this study investigated the expression of SK-29, KR-20, LL-29, and LL-23 which are related as shown.

```
LL-37 LLGDFFRKSKEKIGKEFKRIVQRIKDFLRNLVPRTES
LL-23 LLGDFFRKSKEKIGKEFKRIVQR
LL-29 LLGDFFRKSKEKIGKEFKRIVQRIKDFLR
```

```
KR-20 KRIVQRIKDFLRNLVPRTES
SK-29 SKEKIGKEFKRIVQRIKDFLRNLVPRTES
```

In this study, Li et al. (2007b) used an *E. coli* BL21(DE3) host in conjunction with the vector pET32a to express a thioredoxin fusion of the format:

```
Trx-Histag-LVPR/GS-Stag-D/P-Peptide
```

Following expression, the cells were harvested and lysed, and to the cell lysate 1% v/v Triton was added and the soluble fusion protein purified by batch binding to cobalt-affinity resin. The collected resin was incubated with thrombin for 16 h to release the Stag-peptide segment which was then subjected to formic acid cleavage and HPLC to isolate the peptide. The yields were SK-29 (1.7 mg/L), KR-20 (0.7 mg/L), LL-29 (2.1 mg/L), and LL-23 (5.4 mg/L), and the peptides were shown to inhibit *E. coli*, although the least inhibitory, LL-23 was expressed to the highest level suggesting the peptides affected the host cells.

5.3.5. Piscidins

Piscidins are 22-amino acid long peptides with conserved N-termini, which represent a new family of antimicrobial peptides isolated from the mast cells of fish. Piscidin 1, derived from hybrid-striped bass, displays good inhibitory properties and has the amino acid sequence

```
FFHHIFRGIVHVGKTIHRLVTG
```

(Moon et al., 2007). This peptide is proposed to adopt an amphipathic, helical conformation (Silphaduang and Noga, 2001). The peptide coding region was fused to the C-terminus of a His-tag-ubiquitin-coding sequence. As the peptide codon usage was not optimized for *E. coli*, the construct was expressed in Rosetta (DE3)pLysS, a strain that produces normally poorly represented tRNAs. Following IPTG induction in a rich medium the protein was found in a soluble form, so the cells were lysed by a freeze-thaw method and the fusion protein isolated by nickel-affinity chromatography. In contrast, when growth was on minimal medium the protein aggregated into inclusion bodies, requiring the cells to be lysed by whole cell dissolution in 8 M urea. Following nickel-affinity binding, the fusion protein was washed sequentially in 6, 4, 2, 1, and finally 0 molar urea solutions to allow refolding of the ubiquitin. The eluted protein was then subject to cleavage using a recombinant form of yeast ubiquitin hydrolase to release the piscidin, which was then purified by RP-HPLC. The yields of fusion protein and purified peptide for the rich and minimal media were 15 and 1.5 mg/L, respectively. The isolation of this material was designed to allow NMR structural studies through isotopic labeling during the minimal medium growth and so no bioactivity assays were performed (Moon et al., 2007).

5.3.6. Cationic peptide indolicidin derivative

To produce high levels of cationic antimicrobial peptides, a strategy was used by Metlitskaia et al. (2004) involving the generation of multimer repeats of the coding region, interspersed by a short anionic spacer, to counter the positive charge on the peptide. Additionally, between each coding sequence Met residues were incorporated to allow CNBr cleavage. The fusion partner in this case was a 96-amino acid region derived from the cellulose-binding domain (CBD) of *Clostridium celliulovarans*, termed C96. This domain was unable to bind cellulose, but was effective at targeting the fusion protein to inclusion bodies. The repeat unit of peptide-M-spacer-M therefore had the sequence

ILRWPWWPWRRK**M**AEAEPEAEPI**M**

where the cleaved Met residues (bold) would be converted to homoserine lactone. The most efficiently expressed construct in *E. coli* had 15 such repeat units associated with the C96 fusion and gave up to 65% total cellular protein. The inclusion bodies were recovered by centrifugation and then directly solubilized in 70% formic acid to which CNBr was added to cleave at the Met residues. Cation exchange followed by RP-HPLC chromatography steps were used to purify the peptide resulting in recovery of 100 mg/L, which is a good level of recovery. The recombinant peptide was shown to have similar biological activity to a synthetic version suggesting the C-terminal homoserine lactone (hsl) modification was not important for activity. This example highlights the benefits of targeting peptides to inclusion bodies where they can be protected from degradation and isolated from cellular components against which they may exhibit toxic effects.

5.3.7. Histonin

Another example of mutlimeric constructs giving high yields of antimicrobial peptides in *E. coli* was provided by Kim et al. (2008a) who were also keen to ensure the absence of any additional amino acid residues derived from fusion partners. Multimeric repeats of the histonin coding region

RAGLQFPVGKLLKKLLK**RLKR**

were fused to a truncated segment of the PurF protein, termed F4. Constructs containing between 2 and 32 repeats attached to F4 were generated with maximal levels of protein expression achieved with 12 repeats. As a result of the endogenous RLKR sequence (bold) at the C terminus of histonin, a furin cleavage step was possible allowing the recovery of native histonin monomers with no additional sequence. Inclusion bodies were recovered, solubilized in 8 M urea, and following dialysis, furin cleavage was performed over a period of 24 h at 30 °C prior to cation exchange chromatography resulting in histonin yields of 167 mg/L of culture. The recombinant peptide displayed identical antimicrobial activity to a synthetic version.

5.3.8. Microbial peptide expression in plant systems

Several examples of production of antimicrobial peptides in plants have been reported, but these are predominantly from the perspective of protecting the plant against microbial insult, rather than use of the plant as a biofactory for the large-scale production of peptide.

An early attempt to produce a recombinant human 29-amino acid antimicrobial peptide SMAP29 in tobacco was reported by Morassutti et al. (2002) who used a chitin-binding domain for purification and an associated intein cleavage system to release the peptide. Expression levels appeared to be low, with the peptide associated with plant proteins. Upon electrophoretic separation, the peptide displayed the expected immunological and biological activites.

A further example is the production of proctolin, a neuropeptide with myotropic properties, in tobacco (Rao et al., 2004). The peptide has the sequence RYLPT and so a poly-proctolin gene was generated that had 10 repeats with dibasic cleavage sites between the repeats. These dibasic sites comprised two Arg residues (bold) so the repeats had the format

$$\text{RYLPT}\textbf{RR}\text{YLPTR}(\textbf{R}\text{YLPTR})_n\textbf{R}$$

Endogenous serine protease activity resulted in cleavage between the RR residues (bold) to give the sequence RYLPTR, from which the C-terminal Arg would then be removed by endogenous carboxypeptidase D activity to yield native peptide, which was detectable by mass spectrometry.

The examples outlined in this section serve to provide an indication that the production of peptides is not trivial, but that approaches are available for reasonable levels of recovery of some peptides. The challenge is to apply appropriate approaches for the production of self-assembling peptides.

6. RECOMBINANT EXPRESSION OF SELF-ASSEMBLING PEPTIDES

6.1. Overview

The production of short self-assembling peptides using recombinant technology is still in its infancy. Yields should be sufficiently high to make it cost effective, and one must be able to recover functional peptides from the fusion partner. As with any peptide, it is often difficult to produce a peptide monomer with no additional amino acids following cleavage from its fusion partner (as shown by the cleavage sites in Table 2). Fortunately, additional residues are more likely to be acceptable with self-assembling peptides than they are for medically important bioactive peptides. Self-assembly is a process which is dependent on many physical and chemical properties, one important example being charge. For example, Aggeli et al. (2003) have designed peptides that exploit the pKs of various amino acids in order to promote pH responsiveness. In this instance, the distribution of charged residues and the overall net charge of the peptide are important considerations. With careful consideration of these factors, it has been possible to design peptides which reliably form antiparallel beta-sheet structures at defined pH. This is significant in terms of recombinant production as both enzymatic and chemical cleavage can alter charge. For example, the addition of a homoserine lactone following CNBr cleavage results in the removal of the negative charge at the C-terminus. This could be counteracted by blocking the N-terminus of the resulting peptide, but it is equally possible that the homoserine lactone could prove useful, for example, in surface immobilization of peptides.

6.2. Recombinant self-assembling peptide examples

6.2.1. Aβ_{11-26} peptide

Using the KSI fusion, inclusion body approach, Sharpe et al. (2005) generated tandem repeats of the Aβ_{11-26} peptide region by expression in *E. coli* under conditions where they could achieve uniform labeling with 13C and 15N for NMR structural studies of the aggregates formed by this peptide. They found that greatest yields of peptide were achieved with a triplet repeat. After CNBr cleavage and RP-HPLC purification they achieved reasonable levels of recovery, but not as high as expected. From the 80 mg/L fusion protein they expected to recover 27 mg/L peptide, but only achieved 10 mg/L. As a consequence of its pH responsiveness, the acidic conditions to which the Alzheimer's peptide Aβ_{11-26} peptide was subjected prior to monomer purification probably contributed to the significant loss of peptide due to aggregation or fibril formation. This potential for aggregation is likely to be an issue for most self-assembling peptides, and so conditions will need to be discovered, which allow quantitiative recovery.

6.2.2. AG repeats

Panitch et al. (1997) produced impressive yields of 700 mg/L by *E. coli* fermentation of a simple small artificial protein comprising 60 repeats of an (AG)$_4$ motif with 23 and 33 amino acid N- and C-terminal fusions, respectively. The (AG)240 repeat units were not separated after expression, and the fusion sequences comprised just 10% of the whole fusion protein. This meant that loss of polypeptide material due to inefficient cleavage by

CNBr was negligible and resulted in around 78% yield recovery. The fed batch fermentation allowed the control of oxygen, nitrogen, and glucose sources, improving yields from tens of milligrams per liter previously obtained in standard batch fermentation.

6.2.3. RAD16

Reed et al. (2006) reported on the production of a self-assembling peptide using *R. eutropha*. This peptide, RAD16, was 16 amino acids in length and produced as tandem repeats with the sequence

$$(RADARADARADARADA\mathbf{E})_n$$

The introduction of a glutamic acid (E; bold), after each RAD16 repeat was made to allow cleavage by endoproteinase Glu-C, which cleaves C-terminal to glutamate. A cellulose binding domain (CBD) (Table 1) was selected as the affinity tag, as CBDs bind strongly and specifically to cellulose which is a relatively cheap and abundant purification matrix. The main problem encountered was the low level of peptide recovered. Theoretically, 1 g of fusion protein should give 267 mg of peptide, but only 10.1 mg of peptide was recovered after RP-HPLC.

6.2.4. Vesicle-forming peptides

The ability to generate peptide-derived vesicles may be useful for a range of applications such as the encapsulation and delivery of various pharmaceutical reagents including chemically synthesized drugs and biomolecules such as proteins and nucleic acids. van Hell et al. (2007) developed two peptides designed to self-assemble into vesicles in an aqueous solution. These peptides were amphiphilic possessing the same N-terminal hydrophobic tail, which was also acetylated after purification, but with alternative hydrophilic C-terminal regions comprising either 2 or 7 Glu residues. These Glu residues were intended to provide negatively charged domains at neutral pH which should have the property of forming a relatively large interfacial surface area favoring a vesicular structure. The sequences of the peptides were

```
SA2  Ac-Ala-Ala-Val-Val-Leu-Leu-Leu-Trp-Glu-
     Glu-COOH
SA7  Ac-Ala-Ala-Val-Val-Leu-Leu-Leu-Trp-Glu-
     Glu-Glu-Glu-Glu-Glu-Glu-COOH
```

The recombinant system for peptide production involved designing synthetic oligonucleotides, which could be annealed, strand extended, and then cloned into the pET-SUMO vector to generate a fusion protein. This fusion protein was expressed in *E. coli* BL21(DE3) cells by IPTG induction and then cells lysed by freeze-thaw lysis followed by sonication. The lysate was passed through a Hi-Trap HP column to capture the His-tagged protein which was then eluted in 400 mM imidazole-containing buffer. Following cleavage of the SUMO from the peptide using a His-tagged SUMO protease for a period of 6 h, the samples were passed through another Ni-NTA matrix to separate the peptide from the SUMO protease and SUMO fusion protein. A 5-L fermentation yielded 300 mg of fusion protein and 30 mg of purified peptide.

The peptide was then N-terminally acetylated to increase hydrophobicity and to prevent any charge interactions with the Glu residues. The Trp was used as an integral probe for fluorescence anisotropy measurements to assess the critical aggregate concentration (CAC) of peptide. Vesicle formation was observed by electron microscopy studies, and the efficiency of vesicular structure was probed by entrapment of a small water-soluble fluorophore, calcein. The SA2 vesicles entrapped larger quantities of calcein than did the SA7 vesicles and the CAC of the former was also lower. It was suggested that the SA2 peptides may pack more densely within the vesicular structure enhancing their ability to retain the water-soluble probe. It was also demonstrated that the assembly of vesicles can be controlled by the pH of the surrounding solvent which may be useful for drug release regimes (van Hell et al., 2007).

6.2.5. β-sheet-forming peptides

(a) *mLac21*. Middelberg and colleagues described the expression of an oil–water interfacially active peptide, mLac21, a 21-amino acid peptide derived from the lac repressor. They expressed a single copy of this peptide in the pET31b system as a fusion to the inclusion body-forming protein KSI. Rather than using flanking Met residues for CNBr recovery, they used Cys residues with 1-cyano-4-dimethylaminopyridium tetrafluoroborate (CDAP) cleavage. Following ion exchange and RP-HPLC purification they recovered 2.92 mg peptide/L culture (Morreale et al., 2004).

(b) P_{11}-2. Subsequently, Hartmann et al. (2008) have used a similar approach to express tandem repeats of the peptide P_{11}-2, an 11-amino acid peptide CH_3CO-QQRFQWQFEQQ-NH_2 (Aggeli et al., 1997) that forms interpeptide hydrogen bonding in a pH- and concentration-dependent manner resulting in the formation of peptide hydrogels (Aggeli et al., 2003). Of course in a recombinant system the N- and C-terminal modifications of the synthetic peptide cannot be reproduced without post-synthesis manipulation. There are various applications of this peptide including bone defect treatment (Firth et al., 2006), surface coatings (Whitehouse et al., 2005), and even antimicrobial properties (Protopapa et al., 2006). Hartmann et al. (2008) generated KSI-fusions carrying 1, 2, 4, and 9 repeats of P_{11}-2 each separated by flanking Cys residues. The cultures were grown in minimal medium and induced by

IPTG addition, and interestingly, the level of production of fusion protein decreased with repeat number such that the 9-mer produced very low levels. The most efficient level of peptide could be recovered from the single repeat construct which produced some 85 mg fusion per litre culture. The cell pellets were dissolved in urea and the fusion protein captured on a nickel affinity column resulting in around 92% purity of fusion protein with a yield of 73% for the single repeat construct. The cleavage step with CDAP was not particularly efficient at around 49%, but an acetone precipitation step after cleavage resulted in efficient removal of contaminating proteins including KSI. The peptide was purified by RP-HPLC with a yield of 2.63 mg peptide per litre culture being obtained.

(c) *P_{11}-4*. We used a similar approach (Riley et al., submitted; Figure 1) to express tandem repeats of the related self-assembling peptide P_{11}-4 CH_3CO-QQRFEWEFEQQ-NH_2 that was designed to form β-sheets and nematic gels at low pH (Aggeli et al., 2003). Details of the properties of P_{11}-4 are provided by Carrick et al. (2007). This peptide has applications in hard tissue treatment (Firth et al., 2006) and bone/dental remineralization (Kirkham et al., 2007). P_{11}-4 was expressed as 1–6 tandem repeats with Met residues between the KSI fusion and each peptide repeat. An important aspect of this work was the use of the autoinduction system (Section 3.1.1) developed by Studier (2005) for the high-level production of recombinant proteins from the T7 promoter in pET31b. This system allows high cell densities to be achieved before induction. As was observed by Hartmann et al. (2008) for P_{11}-2 fusions, a reduction in yield was observed with the longer repeats of 4–6 copies. However, in this case the more productive construct contained three copies of P_{11}-4, in keeping with the finding of Sharpe et al. (2005) with $Aβ_{11-26}$. A strategy was used of isolating inclusion bodies by detergent-based cell lysis followed by centrifugation to recover the inclusion bodies which were then dissolved in urea then dialysed against water before CNBr cleavage. The most inefficient step in the process was the RP-HPLC purification. Nonetheless, the very high cell densities and efficient production of recombinant protein allowed recovery of 2.5 g fusion protein/L culture for the $(P_{11}-4)_3$ version which should have provided a theoretical yield of 530 mg peptide/L. The actual yield was around 90 mg/L which is still a further improvement on many other reports and further optimization should provide substantial improvement. The recovered peptide identity was confirmed by mass spectrometry and MS sequencing. The recombinant form demonstrated very similar properties to the chemically synthesized version, with characterization by circular dichroism, atomic force microscopy, and transmission electron microscopy revealing β-sheet structure and the formation of fibrillar structures as expected (Riley et al., submitted).

7. PERSPECTIVE

The examples of recombinant peptide production in bacterial and yeast systems indicate that improvements are being made in the levels of peptide that can be purified. With the use of autoinduction approaches and fermentation these systems have the potential for producing substantial quantities of peptides. With some bacteria there is the issue of potential endotoxin contamination. It seems likely that research will lead to an ever-increasing range of applications of self-assembling peptides such as new designed sequences with potential to replace many oil-based materials. In turn this will generate a demand for large-scale, inexpensive, and safe sources of recombinant peptides. Potentially, one of the most promising routes to meet such demand will be through a molecular farming approach in which transgenic crops are used as the production vehicle. Currently, there appear to be no reports of the development of such systems, although this is an area of active investigation in our laboratory.

ACKNOWLEDGMENTS

KARG, SP, and JMR acknowledge studentship support from the Biotechnology and Biological Sciences Research Council, SK achnowledges support from the Wellcome Trust. We thank The Dow Chemical Company for CASE studentship support (SP and JMR).

REFERENCES

Aggeli, A., Bell, M., Boden, N., Keen, J.N., Knowles, P.F., McLeish, T.C.B., Pitkeathly, M., and Radford, S.E. *Nature* **386**, 259 (1997).
Aggeli, A., Bell, M., Carrick, L.M., Fishwick, C.W.G., Harding, R., Mawer, P.J., Radford, S.E., Strong, A.E., and Boden, N. *J. Am. Chem. Soc.* **125**, 9619 (2003).
Arst, H.N., and Penalva M.A. *Fungal Gen. Biol.* **40**, 1 (2003).
Ayed, A., Rabhi, I., Dellagi, K., and Kallel, H. *Enz. Microb. Tech.* **42**, 173 (2008).
Barnard, G.C., Henderson, G.E., Srinivasan, S., and Gerngross, T.U. *Protein Expr. Purif.* **38**, 264 (2004).
Becerra-Arteaga, A., Mason, H.S., and Shuler, M.L. *Biotechnol. Prog.* **22**, 1643 (2006).
Bellingham, C.M., Woodhouse, K.A., Robson, P., Rothstein, S.J., and Keeley, F.W. *Biochim. Biophys. Acta* **1550**, 6 (2001).
Blommel, P.G., Becker, K.J., Duvnjak, P., and Fox, B.G. *Biotechnol. Prog.* **23**, 585 (2007).
Blommel, P.G., and Fox, B.G. *Protein Expr. Purif.* **55**, 53 (2007).
Boehm, R. *Biol. Emerg. Viruses* **1102**, 121 (2007).
Caddick, M.X., Greenland, A.J., Jepson, I., Krause, K.P., Qu, N., Riddell, K.V., Salter, M.G., Schuch, W., Sonnewald, U., and Tomsett, A.B. *Nat. Biotechnol.* **16**, 177 (1998).
Cappello, J., Crissman, J., Dorman, M., Mikolajczak, M., Textor, G., Marquet, M., and Ferrari, F. *Biotechnol. Prog.* **6**, 198 (1990).
Carrick, L.M., Aggeli, A., Boden, N., Fisher, J., Ingham, E., and Waigh, T.A. *Tetrahedron* **63**, 7457 (2007).
Carrington, E. *Trends Biotech.* **26**, 55 (2008).
Cereghino, J.L., and Cregg, J.M. *FEMS Microbiol. Rev.* **24**, 45 (2000).

Chen, Y.Q., Zhang, S.Q., Li, B.C., Qiu, W., Jiao, B., Zhang, J., and Diao, Z.Y. *Protein Expr.Purif.* **57**, 303 (2008).
Choe, W.S., and Middelberg, A.P.J. *Biotech. Bioeng.* **75**, 451 (2001).
Cipakova, I., Gasperik, J., and Hostinova, E. *Protein Expr. Purif.* **45**, 269 (2006).
Cramer, C.L., Boothe, J.G., and Oishi, K.K. *Plant Biotechnol.* **240**, 95 (1999).
Cregg, J.M., Cereghino, J.L., Shi, J.Y., and Higgins, D.R. *Molec. Biotech.* **16**, 23 (2000).
Daniell, H. *Biotechnol. J.* **1**, 1071 (2006).
Daniell, H., Streatfield, S.J., and Wycoff, K. *Trends Plant Sci.* **6**, 219 (2001).
De Cosa, B., Moar, W., Lee, S.B., Miller, M., and Daniell, H. *Nat. Biotechnol.* **19**, 71 (2001).
Deckor, E.L., and Reski, R. *Curr. Opin. Biotechnol.* **18**, 393 (2007).
Denison, S.H. *Fungal Genet. Biol.* **29**, 61 (2000).
Dougherty, W.G., Cary, S.M., and Parks, T.D. *Virology* **171**, 356 (1989).
Dougherty, W.G., and Parks, T.D. *Virology* **172**, 145 (1989).
During, K., Hippe, S., Kreuzaler, F., and Schell, J. *Plant Molec. Biol.* **15**, 281 (1990).
Fahnestock, S.R., and Bedzyk, L.A. *Appl. Micro. Biotechnol.* **47**, 33 (1997).
Falconer, R.J., O'Neill, B.K., and Middelberg, A.P.J. *Biotechnol. Bioeng.* **62**, 455 (1999).
Fang, L., Jia, K.Z., Tang, Y.L., Ma, D.Y., Yu, M., and Hua, Z.C. *Protein Expr. Purif.* **51**, 102 (2007).
Firth, A., Aggeli, A., Burke, J.L., Yang, X.B., and Kirkham, J. *Nanomedicine* **1**, 189 (2006).
Florack, D., Allefs, S., Bollen, R., Bosch, D., Visser, B., and Stiekema, W. *Transgenic Res.* **4**, 132 (1995).
Foo, C.W.P., and Kaplan, D.L. *Adv. Drug Deliv. Rev.* **54**, 1131 (2002).
Fox, J.L. *Nat. Biotechnol.* **21**, 3 (2003).
Giddings, G. *Curr. Opin. Biotechnol.* **12**, 450 (2001).
Girard, L.S., Fabis, M.J., Bastin, M., Courtois, D., Petiard, V., and Koprowski, H. *Biochem. Biophys. Res. Comm.* **345**, 602 (2006).
Giraud-Guille, M.M. *Calcified Tissue Internat.* **42**, 167 (1988).
Girotti, A., Reguera, J., Rodriguez-Cabello, J.C., Arias, F.J., Alonso, M., and Testera, A.M. *J. Mater. Sci.* **15**, 479 (2004).
Gleba, Y., Klimyuk, V., and Marillonnet, S. *Curr. Opin. Biotechnol.* **18**, 134 (2007).
Gomord, V., Denmat, L.A., Fitchette-Laine, A.C., Satiat-Jeunemaitre, B., Hawes, C., and Faye, L. *Plant J.* **11**, 313 (1997).
Guda, C., Zhang, X., McPherson, D.T., Xu, J., Cherry, J.H., Urry, D.W., and Daniell, H. *Biotechnol. Lett.* **17**, 745 (1995).
Hartmann, B.M., Kaar, W., Falconer, R.J., Zeng, B., and Middelberg, A.P.J. *J. Biotechnol.* **135**, 85 (2008).
Hasslacher, M., Schall, M., Hayn, M., Bona, R., Rumbold, K., Luckl, J., Griengl, H., Kohlwein, S.D., and Schwab, H. *Protein Expr. Purif.* **11**, 61 (1997).
Hellwig, S., Drossard, J., Twyman, R.M., and Fischer, R. *Nat. Biotechnol.* **22**, 1415 (2004).
Higgins, D.R., and Cregg, J.M. In "Pichia Protocols. Methods in Molecular Biology" (D.R. Higgins and J.M. Cregg, Eds.), p. 1. Humana Press, Totowa, NJ (1998).
Hood, E.E., Woodard, S.L., and Horn, M.E. *Curr. Opin. Biotechnol.* **13**, 630 (2002).
Huang, Z., LePore, K., Elkin, G., Thanavala, Y., and Mason, H.S. *Plant Biotechnol. J.* **6**, 202 (2008).
Hyunjong, B., Lee, D.S., and Hwang, I.W. *J. Exper. Botany* **57**, 161 (2006).
Ivanovski, G., Gubensek, F., and Pungercar, J. *Toxicon* **40**, 543 (2002).
Iwashita, K. *J. Biosci. Bioeng.* **94**, 530 (2002).
Johansen, L.K., and Carrington, J.C. *Plant Physiol.* **126**, 930 (2001).
Kaiser, R., and Metzka, L. *Anal. Biochem.* **266**, 1 (1999).
Keeley, F.W., Bellingham, C.M., and Woodhouse, K.A. *Phil. Trans. Royal Soc. Series B* **357**, 185 (2002).
Kempe, T., Kent, S.B.H., Chow, F., Peterson, S.M., Sundquist, W.I., Litalien, J.J., Harbrecht, D., Plunkett, D., and Delorbe, W.J. *Gene* **39**, 239 (1985).
Kim, J.M., Jang, S.A., Yu, B.J., Sung, B.H., Cho, J.H., and Kim, S.C. *Appl. Microbiol. Biotechnol.* **78**, 123 (2008a).

Kim, T.G., Baek, M.Y., Lee, E.K., Kwon, T.H., and Yang, M.S. *Plant Cell Rep.* **27**, 885 (2008b).
Kirkham, J., Firth, A., Vernals, D., Boden, N., Robinson, C., Shore, R.C., Brookes, S.J., and Aggeli, A. *J. Dental Res.* **86**, 426 (2007).
Kluge, J.A., Rabotyagova, U., Leisk, G.G., and Kaplan, D.L. *Trends Biotechnol.* **26**, 244 (2008).
Kohli, A., Twyman, R.M., Abranches, R., Wegel, E., Stoger, E., and Christou, P. *Plant Molec. Biol.* **52**, 247 (2003).
Komarnytsky, S., Borisjuk, N.V., Borisjuk, L.G., Alam, M.Z., and Raskin, I. *Plant Physiol.* **124**, 927 (2000).
Kukuruzinska, M.A., Bergh, M.L.E., and Jackson, B.J. *Annu. Rev. Biochem.* **56**, 915 (1987).
Kuliopulos, A., and Walsh, C.T. *J. Am. Chem. Soc.* **116**, 4599 (1994).
Kumar, G.B.S., Srinivas, L., Ganapathi, T.R., and Bapat, V.A. *Plant Cell Tissue and Organ Cult* **84**, 315 (2006).
Kusnadi, A.R., Nikolov, Z.L., and Howard, J.A. *Biotechnol. Bioeng.* **56**, 473 (1997).
Lazaris, A., Arcidiacono, S., Huang, Y., Zhou, J.F., Duguay, F., Chretien, N., Welsh, E.A., Soares, J.W., and Karatzas, C.N. *Science* **295**, 472 (2002).
Lee, C.T., Morreale, G., and Middelberg, A.P.J. *Biotechnol. Bioeng.* **85**, 103 (2004).
Lee, S.J., Park, C.I., Park, M.Y., Jung, H.S., Ryu, W.S., Lim, S.M., Tan, H.K., Kwon, T.H., Yang, M.S., and Kim, D.I. *Protein Expr. Purif.* **51**, 293 (2007).
Li, B.C., Zhang, S.Q., Dan, W.B., Chen, Y.Q., and Cao, P. *Biotechnol. Lett.* **29**, 1031 (2007a).
Li, Y.F., Li, X., and Wang, G.S. *Protein Expr. Purif.* **55**, 395 (2007b).
Ma, J.K.C., Barros, E., Bock, R., Christou, P., Dale, P.J., Dix, P.J., Fischer, R., Irwin, J., Mahoney, R., Pezzotti, M., Schillberg, S., Sparrow, P., Stoger, E., and Twyman, R.M. *EMBO Reports* **6**, 593 (2005a).
Ma, J.K.C., Chikwarnba, R., Sparrow, P., Fischer, R., Mahoney, R., and Twyman, R.M. *Trends Plant Sci.* **10**, 580 (2005b).
Ma, J.K.C., Drake, P.M.W., Chargelegue, D., Obregon, P., and Prada, A. *Vaccine* **23**, 1814 (2005c).
Marston, F.A.O., and Hartley, D.L. *Meth. Enzymol.* **182**, 264 (1990).
Matzura, O., and Wennborg, A. *Computer Applications in the Biosciences* **12**, 247 (1996).
McDonald, K.A., Hong, L.M., Trombly, D.M., Xie, Q., and Jackman, A.P. *Biotechnol. Prog.* **21**, 728 (2005).
McPherson, D.T., Morrow, C., Minehan, D.S., Wu, J.G., Hunter, E., and Urry, D.W. *Biotechnol. Prog.* **8**, 347 (1992).
Metlitskaia, L., Cabralda, J.E., Suleman, D., Kerry, C., Brinkman, J., Bartfeld, D., and Guarna, M.M. *Biotechnol. Appl. Biochem.* **39**, 339 (2004).
Meyer, P. *Biol. Chem. Hoppe-Seyler* **377**, 87 (1996).
Mitraki, A., and van Raaij, M.J. *Protein Nanotech.* **300**, 125 (2005).
Moon, W.J., Hwang, D.K., Park, E.J., Kim, Y.M., and Chae, Y.K., *Protein Expr. Purif.* **51**, 141 (2007).
Moore, I., Samalova, M., and Kurup, S. *Plant J.* **45**, 651 (2006).
Morassutti, C., De Amicis, F., Skerlavaj, B., Zanetti, M., and Marchetti, S. *FEBS Lett.* **519**, 141 (2002).
Morreale, G., Lee, E.G., Jones, D.B., and Middelberg, A.P.J. *Biotechnol. Bioeng.* **87**, 912 (2004).
Neumann, K., Stephan, D.P., Ziegler, K., Huhns, M., Broer, I., Lockau, W., and Pistorius, E.K. *Plant Biotechnol. J.* **3**, 249 (2005).
Nicol, A., Gowda, C., and Urry, D.W. *J. Vascular Surg.* **13**, 746 (1991).
Nicol, A., Gowda, D.C., and Urry, D.W. *J. Biomed. Mater. Res.* **26**, 393 (1992).
Niu, M.F., Li, X., Wei, J.C., Cao, R.B., Zhou, B., and Chen, P.Y. *Protein Expr. Purif.* **57**, 95 (2008).
Nugent, J.M., and Joyce, S.M. *Curr. Pharmaceut. Design.* **11**, 2459 (2005).
Odell, J.T., Nagy, F., and Chua, N.H. *Nature* **313**, 810 (1985).
Olmos-Soto, J., and Contreras-Flores, R. *Appl. Microbiol. Biotechnol.* **62**, 369 (2003).
Padidam, M. *Curr. Opin. Plant Biol.* **6**, 169 (2003).
Padidam, M., Gore, M., Lu, D.L., and Smirnova, O. *Transgenic Res.* **12**, 101 (2003).

Paine, M.L., and Snead, M.L. *J. Bone Mineral Res.* **12**, 221 (1997).
Panitch, A., Matsuki, K., Cantor, E.J., Cooper, S.J., Atkins, E.D.T., Fournier, M.J., Mason, T.L., and Tirrell, D.A. *Macromolec.* **30**, 42 (1997).
Plunkett, E.R. "Handbook of Industrial Toxicology". Chemical Publishing Co., Inc., New York (1976).
Prince, J.T., McGrath, K.P., Digirolamo, C.M., and Kaplan, D.L. *Biochemistry* **34**, 10879 (1995).
Protopapa, E., Aggeli, A., Boden, N., Knowles, P.F., Salay, L.C., and Nelson, A. *Med. Eng. Phys.* **28**, 944 (2006).
Punt, P.J., van Biezen, N., Conesa, A., Albers, A., Mangnus, J., and van den Hondel, C. *Trends Biotechnol.* **20**, 200 (2002).
Rai, M., and Padh, H. *Curr. Sci.* **80**, 1121 (2001).
Ramessar, K., Sabalza, M., Capell, T., and Christou, P. *Plant Sci.* **174**, 409 (2008).
Rao, R., Breuer, M., Tortiglione, C., Malva, C., Baggerman, G., Corrado, G., De Loof, A., and Pennacchio, F. *Biotechnol. Lett.* **26**, 1413 (2004).
Reed, D.C., Barnard, G.C., Anderson, E.B., Klein, L.T., and Gerngross, T.U. *Protein Expr. Purif.* **46**, 179 (2006).
Rodriguez, B., Kavoosi, M., Koska, J., Creagh, A.L., Kilburn, D.G., and Haynes, C.A. *Biotechnol. Prog.* **20**, 1479 (2004).
Ruf, S., Karcher, D., and Bock, R. *Proc. Nat. Acad. Sci USA* **104**, 6998 (2007).
Salter, M.G., Paine, J.A., Riddell, K.V., Jepson, I., Greenland, A.J., Caddick, M.X., and Tomsett, A.B. *Plant J.* **16**, 127 (1998).
Scheller, J., Guhrs, K.H., Grosse, F., and Conrad, U. *Nat. Biotechnol.* **19**, 573 (2001).
Schillberg, S., Zimmermann, S., Voss, A., and Fischer, R. *Transgenic Res.* **8**, 255 (1999).
Schmidt, F.R. *Appl. Microbiol. Biotechnol.* **65**, 363 (2004).
Sharma, N., Kim, T.G., and Yang, M.S. *Biotechnol. Bioprocess Eng.* **11**, 154 (2006).
Sharpe, S., Yau, W.M., and Tycko, R. *Protein Expr. Purif.* **42**, 200 (2005).
Shen, S.H. *Proc. Nat. Acad. Sci USA* **81**, 4627 (1984).
Silphaduang, U., and Noga, E.J. *Nature* **414**, 268 (2001).
Sivakumar, G. *Biotechnol. J.* **1**, 1419 (2006).
Sparkes, I.A., Runions, J., Kearns, A., and Hawes, C. *Nat. Protocols* **1**, 2019 (2006).
Sparrow, P.A.C., Irwin, J.A., Dale, P.J., Twyman, R.M., and Ma, J.K.C. *Transgenic Res.* **16**, 147 (2007).
Spiegel, H., Schillberg, S., Sack, M., Holzem, A., Nahring, J., Monecke, M., Liao, Y.C., and Fischer, R. *Plant Sci.* **149**, 63 (1999).
Spok, A. *Trends Biotechnol.* **25**, 74 (2007).
Strausberg, R.L., and Link, R.P. *Trends Biotechnol.* **8**, 53 (1990).
Studier, F.W. *Protein Expr. Purif.* **41**, 207 (2005).
Tan, S.H., Wu, W.T., Liu, J.J., Kong, Y., Pu, Y.H., and Yuan, R.Y. *Protein Expr. Purif.* **25**, 430 (2002).
Terpe, K. *Appl. Microbiol. Biotechnol.* **72**, 211 (2006).
Twyman, R.M., Stoger, E., Schillberg, S., Christou, P., and Fischer, R. *Trends Biotechnol.* **21**, 570 (2003).
Urry, D.W. *Angew. Chem. Int. Ed.* **32**, 819 (1993).
van Hell, A.J., Costa, C., Flesch, F.M., Sutter, M., Jiskoot, W., Crommelin, D.J.A., Hennink, W.E., and Mastrobattista, E. *Biomacromol.* **8**, 2753 (2007).
van Ooyen, A.J.J., Dekker, P., Huang, M., Olsthoorn, M.M.A., Jacobs, D.I., Colussi, P.A., and Taron, C.H. *FEMS Yeast Res.* **6**, 381 (2006).
Vandermeulen, G.W.M., and Klok, H.A. *Macromolec. Biosci.* **4**, 383 (2004).
Vaquero, C., Sack, M., Chandler, J., Drossard, J., Schuster, F., Monecke, M., Schillberg, S., and Fischer, R. *Proc. Nat. Acad. Sci USA* **96**, 11128 (1999).
Voinnet, O., Rivas, S., Mestre, P., and Baulcombe, D. *Plant J.* **33**, 949 (2003).
Whitehouse, C., Fang, J.Y., Aggeli, A., Bell, M., Brydson, R., Fishwick, C.W.G., Henderson, J.R., Knobler, C.M., Owens, R.W., Thomson, N.H., Smith, D.A., and Boden, N. *Angew. Chem. Int. Ed.* **44**, 1965 (2005).
Williams, D. *Med. Device Technol.* **14**, 9 (2003).

Xu, H.T., Fan, B.L., Yu, S.Y., Huang, Y.H., Zhao, Z.H., Lian, Z.X., Dai, Y.P., Wang, L.L., Liu, Z.L., Fei, J., and Li, N. *Animal Biotechnol.* **18**, 1 (2007a).

Xu, X.X., Jin, F.L., Yu, X.Q., Ji, S.X., Wang, J., Cheng, H.X., Wang, C., and Zhang, W.Q. *Protein Expr. Purif.* **53**, 293 (2007b).

Yamawaki, S., Matsumoto, T., Ohnishi, Y., Kumada, Y., Shiomi, N., Katsuda, T., Lee, E.K., and Katoh, S. *J. Biosci. Bioeng.* **104**, 403 (2007).

Zhang, J., Zhang, S.Q., Wu, X., Chen, Y.Q., and Diao, Z.Y. *Process Biochem.* **41**, 251 (2006).

Zhang, S.G. *Biotechnol. Adv.* **20**, 321 (2002).

Zhang, S.G. *Nat. Biotechnol.* **21**, 1171 (2003).

Zhang, Y.W., Liu, R.J., and Wu, X.Y. *Annals Microbiol.* **57**, 553 (2007).

Zuo, J.R., and Chua, N.H. *Curr. Opin. Biotechnol.* **11**, 146 (2000).

Zuo, J.R., Niu, Q.W., and Chua, N.H. *Plant J.* **24**, 265 (2000).

CHAPTER **5**

Inspiration from Natural Silks and Their Proteins

Boxun Leng, Lei Huang, and **Zhengzhong Shao**[*]

Contents		
	1. Introduction	119
	2. Structure and Properties of Proteins and Silks	120
	2.1 Structure of the proteins	121
	2.2 Mechanical properties of silk materials	125
	3. Artificial Spinning of Silk Fibroin	133
	3.1 Natural silk fabrication	134
	3.2 Solution spinning	136
	3.3 Electrospining	140
	4. Bioapplication of Silk Fibroin	142
	5. Biomineralization Regulated by Silk Proteins	144
	6. Conclusion and Perspective	148
	Acknowledgment	148
	References	148

1. INTRODUCTION

Silk, a kind of fine, strong, and lustrous fiber, consists of a series of natural polymers spun by any of various caterpillars and spiders. The source of commercial silk is mainly from the cocoons produced by the domestic silkworm larvae, *Bombyx mori*. In textile and fashion industry, natural silk is unfailingly popular for its smooth feeling and shimmering luster. Nowadays, however, silk and its component protein have received much attention as a kind of reproducible natural biomaterial. The outstanding mechanical properties and specific structure of silk provide a good model for researching and gaining better

[*] Department of Macromolecular Science, Key Laboratory of Molecular Engineering of Polymers of Ministry of Education, Advanced Materials Laboratory, Fudan University, Shanghai 200433, People's Republic of China

[*] Corresponding author.
 E-mail address: zzshao@fudan.edu.cn

Advances in Chemical Engineering, Volume 35　　　　　　　　　　© 2009 Elsevier Inc.
ISSN 0065-2377, DOI: 10.1016/S0065-2377(08)00205-6　　　　　All rights reserved.

understanding of the relationship between molecular compositions, the associated structure forming capacity, and the ultimate property of final biomaterials. Moreover, silk and its protein demonstrated to have great potential in the biomedical field, for its biocompatibility, biodegradability, and excellent mechanical properties.

From the viewpoint of zootaxa, the silkworm and the spider belong to insect and arachnid of arthropod, respectively. Their silk proteins (fibroin for silkworm silk and spidroin for spider major ampullate silk) do not have any genetic heritage in common and their amino acids sequence compositions are different too. However, the silkworm and spider employ a similar spinning process to produce silk. Furthermore, the silkworm silk and the major ampullate silk have a number of similar structural characteristics, both at the level of the secondary protein structure and the condensed silk morphology. Therefore, for the sake of convenience, they are discussed together in some parts of this text.

First, a review of the properties and the structure of silk is given followed by a discussion on the relationship between molecular composition, assembled protein structure, and mechanical properties. Second, artificial spinning of silk proteins and their bioapplications are emphasized. Finally, the potential role of silk proteins in biomineralization is introduced and discussed.

2. STRUCTURE AND PROPERTIES OF PROTEINS AND SILKS

The *B. mori* silk fiber is made up of two kinds of protein. One is called sericin, a water-soluble protein responsible for the gum-like, sticky coating covering the fiber, and the other is referred to as fibroin, the core filament of silk. The inner part of silk fiber is composed of two monofilaments called brins (Figure 1a) (Poza et al., 2002; Shao and Vollrath, 2002).

Figure 1 (a) the sericin (outer layer) and fibroin filaments of *Bombyx mori* silkworm; (b) the major ampullate silk of *Nephila edulis* spider.

Sericin which constitutes 25–30% of the weight of fiber can glue the fibers together to form a cocoon. In addition, it has several extraordinary properties aimed at protecting the silk, cocoon, and pupa inside, such as oxidation resistance, antibacterial, and regulating the moisture content (Zhang, 2002). Sericin is also responsible for sensitization and has some surprising properties in bioapplications as well (Zhaorigetu, 2001). However, most research focusses on the fibroin and therefore the outer sericin is removed by washing the cocoon in weakly alkaline solution or hot water (degumming).

The core filament of *B. mori* silk, fibroin, is composed of a heavy chain fibroin (H-fibroin, 391 kDa) and a light chain fibroin (L-fibroin, 28 kDa), as well as P25 protein (25 kDa). These three constituents assemble into the secretory units in the ratio of 6:6:1 (H:L:P25) (Inoue et al., 2000; Shimura et al., 1976). H-fibroin and L-fibroin link together via a disulfide bond and the P25 is thought to act as a kind of chaperon to assist the transport and secretion of the insoluble H-fibroin (Sehnal and Zurovec, 2004; Tanaka et al., 1999; Zhou et al., 2000). H-fibroin has a much higher molecular weight and takes up 90% weight of the core filament. The properties of the core filament are mainly attributed to H-fibroin, which is often referred to as fibroin.

In comparison to the silkworm which produces silk fibroin from a single silk gland, spiders have a set of highly differentiated glands that can produce several kinds of silk. Generally, an orb web spider has at least seven separate and morphologically distinct protein glands that can produce six different kinds of silk, as well as a glue substance. The different kinds of silk and glue substance have distinct amino acid compositions, structures, and properties that serve various biological functions (Lewis, 2006; Hu et al., 2006b; Vollrath and Porter, 2006). The dragline silk produced by the major ampullate gland has drawn the majority of research interests for the gland is relatively big and easy to study, while the fiber has outstanding mechanical properties. Therefore, further reference to spider silk and spidroin implies dragline silk and corresponding spidroin (Figure 1b).

Spider silk only has one protein monofilament, and the core-skin structure has been observed in some of them (Frische et al., 1998; Poza et al., 2002). It is thought that both the skin and the core are mainly composed of spidroins, which are differed from the primary structure (spidroin 1 and spidroin 2, >350 kDa calculated from mRNA) (Hinman and Lewis, 1992; Sponner et al., 2005a, b; Xu and Lewis, 1990).

2.1. Structure of the proteins

To some extent, the properties of the protein are mainly determined by its primary structure (i.e., the amino acid sequence). The two kinds of structural protein, fibroin and spidroin have a distinct and highly repetitive primary structure, which results in specific secondary and tertiary

Table 1 The structure elements in fibroin and spidroin (Hakimi et al., 2007).

	Fibroin (*Bombyx mori*) (ExPASy http://www.expasy.ch/)	Spidroin (*Nephila clavipe*) (Shao et al., 2003)
Main amino acids composition (%)	Gly 43.5 Ala 28.0 Ser 12.3 Val 2.3 Tyr 5.0	Gly 40.3 Ala 28.4 Glu/Gln 10.1 Pro 9.4 Arg 2.3
Repeating motifs	GAGAGS, GAGAGY (Asakura et al., 2004; Sehnal and Zurovec, 2004)	GPGGX/GPGQQ, $(A)_n/(GA)_n$, GGX (Hayashi et al., 1999; Hinman et al., 2000)
β-sheet content	40–50%	35%
Size of crystallite	$2 \times 5 \times 7$ nm (Hakimi et al., 2007)	$2 \times 5 \times 6$ nm (Grubb and Jelinski, 1997)
Molecular weight	391 kDa (Shimura et al., 1976)	>350 kDa (Hinman and Lewis, 1992; Xu and Lewis, 1990)

structures. These structural regularities provide fibroin and spidroin with outstanding properties.

Table 1 shows key structural elements in fibroin. The amino acid repeat unit in *B. mori* silk is mostly Gly-X (Glycine, G) [65% of X is Ala (L-alanine, A), 23% Ser (L-Serine, S), and 9% Tyr (L-Tyrosine, Y)], representing 94% of the amino acids in the total sequence (Zhou et al., 2001). The repeating motif GAGAGS – the first-order repeat among three kinds of repeats at different structural levels – is responsible for the crystalline region in silk. Strings of varying numbers of GAGAGS followed by a terminating repeat (usually GAAS) form the second-order repeat. Some Ala or Ser are replaced by Tyr within the second-order repeat, which is thought to disturb the periodicity to a certain extent. About 2–6 second-order repeats and an amorphous sequence of 43 amino acid residues compose the third-order repeat. The amorphous sequences are rich in charged amino acid residues and function as spacers breaking the repetitive region. Fibroin contains 12 third-order repeats with N- and C-termini. Both of the N- and C-termini contain a high proportion of hydrophilic residues (Sehnal and Zurovec, 2004; Zhou et al., 2001). The repetitive structure of fibroin is pictured schematically in Figure 2a. The charged residues in the amorphous sequence and the termini greatly influence the solubility and the assembly behavior of fibroin (Jin and Kaplan, 2003; Shulha et al., 2006). They also represent sites of fibroin that can interact with

(a) (b) (c)

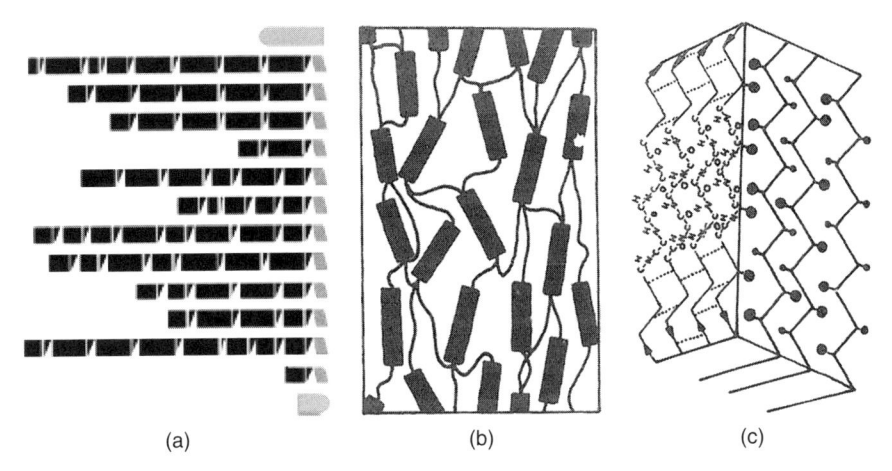

Figure 2 Schematic representation of the repetitive structure of (a) fibroin (Sehnal and Zurovec, 2004), (b) the crystalline and amorphous regions in silk, and (c) the structure of the parallel β-sheet crystals (Zhou et al., 2001). The bullnose symbols in (a) depict the nonrepetitive termini. Each pair of rectangle and triangle represents a second-order repeat unit. Each line on the symbols is the unit of third-order repeat. Crystalline regions are shown in black (Gosline et al., 1986; Heslot, 1998).

ions (Hossain et al., 2003; Peng et al., 2005; Zhou et al., 2003, 2005a, b, 2006b, 2007; Zong, 2004). The gray parts in Figure 2a stand for the hydrophilic regions in fibroin. The hydrophilic and hydrophobic blocks are located alternately along the molecule chain (Jin and Kaplan, 2003; Shulha et al., 2006).

The highly repetitive primary structure leads to the two-phase semicrystalline structure of fibroin (Figure 2b). X-ray diffraction measurements reveal the presence of secondary, β-sheet (Sehnal and Zurovec, 2004) structures (crystal parts) resulting from the repeating motif GAGAGS. In the β-sheet structures, Gly side chains protrude on one side and Ala side chains on the other side of the pleated sheet layer (Figure 2c) (Marsh et al., 1955). The β-sheet structure interlocks with adjacent chains via hydrogen bonds to stack into small 3D crystallites. The size of the crystallites is about $2 \times 5 \times 7$ nm, which is thought to contribute to the strength of silk (Hakimi et al., 2007). As a result, fibroin is traditionally described as rigid with inextensible crystallites embedded in a rubbery matrix (Figure 2b). The matrix is formed by the amorphous parts of the molecules, which have relatively poor orientation and exist as random coils (Mita et al., 1994; Shen et al. 1998). In some cases, they are recognized as 3_1-helices rich in Gly. These helices are somehow oriented parallel to the fiber (van Beek et al., 2002; Kummerlen et al., 1996). The helical region is thought to be related with the extensibility of the silk.

The tertiary structure of fibroin is stabilized by a combination of hydrogen bonds, a high level of crystallinity, and hydrophobic interactions. Such a combination makes the fibroin insoluble to most aqueous solutions including dilute acids and bases. However, several methods have been developed to prepare fibroin solutions. Dissolving silk after degumming in a high-concentrate aqueous lithium salt solution or a $CaCl_2$/ethanol/water solution is widely used (Yamada et al., 2001). Stable fibroin solutions can be prepared with 1,1,1,3,3,3-hexafluoro-2-propanol or hexafluoroacetone (Yao et al., 2002). Recently, preparation of a fibroin solution using an ionic liquid has been reported allowing production of fibroin materials with different morphologies (Gupta et al., 2007; Jiang et al., 2007; Phillips et al., 2004, 2005, 2006). Other strategies allow for high concentration and high-molecular-weight fibroin solutions (Chen et al., 2004; Zhou et al., 2003).

Table 1 shows differences between spidroin and fibroin at the various structure levels. It is noted that spidroin contains high percentages of Ala and Gly, just as fibroin. Spidroin however, contains also Glu (L-Glutamic acid, E), Gln (L-Glutamine, Q), Pro (L-Proline, P), and Arg (L-Argenine, R). Pro mainly exists in spidroin 2 to form repeating motifs, that is, GPGGX/GPGQQ, besides the motifs of $(A)_n$,$(GA)_n$, GGX (Hayashi et al., 1999; Hinman et al., 2000). Amino acids that do not belong to repeating motifs are called "spacers." The $(A)_n/(GA)_n$ module can thus assemble into a β-sheet structure (Hinman et al., 2000). The interactions between the β-sheets are schematically shown in Figure 3. The poly(Gly–Ala) regions have a lower binding energy than poly(Ala) (Hayashi et al., 1999). The GPGGX/GPGQQ repeating units have been suggested to be responsible for β-turn spirals, which account for the elasticity of silk (Liu et al., 2008) and are highly involved in flagelliform silk, an other kind of spider silk with much higher elasticity. The GGX repeating unit is proposed to form helical conformations, 3_{10}-helix (Kummerlen et al., 1996), that could serve as a transition or link between β-sheets and less-rigid protein structures. These motifs are suggested to be structural modules (Figure 3a), and the spacers accordingly contain more charged amino acid residues and separate the repeating units into clusters.

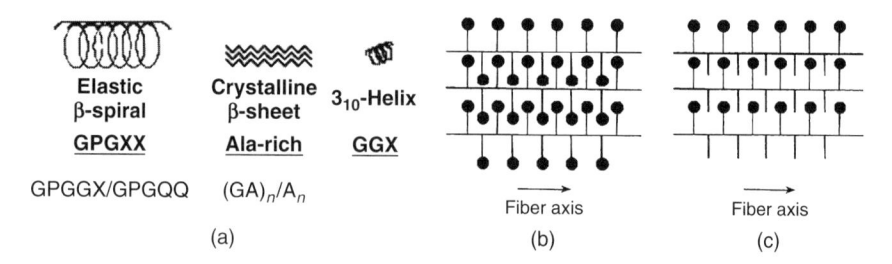

Figure 3 The structural modules as a result from certain amino acid motifs (a) in spidroin (Hayashi et al., 1999; Hinman et al., 2000). The interaction of two different β-sheet forms: poly(Ala) (b) and poly(Gly–Ala) (c) (Hayashi et al., 1999). Poly(Ala) has a tighter structure.

C- and N-terminal regions of spidroin are highly conserved among the spider silk proteins and play an important role in the assembly of spidroin (Huemmerich et al., 2004; Motriuk-Smith et al., 2005).

In addition, one hypothesis for the secondary structure in spidroin suggests that there are amorphous phases, highly oriented crystals, and oriented noncrystalline phases coexisting (Grubb et al., 1997). This structure model has been used to explain the super-contraction of dragline (Liu et al., 2005b).

2.2. Mechanical properties of silk materials

Although the amino acid sequence as well as the secondary structure of fibroin differs from those of spidroin, the fibers spun from these proteins, that is, silkworm silk and spider silk have comparable mechanical properties. These may be attributed to the structural characteristics, both at the molecular and filament level. The superior mechanical properties of silk-based materials, such as films, coatings, scaffolds, and fibers produced using reconstituted or recombinant silk proteins, are determined by their condensed structures.

In view of the various level of structural organization, it is worthwhile to draw a comparison between natural silk fibers (silkworm silk and spider silk) and man-made silk-based materials.

2.2.1. Mechanical properties of natural silk fibers

The significant research attention and the broad number of applications of silk is mainly driven by the prominent mechanical properties. From the mechanical point of view, spider silk (mainly dragline) is regarded as biosteel. The dragline perfectly combines attractive tensile strength with toughness. This composite property is superior to the strongest synthetic fiber such as Kevlar® (Poly-aramide, i.e., poly-paraphenylene-terephtalamide) or highly oriented ultra high-molecular-weight polyethylene. Silks have a better tensile strength than steel considering their much lighter weight. Table 2 compares the mechanical properties of several animal silks and some other natural and man-made fibers (Hinman et al., 2000).

Orb web spiders can produce several kinds of silk. These silks are used for different purposes depending on the different properties of each silk. For example, flagelliform silk is used to capture prey because of its high elasticity. Dragline silk is used for frame and radius building of the web as well as life-saving thread due to its high strength and toughness (Hu et al., 2006b). Interestingly, the dragline also has shape-memory, which prevents an abseiling spider from swinging (Emile et al., 2006).

The properties of natural silk are affected by numerous factors, such as nutrition, temperature, hydration state, extension rate, reeling speed (Knight et al., 2000; Madsen et al., 1999; Riekel et al., 1999; Vollrath and Knight, 1999; Vollrath et al., 2001), and spinning medium during the manufacture (Chen et al.,

Table 2 The mechanical properties of animal silks in comparisons to other man-made and natural materials

Material	Strength (N/m^2)	Elasticity (%)	Energy to break (J/kg)
Silkworm cocoon silk	0.4×10^9	15	6×10^4
Major ampullate silk	1.2×10^9	35	1×10^5
Flagelliform silk	1×10^9	>200	1×10^5
Minor ampullate silk	1×10^9	40–50	3×10^4
Kevlar	4×10^9	5	3×10^4
Rubber	1×10^6	600	8×10^4
Tendon	1×10^9	5	5×10^3
Nylon (type 6)	7×10^7	200	6×10^4
High tensile steel	2×10^9	<1	8×10^4
Collagen	0.9–7.4×10^6	24–68	–
Polylactic acid	2.8–5.0×10^7	2–6	–
Bone	1.6×10^8	3	5×10^3

Modified from literature (Engelberg and Kohn, 1991; Gosline et al., 1999; Hinman et al., 2000; Lewis, 2006; Pins et al., 1997; Vollrath and Porter, 2006; Vepari and Kaplan, 2007).

2002a, b, 2006b; Liu et al., 2005a). Dragline silk is very sensitive to water, which will affect its mechanical properties. The phenomenon is defined as super-contraction when dragline silk shrinks to a certain degree (varies with the species of spider) along its long axis once it contacts with water (Work, 1977). It is conjectured that spiders take advantage of super-contraction to form slack in webs and restore web shape when the silk contacts moist air (Guinea et al., 2003; Savage et al., 2004). It is thought that super-contraction of spider silk results from hydrogen bond breaking by water (or other solvents) and local molecular chains deorientation. The shrinkage depends on the various structure of silk, for example, content of Pro, crystallization of spidroin, orientation of well-defined regular and ill-defined regions, and the ability of solvents to interact with the organization (Eles et al., 2004; Liu et al., 2005b, 2008b). After contraction, silk will increase extensibility and decrease stiffness (Shao and Vollrath, 1999; Shao et al., 1999b, c); however, it can recover its mechanical properties as it can restore the major molecular order (van Beek et al., 1999). The mechanical properties (stress–strain curve) of silk can be determined by its super-contraction ability (Liu et al., 2005b).

Super-contraction, the chaperonage of the special structure of spidroin, is indeed an obstacle to the use of native spider dragline silk, especially in bioapplication (usually wet condition). Recently, it was found that the intrinsic properties of silk fibroin are much better than the data listed in Table 2. The inferior properties are generated by the spinning habit of

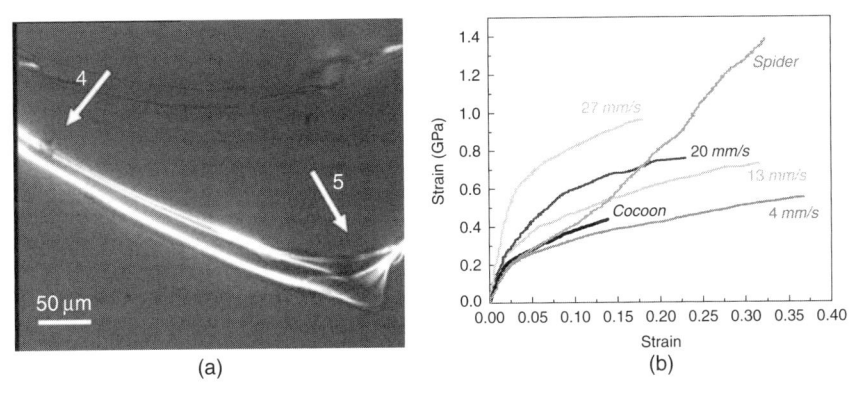

Figure 4 The defects (arrows pointed) on the natural spun silkworm silk (a), and a comparison of stress–strain curves of silks drawn at different speeds from the silkworm *Bombyx mori* (b).

silkworm, which causes defect points in the structure (Figure 4a). The artificial reeling of silk from immobilized silkworms under steady and controlled conditions can produce fibers that are superior to naturally spun ones and approach mechanical properties of *Nephila* spider dragline silk. The silkworm under controlled conditions produces stronger, but more brittle fibers at faster spinning speeds, whereas slower speeds lead to weaker, more extensible fibers (Figure 4b) (Shao and Vollrath, 2002). Consequently, high-performance silkworm silk whose properties are comparable to dragline silk can be produced and used in wet conditions.

The draglines of *N. edulis* have a good (even better) performance at low temperature. This "abnormal" property not only indicates the possibility of such silk to be used as "super-fiber" under severe conditions, but also supplies a model for researchers to investigate the contribution of inter- or intramolecular hydrogen bonds to the physical properties of protein materials (Yang et al., 2005).

2.2.2. Man-made silk materials

The attractive properties of silk fibers as a natural, sustainable product have inspired researchers to look for options to fabricate such fibers without the use of worms or spiders. Furthermore, these natural polymers, silk proteins (both fibroin and spidroin), allow for adjustable mechanical properties, thermal resistance (Drummy et al., 2005; Motta et al., 2002), as well as biomedical compatibility (Vepari and Kaplan, 2007).

The research endeavor to use silk proteins has been a lengthy one. It has generated a lot of insight in the hierarchical organization of proteins into secondary, tertiary, and quaternary structure and is starting to develop industrially viable applications as reflected in multiple spin off companies.

Beyond the well-known textile applications, silk proteins are being proposed for usage especially in the biomedical area (Vepari and Kaplan, 2007). Films, nonwoven mats, sponges, and micro/nanospheres prepared from silk proteins (mainly fibroin) are finding new areas of application. For example, thin films of silk protein are interesting for biomedical (Altman et al., 2003) sutures and optical applications (Putthanarat et al., 2004). A number of approaches aim to synthetically produce silk-like materials using block copolymers or explore recombinant silk proteins by genetic engineering. The mechanical properties of some man-made silk materials are listed in Table 3.

Several methods have been developed to produce silk-based materials with different morphologies, such as wet spinning (fiber) (Zhou et al., 2006a), electrospinning (nonwoven mat) (Jin et al., 2002), casting (film) (Mo et al., 2006), lyophilization (sponge) (Vepari and Kaplan, 2007), and mixing solution (sphere particle) (Cao et al., 2007; Zhang et al., 2007). Of particular interest are wet spinning and electrospinning. However, the complexity of the process requires careful control of multiple parameters such as the composition of solution, the coagulation, and post-treatment of the films or fibers.

Substantially more work is done on the elucidation of properties of regenerated silk, compared to film materials in view of the difficulty to fabricate uniform films. From Table 3, it can be concluded that the mechanical properties of man-made silk materials are inferior to the natural ones. This can be attributed to the fact that the final properties are greatly affected by the structural hierarchies. Typically, the "artificial" materials do not contain the controllable microstructure and supramolecular structure that natural ones possess.

2.2.3. Properties result from structure

The structural differences between natural and man-made silk materials determine the property differences. Even minor changes in the chemical structures give rise to a wide range of variability in mechanical properties (Porter et al., 2005). How this happens is both a challenge to understand and to control when fabricating silk artificially.

For one, the crystalline (or regular) regions mainly define the strength while the amorphous regions allow for the good elasticity in natural silks. Furthermore, the crystallites and most of the amorphous regions are highly aligned along the fiber's long axis. There are, however, some subtle difference between silkworm silk and spider silks.

In *B. mori* silk, the main repeating motif GAGAGS forms the β-sheets, with Gly facing one side and Ala the other side, as depicted in Figure 3c. The small 3D crystallites are stacks of β-sheets. This stacking combined with the hydrophobic interaction between the β-sheets locks the molecules in place. These small crystallites are regarded as the physical cross-links that

Table 3 The mechanical properties of some man-made silk materials

Morphology	Tensile strength (MPa)	Extensibility (%)	References
Film	100	0.5–3.0	Jiang et al. (2007)
Film	9.2	4.0	Li et al. (2003)
Film	58.8	2.1	Jin et al. (2004)
Film (wet)	28.5	51.3	Lv et al. (2005b)
Film (wet)	29.8	59.6	Lv et al. (2005a)
Film (wet)	4.5	10	Kweon et al. (2001)
Film	21	0.7	Freddi et al. (1999)
Fiber	80	11	Matsumoto et al. (1996)
Fiber	very weak	1.5	Ha et al. (2003)
Fiber	630	29.3	Ha et al. (2005)
Fiber	160	17	Um et al. (2004a)
Fiber	190	18	Zhao et al. (2003)
Fiber	180	16	Yao et al. (2002)
Fiber	500	20	Lin et al. (2007)
Fiber	127	12.7	Corsini et al. (2007)
Fiber	0.82 (cN/dtex)	25	Zuo et al. (2007)
Fiber	1.44 (gf/d)	8.4	Ki et al. (2007b)
Fiber	2.0 (gf/d)	20	Ki et al. (2007c)
Fiber	120	35	Marsano et al. (2005)
Fiber (spidroin)	320	4–8	Seidel et al. (2000)
Fiber (spidroin)	110–140	10–27	Shao et al. (2003)
Fiber (recombinant DP-1B)	140	8–10	O'Brien et al. (1998)
Fiber (recombinant ADF-3)	170	43.4	Lazaris et al. (2002)
Fiber (chemical synthesis)	13	22.9	Rathore and Sogah (2001)
Nonwoven mat	2.5	2	Jin et al. (2004)
Nonwoven mat	7.25	3.2	Ayutsede et al. (2005)
Nonwoven mat	1.5	1.6	Chen et al. (2006a)

provide the strength of silk. The amorphous sequences of 43 residues are rich in charged amino acid sequences. They rarely interact with the motifs and function as flexible links between the crystalline regions. Accordingly, a model of crystallites embedded into an amorphous matrix (Figure 2b) explains the high tensile strength and good fiber flexibility (Sehnal and Zurovec, 2004; Termonia, 1994).

Some designing details are conserved in the fibroin to prevent excessive crystallization. GAGAGS occupy most of the repeat length, but the extent of crystallization is about 40–50% (Table 1). If all the GAGAGS formed β-sheets and crystallites, the silk fiber would become a rigid rod. NMR studies on the conformation of the synthetic peptide $(GA)_nG$ found that β-sheets are preferred when n is larger than 2 and that some β-turns appear when n is larger than 9 (Asakura and Yao, 2002). This indicates that long strings of GAGAGS may limit the degree of crystallization. The presences of quantities of Ser and Tyr in the repeating sequence can disturb the homogeneous motif catenations. Tyr partly incorporates into the β-sheets which causes local disorder in the crystal and increases the distance between β-sheets. Such defects are believed to split up the β-sheets under a smaller force, allowing some increase of the extensibility (Asakura and Yao, 2002; Asakura et al., 2002). Sehnal and Zurovec (2004) also suggested that the irregular length at the third-level repeats separated by spacer is important for defining the precise nature of the secondary protein structure and hence for defining properties.

In spider dragline silk, the repeating motif which is responsible for β-sheets is $(A)_n$. Successive Ala residues placed on alternate sides of the backbone (Figure 3b) induce an hydrophobic interaction between molecules that is much stronger than in $(GA)_n$ β-sheets. Also the $(A)_n$ β-sheets have a more compact structure than $(GA)_n$ β-sheets giving rise to $(A)_n$ β-sheet regions with stronger binding energy than $(GA)_n$ region. This structural feature explains why minor ampullate silk [with $(GA)_n$] has a lower tensile strength than major ampullate silk [with $(A)_n$] (Hayashi et al., 1999). This also supports the understanding that tensile strength results from the β-sheet (crystalline) region.

As mentioned above, the Gly-rich domains (GGX) is found to form a helical conformation, 3_{10}-helix (Kummerlen et al., 1996), which serves as a transition between β-sheets and less rigid protein structures. Hydrogen bonds form between these helices and will interlock molecules to some extent, while it also keeps the molecules aligned (Hayashi et al., 1999). Therefore, the GGX sequence may contribute to both the fiber's tensile strength and extensibility.

The repeating motif GPGGX/GPGQQ only exists in major ampullate and flagelliform silk known to have the highest extensibility. This motif has been proposed to conform to a spiral having a similar function as the β-turn spiral (Urry et al., 1975). The Pro residues serve as the points for the retraction after stretching while the side chains of Tyr and Ser stabilize the β-turn spiral through hydrogen bonds. GPGGX/GPGQQ acts as a spring and provides super-elasticity (Figure 5d) (Becker et al., 2003; Hayashi and

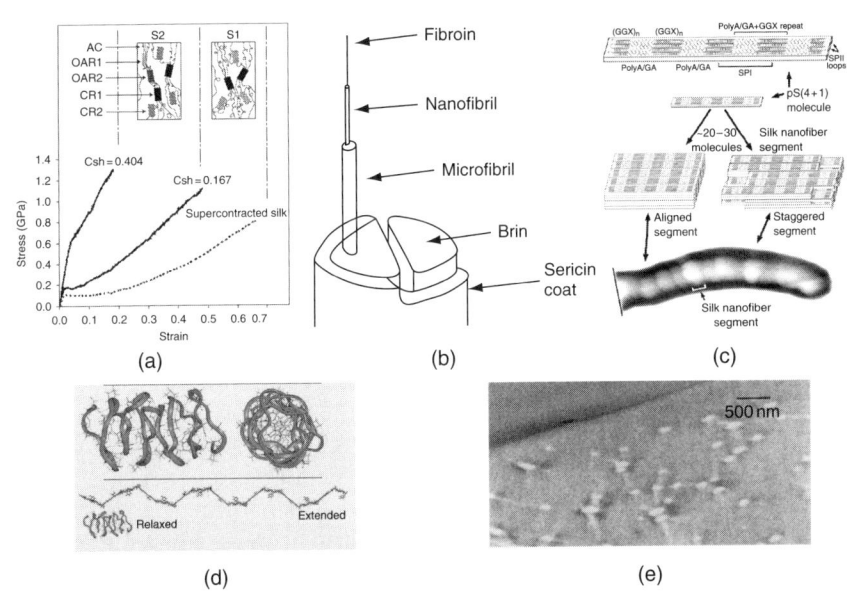

Figure 5 Schematic representation of various molecular structures and supramolecular structure encountered in silk. (a) The two states and the stress–strain curves of their representative spider silks under different degrees of contraction. The supercontraction is induced by loss of alignment with the destruction of hydrogen bonds by a solvent. AC, amorphous molecular chains; OAR1, oriented amorphous region where weak polar solvents such as alcohols can penetrate; OAR2, oriented amorphous region where only strong polar solvents such as water can penetrate; CR1, crystalline region (well defined); CR2, crystalline region (poorly defined) with S1–S2 having increased degree of orientation of this region (Liu et al., 2005b). (b) the schematically hierarchical structure of *B. mori* silk thread (Hakimi et al., 2007). (c) A model for silk nanofibrils organization: well-defined single molecules stacked together to form a nanofibrillar substructure (Oroudjev et al., 2002). (d) Molecular models for relaxed and extended (GPGGX)$_{15}$, which indicates the extensibility of such repeating motif (Becker et al., 2003). (e) the TEM of the silk that was stretched until rupture. Cracks are formed between the light domains. The crack can take up strain energy (Shao et al., 1999a).

Lewis, 1998). The flagelliform silk contains much more GPGGX/GPGQQ than dragline silk, so it is surprisingly elastic (Table 2).

Besides the structural elements mentioned above, mechanical properties of silks are influenced by many other factors. A major one is the aggregation states of the molecular chains (Chen et al., 2006b). Silk properties are strongly influenced by the spinning conditions (Chen et al., 2006b; Liu et al., 2005a). It is theorized that the fabrication conditions have little impact on the crystallites but greatly affect the alignment and orientation of the molecular chains in the amorphous region (Liu et al., 2005b). The combination of these structural elements, alignment of crystallites, and alignment of amorphous region results in the outstanding mechanical properties of silk. Also the alignment of amorphous regions could explain the different capacities to shrink (super-contraction) of the different fibers (Figure 5a) (Liu et al., 2005b). The observation that the spider silk has different properties under different degree of contraction suggests that the overall morphology of silk is very important to the mechanical properties.

The mechanical superiority of silks is also related to the thread substructure, namely the supramolecular structure (Riekel et al., 2001; Sapede et al., 2005). *B. mori* silk and spider dragline silk are made up of bundles of nanofibrils, about 5 nm in diameter (Poza et al., 2002). The nanofibrils align parallel to the axis of the fiber and have a strong interaction (Figure 5b and 5c) (Miller et al., 1999; Oroudjev et al., 2002). This hierarchical structure, from the molecular organization to the nanofibrillar bundles, demonstrates to have a high resistance to the crack spread under tension in view of the heterogeneous morphology of dragline silk. It is suggested that the elongated canaliculi may contain fluid that acts as shock absorbers (Figure 5e) (Shao et al., 1999a; Sponner et al., 2005b).

In view of the intricate structural features in natural silk, it becomes understandable that man-made silks in order to achieve satisfying mechanical properties will require a significant step-up in fabrication control to manage for example a proper degree of crystallinity and crystal size, as well as molecular chains organization and orientation.

Regenerated fibroin films usually have poor physical properties, mainly because the molecules do not align well, and parts of the protein sequence adopt coil conformations. The poor alignment of molecular chains causes insufficient intermolecular interaction to induce the sliding between the molecules at low tensile stress. Although the β-sheet content in artificial silk materials can be increased by methanol or ethanol treatment and heating, allowing some enhancement of the film strength, the elongation at break remains poor. It should be noted that wet films usually have a large extensibility (Table 3), suggesting that molecular water acts as plasticizer, easing the molecular realignment and disentanglement under stress (Kweon et al., 2001; Li et al., 2003; Lv et al., 2005a, b).

Jiang et al. (2007) have investigated the structure and mechanical properties of ultrathin fibroin films prepared by spin-assisted layer-by-layer (SA-LBL) method. It was found that the tensile strength of such film is relatively high but its elasticity is low (Table 3). The SA-LBL process increased the β-sheet content and enhanced intermolecular interactions. The ultrathin film may confine the molecules into a 2D plane, but the crystalline structure is poorly ordered. The very high β-sheet content and the poor molecular alignment are associated to the high tensile strength and low elasticity.

These insights indicate that for artificial wet spinning of silk, there are a number of important processing parameters to consider for optimizing performance, including composition of spinning dope, spinning speed, temperature, coagulation bath, and post-treatment.

3. ARTIFICIAL SPINNING OF SILK FIBROIN

Although spider dragline silk has superior mechanical properties, most suited spider species are cannibals and not domesticated, making daily silk production difficult and with very low yield. To date, it is clearly not realistic to even consider spiders as an economically viable option for the commercial silk production similar to the silkworm silk. Still research on dragline silk has led to understand the hierarchical structure–property relationships of such nonbioactive proteins. Recently, genetic engineering has been carried out to clone and express recombinant spider silk protein in different host systems, such as transgenic tobacco, yeast, and mammalian cells (Fahnestock and Steinbuchel, 2003; Lazaris et al., 2002). The recombinant proteins were spun into fibers with fair mechanical properties. However, the high production cost and complicated process make it difficult for such a strategy to produce spider protein viable for commercialization and wide usage, at least for the time being (Chen et al., 2006b; Vepari and Kaplan, 2007).

Accordingly, major hurdles will need to be overcome to achieve required performance including overcoming inherent weaknesses of natural silk such as super-contraction and fabrication consistency (Hakimi et al., 2007). To date, film, sponge-like silk, and nonwoven mat materials can artificially be made from silk solution, as mentioned in previous section (Vepari and Kaplan, 2007).

As to fibers, it was reported that the inferior mechanical properties of silk from cocoons compared to spider silk result from the silkworm spinning process. If silkworm silk is processed at a constant pulling speed rather than constant force pulling, it possesses excellent properties, approaching the spider dragline silk (Shao and Vollrath, 2002). This suggests that the silkworm silk has the potential to produce better fibers, and the regenerated fibroin, which is easy to harvest, has the possibility to be fabricated into a reconstituted super-fiber.

More than 80 years ago, artificial spinning was carried out using regenerated spidroin and fibroin as spinning dope. Although the properties of regenerated silk then and now are still not good enough compared to those of natural silk, much progress has been made on the fabrication methodology.

It is therefore of interest to review the mechanism of natural silk processing first. In addition, some promising artificial spinning methodologies of silk-like materials are discussed, including wet solution spinning and electrospinning. It is believed that understanding the mechanism and the factors involved in the process will help researchers to produce high-performance artificial silk successfully via a biomimetic method.

3.1. Natural silk fabrication

Silkworms and spiders have developed a set of complicated but efficient spinning systems. They can produce silks with different properties under mild, ambient conditions in an aqueous solution. Considering the supreme properties, they really employ an efficient procedure with minimum energy consumption. Only the conformation transitions happen and no active enzymes work in the solidifying process. As a result, there is a great deal of interest in understanding the precise details of how silk forms from silk proteins – whether in vivo or in artificial circumstances (Fahnestock and Steinbuchel, 2003; Vollrath and Knight, 2001).

Here the production of dragline silk of *Nephila* is considered as an example. The dragline silk is produced in the major ampullate gland which mainly contains four parts, namely tail, lumen (sac), spinning duct, and spinneret (valve and spigot). The schematic structure of the gland is shown in Figure 6a (Lewis, 2006). The tail is the major part of the "spinning dope," serving as a reservoir for the aqueous solution of silk proteins. The sac is responsible for the storage of such spinning dope. The duct has three loops (limbs) and has a tapered shape. Spiders exude the protein solution in a duct where much of the fiber formation occurs. The duct terminates in the valve which has a muscular function that controls the flow rate and the fiber diameter. The final silk thread exits at the spigot (Vollrath and Knight, 2001). The special morphology of duct (three loops, decreasing in diameter with length) favors the formation of the silk. When the silk solution is pulled down a narrow duct, shear and elongational forces can be generated, which cause the alignment of the molecules. The dehydration of the solution also happens in the duct. There is a so-called internal drawdown in the third limb of the duct, in which the forming thread is suddenly stretched. The high elongational stress generated during this drawdown brings the molecules into a more extended conformation, so that they are able to join together forming intermolecular hydrogen bonds. The flexible lips of the spigot can fit tightly around the silk, which is suggested to assist in the water retention (Vollrath and Knight, 1999). Also the pulling force exerted by the spider legs in combination with the holding force by the muscular valve induce an external drawdown on the silk (Vollrath and Knight, 2001).

Vollrath and Knight (2001) have studied and reviewed the spinning process of spiders in detail. The spinning process is believed to employ a liquid crystal strategy. The state of the protein in the sac and in the first and second loop of the duct is a liquid crystal (Knight and Vollrath, 1999). The liquid crystal is considered to be in the nematic phase. The liquid crystal order (as well as the ordered structure observed in the solid silk) seems to form progressively in the duct because silk becomes increasingly birefringent as it passes along the duct while the solution in the sac is not birefringent (Work, 1977). This is attributed to the shear and elongational

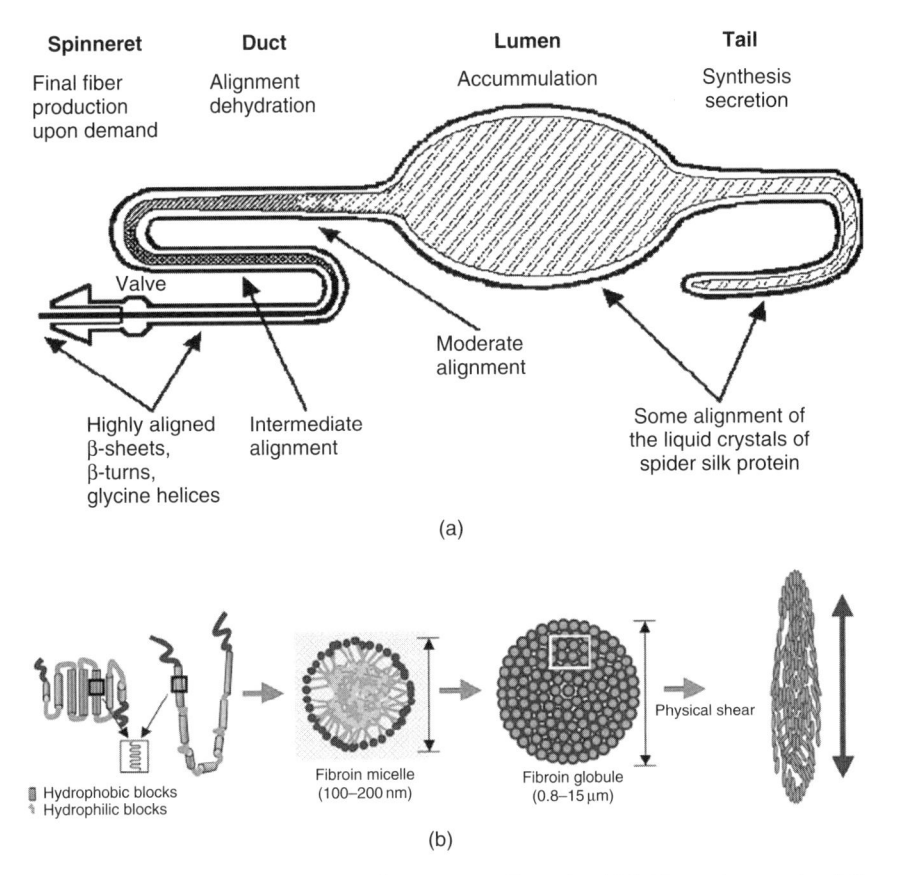

Figure 6 Schematic representation of major ampullate gland (a). The various parts of the gland and their function, as well as the structure of silk under each part, are shown (Lewis, 2006). Also shown is a model of chain folding, micelle formation, globule formation and curing, and shear processing of silk proteins. This model is motivated by experiments in vitro (Jin and Kaplan, 2003) (b).

forces aligning the protein molecules and altering their secondary structure (Lewis, 2006). It is thought that there are several advantages for using the liquid crystal strategy. First, it will prevent fiber formation in the sac. Second, it will form a compact conformation that allows the silk to be processed at high concentrations without aggregation. Third, it allows the viscous solution to flow easily allowing the spider to draw silk using minimal forces. Fourth, the prealignment will prevent defects from forming in the final fiber. Finally, molecular chains can still reorient with the plasticization of water.

Besides the shear force and internal or external drawdown, that is, physical boundary conditions, the chemical environment for fiber formation, that is, presence of metallic ions and pH value of the solutions, defines the conformational transition and final properties of the fiber (Chen et al., 2002a, b, c; Dicko et al., 2004a, b; Knight and Vollrath, 2001; Zhou et al., 2005b; Zong et al., 2004). For example, there is a gradual decreasing pH from 7.2 in tail part to 6.3 in the third limb of the duct (Dicko et al., 2004b). This acidification of the spinning dope aids the conformational transition of the proteins from the random coil and/or helix into the final conformation of a mix of α-helix, β-spiral, and β-sheet in the final fiber (Chen et al., 2002b, 2006b). Metallic ions, such as sodium (Na^+), whose concentration decreases, and potassium (K^+), whose concentration increases along the duct, induce the formation of β-sheet and nanofibrils (Chen et al., 2002b, c).

The combination of physical and chemical boundary conditions defines how the silk protein solution turns into a solid silk thread consisting of highly oriented molecules and hierarchically organized structures.

Further insight into early steps in fiber production and the phase separation of the fibroin was provided by Kaplan and his coworkers describing the behavior of fibroin and polyethyleneoxide (PEO) blends (Jin and Kaplan, 2003). They proposed a model of molecular folding, micelle and globule formation, and thread development under shearing (Figure 6b). Gradually increasing the amounts of high-molecular-weight PEO to the fibroin aqueous solution induces competition for water between PEO and fibroin. It simulates the state change of fibroin into forming micelles and globules when water becomes less available inducing an assembly process (Shulha et al., 2006). These authors also suggested that the role of sericin is to adsorb water and increase the fibroin concentration.

These insights are forming the basis for advancing artificial spinning through mimicking of the natural process.

3.2. Solution spinning

The overall performance difference between the artificial fibroin silk and natural silk is induced by many factors. Composition of the spinning dope is critical but not the only factor. Important to understand is that the spinning process which determines the condensed structure of silk is crucial. It suggest that knowing the spinning process details it should be feasible to produce high-performance silk artificially and "design" silk.

The natural spinning by silkworm and spider is a "dry-wet spinning." It implies that an aqueous silk protein solution with high concentration is spun into a solid thread through without a coagulation bath. Current artificial spinning of silk protein are nearly all operating using the wet spinning process which needs a coagulation bath.

Regenerated spidroin and fibroin dissolved in various solvents are used as spinning dope while the coagulation baths are mainly alcohol (Table 4). In addition certain fiber post-treatments, such as drawing, are used as well.

Table 4 Examples of wet spinning processes using regenerated fibroin, spidroin, and recombinant spidroin, with different solvents and coagulation baths (Zhou et al., 2006a)

Spinning dope	Coagulation bath	References
Spidroin in HFIP	Acetone	Seidel et al. (2000)
Spidroin aqueous solution	Air	Shao et al. (2003)
Recombinant spidroin (DP-1B) in HFIP	Methanol	O'Brien et al. (1998)
Recombinant spidroin (ADF-3) in water	Methanol/water	Lazaris et al. (2002)
Fibroin in LiBr·H$_2$O—EtOH—H$_2$O	Methanol	Matsumoto et al. (1996)
Fibroin aqueous solution (dialysis after dissolving in Ca(NO$_3$)$_2$·4H$_2$O$_2$MeOH)	Methanol	Ha et al. (2003)
Fibroin in 95% formic acid and anhydrous TFA[a]	Methanol	Ha et al. (2005)
Fibroin in 98% formic acid[a]	Methanol	Um et al. (2004a)
Fibroin in HFIP[a]	Methanol	Zhao et al. (2003)
Fibroin in HFA[a]	Methanol	Yao et al. (2002)
Fibroin in mixture of formic acid and LiCl	Methanol	Lock (1992)
Fibroin aqueous solution with high concentration	(NH$_4$)$_2$SO$_4$	Lin et al. (2007)
Fibroin in NMMO	Air gap and ethanol	Corsini et al. (2007)
Fibroin in HFIP[a]	Ethanol and methanol	Zuo et al. (2007)
Fibroin in 98% formic acid[a]	Methanol	Ki et al. (2007b)
Fibroin in mixture of phosphoric and formic acid	Methanol	Ki et al. (2007c)
Fibroin in NMMO containing n-propyl	Ethanol	Marsano et al. (2005)

The properties of some production are listed in Table 3.
[a]The corresponding solutions were prepared by dissolving fibroin films.

The wet spinning of regenerated spidroin was reported in the early 1990s by Jelinski et al. They dissolved spider silk in hexafluoroisopropanol (HFIP) at a concentration of 2.5 wt% to produce an artificial fiber using water, methanol, isopropanol, and acetone as coagulation bath. The reconstituted silk could only be shaped in acetone but the structure

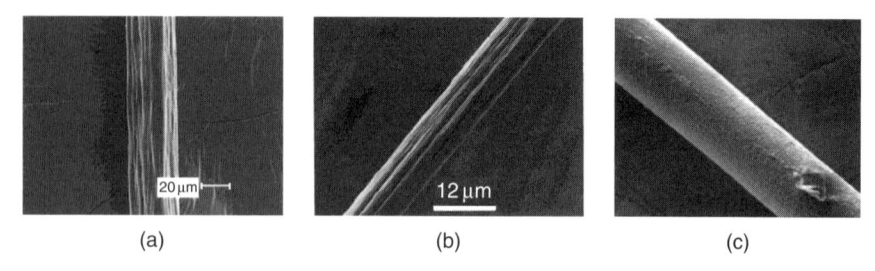

Figure 7 The morphologies of regenerated spidroin silks. (a) The silk formed in acetone and subsequently stretched (Seidel et al., 2000). (b) Silk drawn out of solution into air (Shao et al., 2003). (c) Silk produced from the recombinant ADF-3 (Lazaris et al., 2002).

as well as the properties were of low quality (Figure 7a) (Seidel et al., 1998, 2000). Shao et al. (2003) also used a regenerated spidroin aqueous solution to obtain a thread in air but again the properties and structure of the silk was not satisfactory. Poor structural regularity and lower β-sheet content were considered as causes, as the amino acids sequence was nearly identical with the originating natural silk (Figure 7b).

Recombinant spidroins have also been synthesized and spun into fibers (Hinman and Lewis, 1992; Lazaris et al., 2002; O'Brien et al., 1998; Xu and Lewis, 1990). Among these reports, Lazaris et al. (2002) obtained the best performing artificial spider silk. The spinning dope is the ADF-3 aqueous solution (23 wt%) and the resulting silk has higher elasticity but lower tensile strength compared to dragline silk (Figure 7c).

Considering the source of raw materials, neither regenerated spidroin nor recombinant spidroin is available in large amounts. This is an obstacle for these two production methods when commercial usage or even research is aimed for. Obtaining fibroin from the silkworm silk is well known and widely used as it is easier to obtain fibroin solution as spinning dope. Accordingly, research has focused more on the wet spinning of fibroin solution.

Pioneering work in fibroin wet spinning can be traced back to 1930s. After that, little work has been done until the late 1980s, when more research was done to investigate the spinning dope systems, and structure and properties of the artificial fibroin silk. The composition of the dope is very important to the properties of the final fiber. Several kinds of solvents, such as LiBr—EtOH, $Ca(NO_3)_2$—MeOH, formic acid, HFIP, hexafluoro acetone (HFA), and so on, are used to prepare the spinning dope (Table 4). Very recently, an ionic liquid was used as dope solvent (Phillips et al., 2005).

Hudson et al. used $Ca(NO_3)_2$—MeOH to dissolve silkworm silk and did not manage to obtain good fibers directly. They also casted film from such a solution and then dissolved the film in formic acid and trifluoroacetate acid

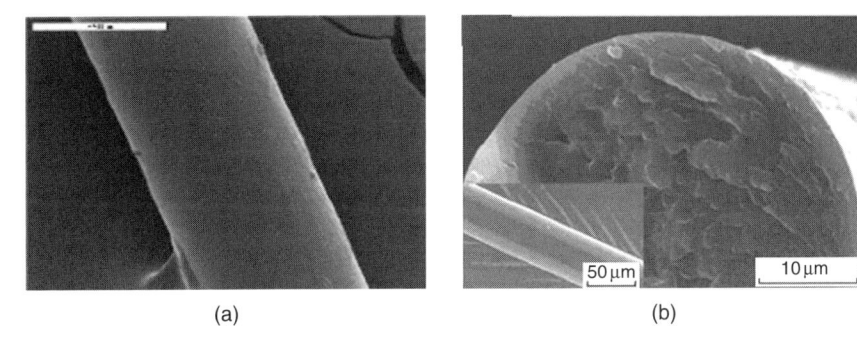

Figure 8 SEM pictures showing the morphology of artificial fibroin fibers from different processings. (a) Drawn fibers from TFA dope solution (Ha et al., 2005). (b) Cross-section and surface (inset) of fiber after post-drawn from aqueous solution (Lin et al., 2007).

(TFA). The reobtained solution could be spun into better fibers (Figure 8a) (Ha et al., 2003, 2005). Park and his coworkers did similar experiments but focused on the formic acid system. They compared the influence of many coagulation solvents and the processing temperature to the morphology of the final fibers. For alcohol systems (ROH), a larger R group will cause nonround fiber cross sections, and increasing the temperature of the coagulation bath will decrease the diameter of fiber (Um et al., 2004a, b). HFIP and HFA are thought to be good solvents for the fibroin wet spinning, as the solvents do not degrade fibroin and the solutions are stable. Although the produced artificial fibroin fibers have very good properties, the applied solvents are very expensive, being prohibitive for further commercial developments (Lock, 1992; Zhao et al., 2003).

Most of the solvent systems mentioned are rather harsh, and some will severely degrade the fibroin. Some will dissolve the fibroin only in relatively low concentration at room temperature. It is also difficult to obtain a spinning dope with a high-molecular-weight fibroin and a high concentration at the same time. Chen et al. (2004), however, have developed a novel method to prepare the fibroin aqueous solution, using high molecular weight and high concentration. Such a solution is good for the conformation transition and for the fabrication. Furthermore, the process of producing the solution is environmental friendly. With this aqueous fibroin solution, Lin et al. (2007) have spun the corresponding fiber with good mechanical properties using $(NH_4)_2SO_4$ as coagulation bath and having a post-treatment, that is, postdrawing and steam annealing.

Overall Table 3 indicates that the properties of artificial silk fibers cannot match those of native fibers (Table 2 and Figure 4).

From a scientific perspective, the artificial silk experiments have provided insight into the morphology of reconstituted silk. In the spinning dope, fibroin molecules adapt a random coil or other less extended conformations.

As the dope is pressed into the coagulation bath, the fibroin precipitates immediately and the random coil state is fixed limiting the intermolecular–protein interactions and producing a brittle fiber. Postspin treatment, such as postdrawn, may extend the molecules and ameliorate some intramolecular interaction into the needed intermolecular ones to the benefit of the orientation of molecules and the formation of β-sheets. These structural changes increase the mechanical properties to a certain degree (Corsini et al., 2007; Ha et al., 2005; Lin et al., 2007). Lin et al. (2007) have obtained an artificial fibroin silk from aqueous dope, whose mechanical properties are close or even better than those of natural silk (cocoon fiber). This encouraging result was obtained from an aqueous fibroin solution with high molecular weight and high concentration (Chen et al., 2004) and from the steam-annealing post-treatment.

Nevertheless, silk spinning remains a very complex process. Spiders and silkworms not only have a set of well-developed spinning glands but also have a set of well-defined and controlled chemical boundary conditions. Besides the composition of the spinning dope, the spinning techniques and the combination of chemical parameters (pH and metallic ions) must be considered and optimized.

3.3. Electrospining

Electrospinning is a direct method to produce nanofibers out of polymer solutions. The formation of fine fibers by electrospinning has been widely explored for applications such as high-performance filters and biomaterial scaffolds for cell growth, vascular grafts, wound dressings, or tissue engineering (Bognitzki et al., 2001; Boland et al., 2006). These applications benefit from the high surface area of the electrospun mat, their good biocompatibility, and fair mechanical properties (Jin et al., 2002). Therefore, electrospinning with fibroin/spidroin solutions is increasingly investigated to broaden the application of silk proteins.

Zarkoob et al. (1998, 2004) were the first to report on the electrospinning of silkworm silk and *Nephila clavipes* dragline protein. They used an HFIP solution of protein as the spinning dope. The resulting fibers had a wide distribution in diameter and the continuity during spinning could be significantly improved.

Since then research on the electrospinning of fibroin has led to improved results (Figure 9a). Most research is focused on improving the electrospinning device and optimizing the processing variables, such as the concentration of the fibroin solution, the strength of electricfield, and distance of collection, all greatly influencing the diameter distribution and the morphology of obtained silk nonwoven mat (Figure 9b). Also, nanofibers of silk protein with different oriented alignment were obtained by changing the shape of the injector electrode defining the Taylor cone geometry (Ayutsede

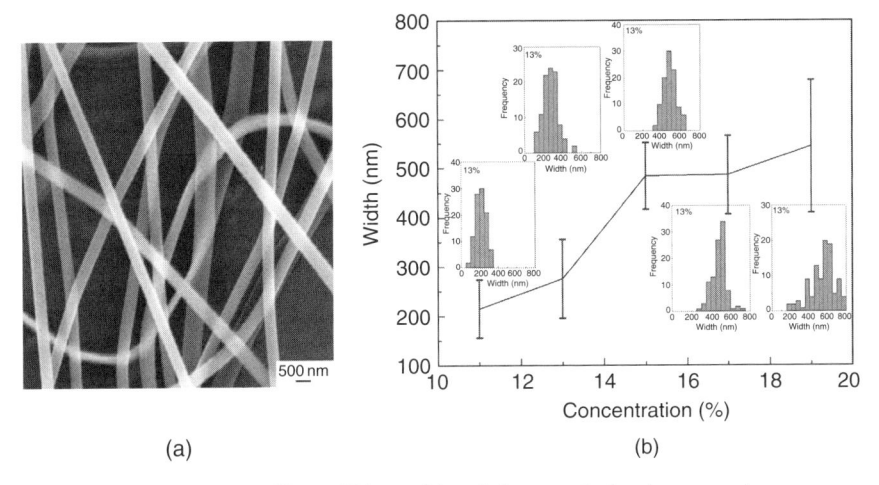

Figure 9 The electrospun fiber of fibroin (a) and the correlation between the concentration of fibroin and the diameter of electrospun fiber (b).

et al., 2006). Similar to the wet spinning process, the spinning dope is important in electrospinning. A number of electrospinning of fibroin experiments are summarized in Table 5.

In order to improve the properties and the spinnability, fibroin sometimes has been electrospun together with other natural or synthetic polymers (Jin et al., 2002; Park et al., 2004, 2006; Wang et al., 2004, 2006). For instance, Jin et al. (2002) developed an aqueous process for silk electrospinning in combination with PEO. More recently, Cao (2008) used PVA/Silk Fibroin (SF), Gelatin/SF, and Hydroxyapatite (HAP)/SF to produce double-layered (core-shell) nanofibers (mats) by coelectrospinning.

Several in vitro experiments have been carried out to examine the biological properties of fibroin nonwoven mats. The results show that the mats are suitable for cell adsorption, differentiation, growth, and propagation. The fibroin nonwoven mats really indicated the great potential for application in wound dressings and tissue engineering (Jin et al., 2004; Kim et al., 2005b; Meechaisue et al., 2007; Min et al., 2004, 2006; Wang et al., 2006).

Most electrospinning of fibroin is performed using formic acid or HFIP as solvent. Considering the biomedical application and environmental issues related to these solvents, electrospinning of fibroin without use of organic solvents is being pursued. All-aqueous processes for regenerated silk fibroin (RSF) electrospinning are reported (Wang et al., 2005, 2006; Zhu et al., 2007), with optimal concentrations of RSF around 30 wt%. In such electrospinning dopes, the RSF can only be considered as "protein fragments" rather than the complete protein in terms of molecular weight. Accordingly, only poor mechanical properties of the prepared nonwoven RSF mat are observed (e.g., breaking stress and strain were 0.82 MPa

Table 5 Some fibroin electrospinning results using different solution system

Solvent	Concentration (%)	Diameter (nm)	References
HFIP	0.23–1.2	6.5–200	Zarkoob et al. (1998, 2004)
Formic acid		30–120	Kim et al. (2003)
HFA	2–10	100–1,000	Ohgo et al. (2003)
Formic acid	5–20	20–400	Sukigara et al. (2003)
Formic acid	3–15	30–120	Min et al. (2004)
Formic acid	10–16.5	<40	Ayutsede et al. (2005); Sukigara (2004)
Formic acid	12	20–150	Ayutsede et al. (2006)
Water	28–37	~1,000	Chen et al. (2006a)
HFIP	5–10	250–550	Min et al. (2006)
Water	17–39	~700	Wang et al. (2005, 2006)
HFIP	7	500–1,200	Jeong et al. (2006)
Formic acid	12–18	100–400	Ki et al. (2007a)
Water	30–38	1,330–2,780	Zhu et al. (2007)

The morphology of the result nano-fibers can be tuned.

and 0.76%, respectively). It should be noted that the mechanism of electrospinning totally differs from that of wet spinning, for determining the final structure and associated performance as previously described.

4. BIOAPPLICATION OF SILK FIBROIN

Fibroin products with different morphologies (such as film, fiber, sponge, and nonwoven mat) have been produced successfully. In view of the overall economics, most applications are in the biomedical field. Silk sutures are used for centuries, and silk fibroin has been applied in clinical repairs and tissue engineering (Vepari and Kaplan, 2007). As a biomaterial, silk fibroin is biocompatible and nonbiotoxic, that is, biologically inert, and possesses some special advantages, namely excellent mechanical properties, controllable proteolytic biodegradability, morphologic flexibility, versatility for sterilization (Karageorgiou et al., 2004; Meinel et al., 2004a; Sugihara et al., 2000), and options for chemical modification (Gotoh et al., 1998; Hakimi et al., 2007; Vepari and Kaplan, 2006, 2007).

Biodegradability and in particular control over the rate of degradation is critical for tissue regeneration. In an optimized design, the rate of scaffold degradation needs to match the rate of tissue growth (Altman et al., 2003). Although silk itself is defined as a nondegradable biomaterial by the United States Pharmacopeia, it is reported that fibroin is susceptible to proteolytic degradation and can be absorbed slowly after implantation in vivo and after a long-term, indicating that silk materials can provide adequate and robust support over a period of time (Arai et al., 2004; Kim et al., 2005c), in particular useful for slow tissue growth. The degradation of fibroin materials depends on many parameters, such as structure (porosity and pore size), morphology, molecular weight, and biological conditions at different implantation locations. Essential is the content of β-sheet which directly affects the rate of degradation (Huang et al., 2007; Minoura et al., 1990).

Many silk-based materials have been investigated in vitro and in vivo because of the perceived morphologic flexibility and the need for biocompatible materials for use in tissue engineering, bone engineering, and ligament repair. Table 6 lists the silk-based materials with different morphologies in different bioapplications. Some evidence suggests that these materials exhibit better biomedical properties than collagen, another protein fiber and a common biomedical material (Altman et al., 2003; Vepari and Kaplan, 2007). For example, the anterior cruciate ligaments (ACLs) have complex structures and require high mechanical properties. Several

Table 6 A listing of some bioapplication of silk-based materials

Morphology	Application	References
Fiber	Ligament tissue	Altman et al. (2002); Chen et al. (2003); Moreau et al. (2005)
	Tendon tissue	Kardestuncer et al. (2006)
Twisted silk rope	Ligament tissue	Altman et al. (2003)
Film	Wound dressings	Sugihara et al. (2000)
	Bone engineering	Karageorgiou et al. (2004); Kino et al. (2006); Sofia et al. (2001); Tanaka et al. (1999)
	Hepatic tissue	Hu et al. (2006a)
	Antithrombogenesis	Lee et al. (1998)
Nonwoven mat	Bone engineering	Kim et al. (2005b); Li et al. (2006)
	Connective tissue	Dal Pra et al. (2005)
	Endothelial and blood vessel	Fuchs et al. (2006); Unger et al. (2004)

Table 6 (*Continued*)

Morphology	Application	References
Porous sponge	Wound dressings	Yeo et al. (2000)
	Bone engineering	Karageorgiou et al. (2006); Kim et al. (2005a); Marolt et al. (2006); Meinel et al. (2004b, 2005, 2006a, b)
	Cartilage engineering	Aoki et al. (2003)
Hydrogel	Cartilage engineering	Aoki et al. (2003)
	Bone engineering	Fini et al. (2005); Motta et al. (2004)
Spherical particles	Drug carrier and release	Cao et al. (2007); Wang et al. (2007a, b)

Modified from literature (Vepari and Kaplan, 2007).

materials have been tried in ACL, but because of their supreme mechanical properties, silk-based materials seem better candidates in this particular area (Altman et al., 2002, 2003; Ge et al., 2006). Moreover, silk proteins have been probed in bone reconstruction because of their ability to deposit minerals.

As the coating protein of cocoon silk, sericin is proposed for bioapplications, although it has been reported to cause hypersensitivity (Altman et al., 2003). It is suggested that sericin could suppress the development of colon tumor by reducing cell proliferation and nitric acid production (Zhaorigetu, 2001). In addition, it can induce the deposit of bone-like apatite (Takeuchi et al., 2003; Terada et al., 2005; Tsubouchi et al., 2005).

5. BIOMINERALIZATION REGULATED BY SILK PROTEINS

Living organisms are capable of using proteins to deposit several kinds of minerals with specific function such as teeth and bone. This phenomenon is termed biomineralization. Typically, the biominerals formed during such a process have exquisite structure and possess outstanding mechanical and optical properties (Mann, 2001). Among these minerals, calcium carbonate, HAP, and silica have been widely investigated. There are several different morphologies of calcium carbonate in organisms, such as the brick-like structure sandwiched by proteins in nacre (Addadi et al., 2006; Kato et al., 2002), the sponge-like and fenestrated structure in sea urchins (Park and

Meldrum, 2002), the elaborately shaped spicules in ascidians, and finely the sculpted coccolith shells in coccolithophores (Mann, 2001). For silica structure, the diatoms produce the most exquisite siliceous cell wall (Sumper and Brunner, 2006). Also the calcium phosphate in bone shows a highly regulated organization and arrangement. Interestingly, avians have two main biomineralization systems, one produces calcium carbonate for egg shell and the other produces calcium phosphate for bone (Bauerlein, 2000; Mann, 2001).

Research related to biomineralization mainly includes three aspects, investigation of the structure of biominerals and the structure–function relationship (Bauerlein, 2003; Mann, 2001), the mechanism of the formation of minerals (in vivo and in vitro) and the exact control over the morphology of minerals by organism (Xu et al., 2007), and the production of new and high-performance materials under the concept of biomineralization (Slocik and Naik, 2007). In the latter field, new high-performance or functional hybrid-materials that are environmental friendly and low energy consuming could be produced by mimicking the process of biomineralization. Calcium carbonate, for example, is commonly used as model mineral. Synthesizing calcium carbonate-rich hybrid materials with controlled morphology is an important research subject (Colfen, 2003). Generally, the calcium carbonate is synthesized with the collaboration of an insoluble matrix and soluble molecules in the organism (mainly mollusk shell) (Addadi et al., 2006; Wei et al., 2007; Ajikumar et al., 2004; Hosoda et al., 2003; Litvin et al., 1997; Nudelman et al., 2006). In the experiments that investigate mechanism in vitro, synthesis of calcium carbonate in the presence of organic templates and soluble inhibitors is an effective technique, namely the template direction and growth inhibition method. Polyelectrolytes (Sugawara et al., 2003; Wang et al., 2005, 2006), double-hydrophilic block copolymers (DHBCs) (Colfen, 2001; Xu et al., 2007; Yu and Colfen, 2004), and biopolymers (Belcher et al., 1996; Butler et al., 2006; Falini et al., 1996; Feng et al., 2000; Hardikar et al., 2001; Leveque et al., 2004; Li et al., 2002; Raz et al., 2000; Shen et al., 2002; Verraest et al., 1996) including dextran, peptides, and proteins, such as collagen and fibroin, have been employed as soluble additives.

It is well known that well-ordered β-chitin (a polysaccharide) associated with a less ordered protein in the β-sheet conformation is the main component of nacreous organic matrix in shell. The amino acid sequence of such proteins is very similar to those of silk fibroins. Indeed, the amino acid sequence of a major protein from the nacreous shell layer of the pearl oyster resembles that of spidroin (Sudo et al., 1997; Weiner and Traub, 1980). The question of whether silk-like proteins play an important role in shell formation is raised. When Falini et al. (1996) did the experiment with the proteins from the shell, they assembled a substrate in vitro that contained β-chitin and natural silk fibroin and concluded that the silk fibroin may influence ion diffusion or the accessibility to the chitin surface or both. Furthermore, cryo-TEM study of the structure of the *Atrina* shell nacreous organic matrix without dehydration suggested that the silk-like shell protein was not in the form of a distinguishable ordered layer but in the form of an disordered hydrated gel (Levi-Kalisman et al., 2001). It meant that silk-like proteins acting as an organizing template require the cooperation of other organic materials, for example, β-chitin and other acidic proteins.

The role of the individual silk-like protein played is unclear, and whether the silk-like protein may dominate the crystallization of calcium carbonate or not is still unknown. To provide experimental insights into the interaction of minerals and proteins, a model system containing RSF or spidroin as templates may be used for the crystallization of calcium carbonate.

Li (2005) used regenerated fibroin film as a substrate for $CaCO_3$ crystallization, in cooperation with soluble acidic peptides. It was found that the growth of $CaCO_3$ was aligned along the oriented silk films. It was suggested that fibroin molecules could cause the acid functional groups to be ordered over a large distance by orientation of molecules, which led to the alignment of $CaCO_3$. Moreover, a conformation transition of silk fibroin happened during the crystallization, which implied that the silk-like proteins in mollusk shell play a pivotal role during the formation of aragonite crystals in the nacre sheets through the interaction between crystal and proteins.

The use of fibroin, as a soluble additive rather than an insoluble matrix, can influence the crystallization of $CaCO_3$ (Cheng et al., in press). It was observed that the inherent (self-assembling) aggregation process of silk fibroin molecules affected both the morphology and crystallographic polymorphism of $CaCO_3$. This combination stimulated the growth of a novel, rice-grain-shaped protein/mineral hybrid with a hollow structure with an aragonite polymorphism (Figure 10a). These observations suggest new hypotheses about the role of silk-like protein in the natural biomineralization process. It also may serve to shed light on the formation process of those ersatz hybrids regulated by artificially selected structural proteins.

As fibroin is considered a good biomaterial, a number of experiments have been performed in bone engineering. Experiments in which fibroin or spidroin is used to modulate the crystallization of calcium phosphate have been carried out (Cao and Mao, 2007; Kino et al., 2006; Kong et al., 2004; Yao et al., 2007; Wang et al., 2007; Takeuchi et al., 2005). The fibroin solution can regulate the crystallization of calcium phosphate through the change of the initial structures, which is controlled by some chemical conditions such as metallic ions and pH value (Kong et al., 2004; Yao et al., 2007). Takeuchi et al. (2003) managed to deposit bone-like apatite on natural silk in solutions that mimic physiological conditions. It was found that the deposition benefited from the existence of sericin in natural silk fiber. Cao and Mao (2007) carried out the mineralization experiment

(a)   (b)

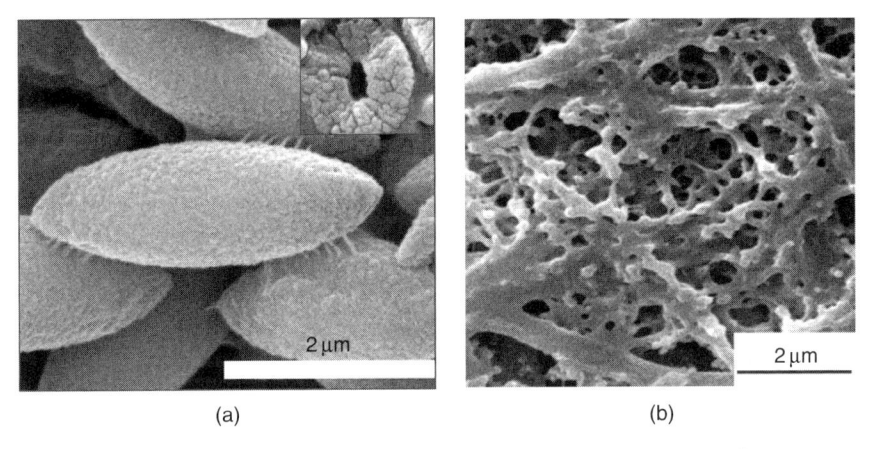

Figure 10 A novel micron-sized hybrid of silk fibroin and CaCO$_3$, with a hollow structure and unusual rice-like shape is synthesized in the presence of regenerated silk fibroin via aggregation-mediated crystallization (a); mineralization of HAP on the electrospun mat of silk fibroin (b).

on the spider dragline silk. As spider dragline silk can be pictured as the oriented organization of protein nanocrystals along the long axis of fiber, the nucleation sites for HAP orient along the fiber, which results in the preferentially c-axis orientation of HAP forming on the surface of silk. The authors suggested that the mineralized silks would combine the good mechanical properties of the spider silks and the biocompatibility of HAP and may be assembled into the ideal biomaterials for bone implants. Furthermore, a preliminary study has indicated that the HAP can mineralize on the surface of electrospun fibroin nanofibers as well as in the interstitial space of the mat (Figure 10b) (Cao, 2008).

Some silk-based materials have been used for bone reconstruction or cell proliferation (Meinel et al., 2004c, 2006b; Sofia et al., 2001; Tanaka et al., 2007; Vepari and Kaplan, 2007; Wang et al., 2007). Both in vivo and in vitro tests show that silk proteins have good affinity to bone cell and good bone conductivity. It is therefore anticipated that silks and their proteins will play more important roles in the field of bone tissue engineering, especially the material blend of silk and HAP or other growth agents (Kirker-Head et al., 2007; Li et al., 2005).

Recently, Foo et al. (2006) produced some novel nanocomposites from spider silk–silica fusion (chimeric) proteins. The composite morphology and structure could be regulated by controlling processing conditions to produce films and fibers. Silk and biomineralization being natural inspiration sources will allow production of numerous new materials in various fields of application.

6. CONCLUSION AND PERSPECTIVE

Silks are a unique group of fibrous proteins with unusually high mechanical strength in fiber form. The impressive mechanical properties result from a well-defined hierarchical structure. The primary amino acid sequence structure and the aggregation state of secondary and tertiary levels of organization of the silk proteins contribute to the fiber performance. Several kinds of silk-based materials such as fibers, films, sponges, and spheres have been produced. Due to their natural origin and good properties, silks and their proteins have been widely investigated in biomedical applications. Different morphologies are used in different tissue engineering applications. The rate of degradability can be adjusted by the β-sheet content and the overall fiber morphology.

Among the man-made silk-based materials, fibroin draws the most interest. Obtaining a "super-fiber" without silkworms or spiders from regenerated fibroin or spidroin is the ultimate goal. To date, artificial fibers still lags behind natural ones in term of mechanical properties. The reason is that the artificial fibers do not have the well-defined structure of natural ones, despite the similar amino acids compositions. Studying the natural spinning process has generated a lot of understanding useful for developing better artificial silk fabrication processes. Key is to optimize the process of artificial spinning by combining the physical (spinning dope, coagulation bath, spinning rate, and post-treatment) and chemical conditions (metallic ions and pH) that allow for good orientation and crystallization of the primary structures.

To extend the application area of silk proteins-based materials, blending the fibroin with other natural macromolecules and synthetic polymers, or even manufacturing composites with silk fibers are a few of the possible strategies.

ACKNOWLEDGMENT

The authors thank the National Natural Science Foundation of China (NSFC 20525414 and 20434010), as well as the Programme for New Century Excellent Talents in University of China (NCET), the Programme for Changjiang Scholars and Innovative Research Team in Fudan University for financial supports.

REFERENCES

Addadi, L., Joester, D., Nudelman, F., and Weiner, S. "Mollusk shell formation: A source of new concepts for understanding biomineralization processes". *Chem. Eur. J.* **12**(4), 981–987 (2006).

Ajikumar, P.K., Lakshminarayanan, R., and Valiyaveettil, S. "Controlled deposition of thin films of calcium carbonate on natural and synthetic templates". *Crystal Growth Des.* **4**(2), 331–335 (2004).

Altman, G.H., Horan, R.L., Lu, H.H., Moreau, J., Martin, I., Richmond, J.C., and Kaplan, D.L. "Silk matrix for tissue engineered anterior cruciate ligaments". *Biomaterials* **23**(20), 4131–4141 (2002).

Altman, G.H., Diaz, F., Jakuba, C., Calabro, T., Horan, R.L., Chen, J.S., Lu, H., Richmond, J., and Kaplan, D.L. "Silk-based biomaterials". *Biomaterials* **24**(3), 401–416 (2003).

Aoki, H., Tomita, N., Morita, Y., Hattori, K., Harada, Y., Sonobe, M., Wakitani, S., and Tamada, Y. "Culture of chondrocytes in fibroin-hydrogel sponge". *Bio. Med. Mater. Eng.* **13**(4), 309–316 (2003).

Arai, T., Freddi, G., Innocenti, R., and Tsukada, M. "Biodegradation of Bombyx mori silk fibroin fibers and films". *J. Appl. Polym. Sci.* **91**(4), 2383–2390 (2004).

Asakura, T., Sugino, R., Okumura, T., and Nakazawa, Y. "The role of irregular unit, GAAS, on the secondary structure of Bombyx mori silk fibroin studied with C-13 CP/MAS NMR and wide-angle X-ray scattering". *Protein Sci.* **11**(8), 1873–1877 (2002).

Asakura, T., Suita, K., Kameda, T., Afonin, S., and Ulrich, A.S. "Structural role of tyrosine in Bombyx mori silk fibroin, studied by solid-state NMR and molecular mechanics on a model peptide prepared as silk I and II". *Magn. Reson. Chem.* **42**(2), 258–266 (2004).

Asakura, T., and Yao, J.M. "C-13 CP/MAS NMR study on structural heterogeneity in Bombyx mori silk fiber and their generation by stretching". *Protein Sci.* **11**(11), 2706–2713 (2002).

Ayutsede, J., Gandhi, M., Sukigara, S., Micklus, M., Chen, H.E., and Ko, F. "Regeneration of Bombyx mori silk by electrospinning. Part 3: Characterization of electrospun nonwoven mat". *Polymer* **46**(5), 1625–1634 (2005).

Ayutsede, J., Gandhi, M., Sukigara, S., Ye, H.H., Hsu, C.M., Gogotsi, Y., and Ko, F. "Carbon nanotube reinforced Bombyx mori silk nanofibers by the electrospinning process". *Biomacromolecules* **7**(1), 208–214 (2006).

Baeuerlein, E. "Biomineralization. From Biology to Biotechnology and Medical Applications". Wiley-VCH: , Weinheim, 2000.

Bauerlein, E. "Biomineralization of unicellular organisms: an unusual membrane biochemistry for the production of inorganic nano- and microstructures". *Angew. Chem. Int. Ed.* **42**(6), 614–641 (2003).

Becker, N., Oroudjev, E., Mutz, S., Cleveland, J.P., Hansma, P.K., Hayashi, C.Y., Makarov, D.E., and Hansma, H.G. "Molecular nanosprings in spider capture-silk threads". *Nat. Mater.* **2**(4), 278–283 (2003).

van Beek, J.D., Hess, S., Vollrath, F., and Meier, B.H. "The molecular structure of spider dragline silk: Folding and orientation of the protein backbone". *Proc. Natl. Acad. Sci. U.S.A* **99**(16), 10266–10271 (2002).

van Beek, J.D., Kummerlen, J., Vollrath, F., and Meier, B.H. "Supercontracted spider dragline silk: a solid-state NMR study of the local structure". *Int. J. Biol. Macromol.* **24**(2–3), 173–178 (1999).

Belcher, A.M., Wu, X.H., Christensen, R.J., Hansma, P.K., Stucky, G.D., and Morse, D.E. "Control of crystal phase switching and orientation by soluble mollusc-shell proteins". *Nature* **381**(6577), 56–58 (1996).

Bognitzki, M., Czado, W., Frese, T., Schaper, A., Hellwig, M., Steinhart, M., Greiner, A., and Wendorff, J.H. "Nanostructured fibers via electrospinning". *Adv. Mater.* **13**(1), 70–72 (2001).

Boland, E.D., Pawlowski, K.J., Barnes, C.P., Simpson, D.G., Wnek, G.E., and Bowlin, G.L. Electrospinning of bioresorbable polymers for tissue engineering scaffolds, in "Polymeric Nanofibers", Vol. 918, pp. 188–204. . American Chemical Society, , Washington, DC (2006).

Butler, M.F., Glaser, N., Weaver, A.C., Kirkland, M., and Heppenstall-Butler, M. "Calcium carbonate crystallization in the presence of biopolymers". *Crystal Growth Des.* **6**(3), 781–794 (2006).

Cao, B., and Mao, C. "Oriented nucleation of hydroxylapatite crystals on spider dragline silks". *Langmuir* **23**(21), 10701–10705 (2007).

Cao, H. "Preparation of Silk Fibroin-based Electrospun Mat and Deposition of Inorganic Mineral on Silk Fibroin Matrix" Ph.D dissertation, Fudan University, 2008

Cao, Z.B., Chen, X., Yao, J.R., Huang, L., and Shao, Z.Z. "The preparation of regenerated silk fibroin microspheres". *Soft Matter* **3**(7), 910–915 (2007).
Chen, C., Cao, C.B., Ma, X.L., Tang, Y., and Zhu, H.S. "Preparation of non-woven mats from all-aqueous silk fibroin solution with electrospinning method". *Polymer* **47**(18), 6322–6327 (2006a).
Chen, J.S., Altman, G.H., Karageorgiou, V., Horan, R., Collette, A., Volloch, V., Colabro, T., and Kaplan, D.L. "Human bone marrow stromal cell and ligament fibroblast responses on RGD-modified silk fibers". *J. Biomed. Mater. Res. Part A* **67A**(2), 559–570 (2003).
Chen, X., Knight, D.P., Shao, Z.Z., and Vollrath, F. "Conformation transition in silk protein films monitored by time-resolved Fourier transform infrared spectroscopy: Effect of potassium ions on Nephila spidroin films". *Biochemistry* **41**(50), 14944–14950 (2002a).
Chen, X., Knight, D.P., and Vollrath, F. "Rheological characterization of Nephila spidroin solution". *Biomacromolecules* **3**(4), 644–648 (2002b).
Chen, X., Shao, Z.Z., Knight, D.P., and Vollrath, F. "Conformation transition of silk protein membranes monitored by time-resolved FTIR spectroscopy: Effect of alkali metal ions on Nephila spidroin membrane". *Acta Chimica Sinica* **60**(12), 2203–2208 (2002c).
Chen, X., Shao, Z.Z., and Vollrath, F. "The spinning processes for spider silk". *Soft Matter* **2**(6), 448–451 (2006).
Chen, X., Shao, Z.Z., and Zhou, L. Regenerated silk protein solutions having high concentrations and preparation methods. Chinese patent 2004, CN1483866 (2004).
Cheng, C., Shao, Z.Z., and Vollrath, F. "Silk fibroin regulating the crystallization process of calcium carbonate". *Adv. Funct. Mater.* 18(15), 2172–2179 (2008).
Colfen, H. "Precipitation of carbonates: Recent progress in controlled production of complex shapes". *Curr. Opin. Colloid Interface Sci.* **8**(1), 23–31 (2003).
Colfen, H. "Double-hydrophilic block copolymers: Synthesis and application as novel surfactants and crystal growth modifiers". *Macromol. Rapid Commun.* **22**(4), 219–252 (2001).
Corsini, P., Perez-Rigueiro, J., Guinea, G.V., Plaza, G.R., Elices, M., Marsano, E., Carnasciali, M.M., and Freddi, G. "Influence of the draw ratio on the tensile and fracture behavior of NMMO regenerated silk fibers". *J. Polym. Sci. B Polym. Phys.* **45**(18), 2568–2579 (2007).
Dal Pra, I., Freddi, G., Minic, J., Chiarini, A., and Armato, U. "De novo engineering of reticular connective tissue in vivo by silk fibroin nonwoven materials". *Biomaterials* **26**(14), 1987–1999 (2005).
Dicko, C., Kenney, J.M., Knight, D., and Vollrath, F. "Transition to a beta-sheet-rich structure in spidroin in vitro: The effects of pH and cations". *Biochemistry* **43**(44), 14080–14087 (2004).
Dicko, C., Vollrath, F., and Kenney, J.M. "Spider silk protein refolding is controlled by changing pH". *Biomacromolecules* **5**(3), 704–710 (2004).
Drummy, L.F., Phillips, D.M., Stone, M.O., Farmer, B.L., and Naik, R.R. "Thermally induced alpha-helix to beta-sheet transition in regenerated silk fibers and films". *Biomacromolecules* **6**(6), 3328–3333 (2005).
Eles, P.T., and Michal, C.A. "Strain dependent local phase transitions observed during controlled supercontraction reveal mechanisms in spider silk". *Macromolecules* **37**(4), 1342–1345 (2004).
Emile, O., Le Floch, A., and Vollrath, F. "Biopolymers: Shape memory in spider draglines". *Nature* **440**(7084), 621 (2006).
Engelberg, I., and Kohn, J. "Physicomechanical properties of degradable polymers used in medical applications – a comparative-study". *Biomaterials* **12**(3), 292–304 (1991).
The ExPASy (Expert Protein Analysis System, http://www.expasy.ch/) proteomics server of the Swiss Institute of Bioinformatics, P05790.
Fahnestock, S. R., and Steinbuchel, A. (Eds), "Biopolymers: Polyamides and Complex Proteinaceous Materials II", Vol. 8, Chapter 1 to Chapter 5, Wiley-VCH GmbH & Co. KGaA, Weinheim, Germany (2003).

Falini, G., Albeck, S., Weiner, S., and Addadi, L. "Control of aragonite or calcite polymorphism by mollusk shell macromolecules". *Science* **271**(5245), 67–69 (1996).

Feng, Q.L., Pu, G., Pei, Y., Cui, F.Z., Li, H.D., and Kim, T.N. "Polymorph and morphology of calcium carbonate crystals induced by proteins extracted from mollusk shell". *J. Crystal Growth* **216**(1–4), 459–465 (2000).

Fini, M., Motta, A., Torricelli, P., Glavaresi, G., Aldini, N.N., Tschon, M., Giardino, R., and Migliaresi, C. "The healing of confined critical size cancellous defects in the presence of silk fibroin hydrogel". *Biomaterials* **26**(17), 3527–3536 (2005).

Foo, C.W.P., Patwardhan, S.V., Belton, D.J., Kitchel, B., Anastasiades, D., Huang, J., Naik, R.R., Perry, C.C., and Kaplan, D.L. "Novel nanocomposites from spider silk-silica fusion (chimeric) proteins". *Proc. Natl. Acad. Sci. U.S.A.* **103**(25), 9428–9433 (2006).

Freddi, G., Tsukada, M., and Beretta, S. "Structure and physical properties of silk fibroin polyacrylamide blend films". *J. Appl. Polym. Sci.* **71**(10), 1563–1571 (1999).

Frische, S., Maunsbach, A.B., and Vollrath, F. "Elongate cavities and skin-core structure in Nephila spider silk observed by electron microscopy". *J. Microsc. Oxford* **189**, 64–70 (1998).

Fuchs, S., Motta, A., Migliaresi, C., and Kirkpatrick, C.J. "Outgrowth endothelial cells isolated and expanded from human peripheral blood progenitor cells for endothelialization as a potential source of autologous of silk fibroin biomaterials". *Biomaterials* **27**(31), 5399–5408 (2006).

Ge, Z.G., Yang, F., Goh, J.C.H., Ramakrishna, S., and Lee, E.H. "Biomaterials and scaffolds for ligament tissue engineering"*J. Biomed. Mater. Res. A* **77A**(3), 639–652 (2006).

Gosline, J.M., Demont, M.E., and Denny, M.W. "The Structure and Properties of Spider Silk". *Endeavour* **10**(1), 37–43 (1986).

Gosline, J.M., Guerette, P.A., Ortlepp, C.S., and Savage, K.N. "The mechanical design of spider silks: from fibroin sequence to mechanical function". *J. Exp. Biol.* **202**(23), 3295–3303 (1999).

Grubb, D.T., and Jelinski, L.W. "Fiber morphology of spider silk: the effects of tensile deformation". *Macromolecules* **30**(10), 2860–2867 (1997).

Guinea, G.V., Elices, M., Perez-Rigueiro, J., and Plaza, G. "Self-tightening of spider silk fibers induced by moisture". *Polymer* **44**(19), 5785–5788 (2003).

Gotoh, Y., Tsukada, M., and Minoura, N. "Effect of the chemical modification of the arginyl residue in Bombyx mori silk fibroin on the attachment and growth of fibroblast cells". *J. Biomed. Mater. Res.* **39**(3), 351–357 (1998).

Gupta, M.K., Khokhar, S.K., Phillips, D.M., Sowards, L.A., Drummy, L.F., Kadakia, M.P., and Naik, R.R. "Patterned silk films cast from ionic liquid solubilized fibroin as scaffolds for cell growth". *Langmuir* **23**(3), 1315–1319 (2007).

Ha, S.W., Park, Y.H., and Hudson, S.M. "Dissolution of Bombyx mori silk fibroin in the calcium nitrate tetrahydrate-methanol system and aspects of wet spinning of fibroin solution". *Biomacromolecules* **4**(3), 488–496 (2003).

Ha, S.W., Tonelli, A.E., and Hudson, S.M. "Structural studies of Bombyx mori silk fibroin during regeneration from solutions and wet fiber spinning". *Biomacromolecules* **6**(3), 1722–1731 (2005).

Hakimi, O., Knight, D.P., Vollrath, F., and Vadgama, P. "Spider and mulberry silkworm silks as compatible biomaterials". *Compos. B Eng.* **38**(3), 324–337 (2007).

Hardikar, V.V., and Matijevic, E. "Influence of ionic and nonionic dextrans on the formation of calcium hydroxide and calcium carbonate particles". *Colloids Surf. A* **186**(1–2), 23–31 (2001).

Hayashi, C.Y., and Lewis, R.V. "Evidence from flagelliform silk cDNA for the structural basis of elasticity and modular nature of spider silks". *J. Mol. Biol.* **275**(5), 773–784 (1998).

Hayashi, C.Y., Shipley, N.H., and Lewis, R.V. "Hypotheses that correlate the sequence, structure, and mechanical properties of spider silk proteins". *Int. J. Biol. Macromol.* **24**(2–3), 271–275 (1999).

Heslot, H. "Artificial fibrous proteins: A review". *Biochimie* **80**(1), 19–31 (1998).
Hinman, M.B., Jones, J.A., and Lewis, R.V. "Synthetic spider silk: A modular fiber". *Trends Biotechnol.* **18**(9), 374–379 (2000).
Hinman, M.B., and Lewis, R.V. "Isolation of a Clone Encoding a 2nd Dragline Silk Fibroin – Nephila-Clavipes Dragline Silk Is a 2-Protein Fiber". *J. Biol. Chem.* **267**(27), 19320–19324 (1992).
Hosoda, N., Sugawara, A., and Kato, T. "Template effect of crystalline poly(vinyl alcohol) for selective formation of aragonite and vaterite CaCO3 thin films". *Macromolecules* **36**(17), 6449–6452 (2003).
Hossain, K.S., Ochi, A., Ooyama, E., Magoshi, J., and Nemoto, N. "Dynamic light scattering of native silk fibroin solution extracted from different parts of the middle division of the silk gland of the Bombyx mori silkworm". *Biomacromolecules* **4**(2), 350–359 (2003).
Hu, K., Lv, Q., Cui, F.Z., Feng, Q.L., Kong, X.D., Wang, H.L., Huang, L.Y., and Li, T. "Biocompatible fibroin blended films with recombinant human-like collagen for hepatic tissue engineering". *J. Bioact. Compat. Polym.* **21**(1), 23–37 (2006a).
Hu, X., Vasanthavada, K., Kohler, K., McNary, S., Moore, A.M.F., and Vierra, C.A. "Molecular mechanisms of spider silk". *Cell. Mol. Life Sci.* **63**(17), 1986–1999 (2006b).
Huang, X.T., Shao, Z.Z., and Chen, X. "Investigation on the biodegradation behavior of Bombyx mori silk and porous regenerated fibroin scaffold". *Acta Chimica Sinica* **65**(22), 2592–2596 (2007).
Huemmerich, D., Helsen, C.W., Quedzuweit, S., Oschmann, J., Rudolph, R., and Scheibel, T. "Primary structure elements of spider dragline silks and their contribution to protein solubility". *Biochemistry* **43**(42), 13604–13612 (2004).
Inoue, S., Tanaka, K., Arisaka, F., Kimura, S., Ohtomo, K., and Mizuno, S. "Silk fibroin of Bombyx mori is secreted, assembling a high molecular mass elementary unit consisting of H-chain, L-chain, and P25, with a 6:6:1 molar ratio". *J. Biol. Chem.* **275**(51), 40517–40528 (2000).
Jeong, L., Lee, K.Y., Liu, J.W., and Park, W.H. "Time-resolved structural investigation of regenerated silk fibroin nanofibers treated with solvent vapor". *Int. J. Biol. Macromol.* **38**(2), 140–144 (2006).
Jiang, C.Y., Wang, X.Y., Gunawidjaja, R., Lin, Y.H., Gupta, M.K., Kaplan, D.L., Naik, R.R., and Tsukruk, V.V. "Mechanical properties of robust ultrathin silk fibroin films". *Adv. Funct. Mater.* **17**(13), 2229–2237 (2007).
Jin, H.J., Chen, J.S., Karageorgiou, V., Altman, G.H., and Kaplan, D.L. "Human bone marrow stromal cell responses on electrospun silk fibroin mats". *Biomaterials* **25**(6), 1039–1047 (2004).
Jin, H.J., Fridrikh, S.V., Rutledge, G.C., and Kaplan, D.L. "Electrospinning Bombyx mori silk with poly(ethylene oxide)". *Biomacromolecules* **3**(6), 1233–1239 (2002).
Jin, H.J., and Kaplan, D.L. "Mechanism of silk processing in insects and spiders". *Nature* **424**(6952), 1057–1061 (2003).
Jin, H.J., Park, J., Valluzzi, R., Cebe, P., and Kaplan, D.L. "Biomaterial films of Bombyx mori silk fibroin with poly(ethylene oxide)". *Biomacromolecules* **5**(3), 711–717 (2004).
Karageorgiou, V., Meinel, L., Hofmann, S., Malhotra, A., Volloch, V., and Kaplan, D. "Bone morphogenetic protein-2 decorated silk fibroin films induce osteogenic differentiation of human bone marrow stromal cells". *J. Biomed. Mater. Res. A* **71A**(3), 528–537 (2004).
Karageorgiou, V., Tomkins, M., Fajardo, R., Meinel, L., Snyder, B., Wade, K., Chen, J., Vunjak-Novakovic, G., and Kaplan, D.L. "Porous silk fibroin 3-D scaffolds for delivery of bone morphogenetic protein-2 in vitro and in vivo". *J. Biomed. Mater. Res. A* **78A**(2), 324–334 (2006).
Kardestuncer, T., McCarthy, M.B., Karageorgiou, V., Kaplan, D., and Gronowicz, G. "RGD-tethered silk substrate stimulates the differentiation of human tendon cells". *Clin. Orthopaedics Relat. Res.* **448**, 234–239 (2006).
Kato, T., Sugawara, A., and Hosoda, N. "Calcium carbonate-organic hybrid materials". *Adv. Mater.* **14**(12), 869–877 (2002).
Ki, C.S., Kim, J.W., Hyun, J.H., Lee, K.H., Hattori, M., Rah, D.K., and Park, Y.H. "Electrospun three-dimensional silk fibroin nanofibrous scaffold". *J. Appl. Polym. Sci.* **106**(6), 3922–3928 (2007a).

Ki, C.S., Kim, J.W., Oh, H.J., Lee, K.H., and Park, Y.H. "The effect of residual silk sericin on the structure and mechanical property of regenerated silk filament". *Int. J. Biol. Macromol.* **41**(3), 346–353 (2007b).

Ki, C.S., Lee, K.H., Baek, D.H., Hattori, M., Urn, I.C., Ihm, D.W., and Park, Y.H. "Dissolution and wet spinning of silk fibroin using phosphoric acid/formic acid mixture solvent system". *J. Appl. Polym. Sci.* **105**(3), 1605–1610 (2007c).

Kim, H.J., Kim, U.J., Vunjak-Novakovic, G., Min, B.H., and Kaplan, D.L. "Influence of macroporous protein scaffolds on bone tissue engineering from bone marrow stem cells". *Biomaterials* **26**(21), 4442–4452 (2005a).

Kim, K.H., Jeong, L., Park, H.N., Shin, S.Y., Park, W.H., Lee, S.C., Kim, T.I., Park, Y.J., Seol, Y.J., Lee, Y.M., Ku, Y., Rhyu, I.C., Han, S.B., and Chung, C.P. "Biological efficacy of silk fibroin nanofiber membranes for guided bone regeneration". *J. Biotechnol.* **120**(3), 327–339 (2005b).

Kim, S.H., Nam, Y.S., Lee, T.S., and Park, W.H. "Silk fibroin nanofiber: electrospinning, properties, and structure". *Polym. J.* **35**(2), 185–190 (2003).

Kim, U.J., Park, J., Kim, H.J., Wada, M., and Kaplan, D.L. "Three-dimensional aqueous-derived biomaterial scaffolds from silk fibroin". *Biomaterials* **26**(15), 2775–2785 (2005c).

Kino, R., Ikoma, T., Monkawa, A., Yunoki, S., Munekata, M., Tanaka, J., and Asakura, T. "Deposition of bone-like apatite on modified silk fibroin films from simulated body fluid". *J. Appl. Polym. Sci.* **99**(5), 2822–2830 (2006).

Kirker-Head, C., Karageorgiou, V., Hofmann, S., Fajardo, R., Betz, O., Merkle, H.P., Hilbe, M., von Rechenberg, B., McCool, J., Abrahamsen, L., Nazarian, A., Cory, E., Curtis, M., Kaplan, D., and Meinel, L. "BMP-silk composite matrices heal critically sized femoral defects". *Bone* **41**(2), 247–255 (2007).

Knight, D.P., Knight, M.M., and Vollrath, F. "Beta transition and stress-induced phase separation in the spinning of spider dragline silk". *Int. J. Biol. Macromol.* **27**(3), 205–210 (2000).

Knight, D.P., and Vollrath, F. "Liquid crystals and flow elongation in a spider's silk production line". *Proc. R. Soc. London Ser. B Biol. Sci.* **266**(1418), 519–523 (1999).

Knight, D.P., and Vollrath, F. "Changes in element composition along the spinning duct in a Nephila spider". *Naturwissenschaften* **88**(4), 179–182 (2001).

Kong, X.D., Cui, F.Z., Wang, X.M., Zhang, M., and Zhang, W. "Silk fibroin regulated mineralization of hydroxyapatite nanocrystals". *J. Cryst. Growth* **270**(1–2), 197–202 (2004).

Kummerlen, J., vanBeek, J.D., Vollrath, F., and Meier, B.H. "Local structure in spider dragline silk investigated by two-dimensional spin-diffusion nuclear magnetic resonance". *Macromolecules* **29**(8), 2920–2928 (1996).

Kweon, H., Ha, H.C., Um, I.C., and Park, Y.H. "Physical properties of silk fibroin/chitosan blend films". *J. Appl. Polym. Sci.* **80**(7), 928–934 (2001).

Lazaris, A., Arcidiacono, S., Huang, Y., Zhou, J.F., Duguay, F., Chretien, N., Welsh, E.A., Soares, J.W., and Karatzas, C.N. "Spider silk fibers spun from soluble recombinant silk produced in mammalian cells". *Science* **295**(5554), 472–476 (2002).

Lee, K.Y., Kong, S.J., Park, W.H., Ha, W.S., and Kwon, I.C. "Effect of surface properties on the antithrombogenicity of silk fibroin/S-carboxymethyl kerateine blend films". *J. Biomater. Sci. Polym. Ed.* **9**(9), 905–914 (1998).

Leveque, I., Cusack, M., Davis, S.A., and Mann, S. "Promotion of fluorapatite crystallization by soluble-matrix proteins from Lingula anatina shells". *Angew. Chem. Int. Ed.* **43**(7), 885–888 (2004).

Levi-Kalisman, Y., Falini, G., Addadi, L., and Weiner, S. "Structure of the nacreous organic matrix of a bivalve mollusk shell examined in the hydrated state using Cryo-TEM". *J. Struct. Biol.* **135**(1), 8–17 (2001).

Lewis, R.V. "Spider silk: Ancient ideas for new biomaterials". *Chem. Rev.* **106**(9), 3762–3774 (2006).

Li, C.M. "Silk polymer templates in biomineralization", Ph.D. Thesis, Tufts University (2005).

Li, C.M., Jin, H.J., Botsaris, G.D., and Kaplan, D.L. "Silk apatite composites from electrospun fibers". *J. Mater. Res.* **20**(12), 3374–3384 (2005).
Li, C.M., Botsaris, G.D., and Kaplan, D.L. "Selective in vitro effect of peptides on calcium carbonate crystallization". *Cryst. Growth Des.* **2**(5), 387–393 (2002).
Li, C.M., Vepari, C., Jin, H.J., Kim, H.J., and Kaplan, D.L. "Electrospun silk-BMP-2 scaffolds for bone tissue engineering". *Biomaterials* **27**(16), 3115–3124 (2006).
Li, M.Z., Tao, W., Lu, S.Z., and Kuga, S. "Compliant film of regenerated Antheraea pernyi silk fibroin by chemical crosslinking". *Int. J. Biol. Macromol.* **32**(3–5), 159–163 (2003).
Lin, J.B., Yao, J.R., Zhou, L., Chen, X., and Shao, Z.Z. "Preliminary exploration of the artificial preparation, structures and properties of regenerated silk fibers". *Chem. J. Chin. Univ. Chin.* **28**(6), 1181–1185 (2007).
Liu, Y., Shao, Z.Z., and Vollrath, F. "Extended wet-spinning can modify spider silk properties". *Chem. Commun.* **19**, 2489–2491 (2005a).
Liu, Y., Shao, Z.Z., and Vollrath, F. "Relationships between supercontraction and mechanical properties of spider silk". *Nat. Mater.* **4**(12), 901–905 (2005b).
Liu, Y., Shao, Z., and Vollrath, F. "Elasticity of Spider Silks". *Biomacromolecules* **9**(7), 1782–1786 (2008a).
Liu, Y., Sponner, A., Porter, D., and Vollrath, F. "Proline and processing of spider silks". *Biomacromolecules* **9**(1), 116–121 (2008b).
Litvin, A.L., Valiyaveettil, S., Kaplan, D.L., and Mann, S. "Template-directed synthesis of aragonite under supramolecular hydrogen-bonded Langmuir monolayers". *Adv. Mater.* **9**(2), 124–127 (1997).
Lock, R.L."Process for Making Silk Fibroin Fibers". US 5171505 [P] (1992).
Lv, Q., Cao, C.B., Zhang, Y., Ma, X.L., and Zhu, H.S. "Preparation of insoluble fibroin films without methanol treatment". *J. Appl. Polym. Sci.* **96**(6), 2168–2173 (2005a).
Lv, Q., Cao, C.B., and Zhu, H.S. "Clotting times and tensile properties of insoluble silk fibroin films containing heparin". *Polym. Int.* **54**(7), 1076–1081 (2005b).
Madsen, B., Shao, Z.Z., and Vollrath, F. "Variability in the mechanical properties of spider silks on three levels: Interspecific, intraspecific and intraindividual". *Int. J. Biol. Macromol.* **24**(2–3), 301–306 (1999).
Mann, S. "Biomineralization Principles and Concepts in Bioinorganic Materials Chemistry". Oxford University Press: , New York (2001).
Marolt, D., Augst, A., Freed, L.E., Vepari, C., Fajardo, R., Patel, N., Gray, M., Farley, M., Kaplan, D., and Vunjak-Novakovic, G. "Bone and cartilage tissue constructs grown using human bone marrow stromal cells, silk scaffolds and rotating bioreactors". *Biomaterials* **27**(36), 6138–6149 (2006).
Marsano, E., Corsini, P., Arosio, C., Boschi, A., Mormino, M., and Freddi, G. "Wet spinning of Bombyx mori silk fibroin dissolved in N-methyl morpholine N-oxide and properties of regenerated fibres". *Int. J. Biol. Macromol.* **37**(4), 179–188 (2005).
Marsh, R.E., Corey, R.B., and Pauling, L. "An investigation of the structure of silk fibroin". *Biochim. Biophys. Acta* **16**(1), 1–34 (1955).
Matsumoto, K., Uejima, H., Iwasaki, T., Sano, Y., and Sumino, H. "Studies on regenerated protein fibers 3. Production of regenerated silk fibroin fiber by the self-dialyzing wet spinning method". *J. Appl. Polym. Sci.* **60**(4), 503–511 (1996).
Meechaisue, C., Wutticharoenmongkol, P., Waraput, R., Huangjing, T., Ketbumrung, N., Pavasant, P., and Supaphol, P. "Preparation of electrospun silk fibroin fiber mats as bone scaffolds: A preliminary study". *Biomed. Mater.* **2**(3), 181–188 (2007).
Meinel, L., Hofmann, S., Karageorgiou, V., Zichner, L., Langer, R., Kaplan, D., and Vunjak-Novakovic, G. "Engineering cartilage-like tissue using human mesenchymal stem cells and silk protein scaffolds". *Biotechnol. Bioeng.* **88**(3), 379–391 (2004a).
Meinel, L., Betz, O., Fajardo, R., Hofmann, S., Nazarian, A., Cory, E., Hilbe, M., McCool, J., Langer, R., Vunjak-Novakovic, G., Merkle, H.P., Rechenberg, B., Kaplan, D.L., and Kirker-Head, C. "Silk based biomaterials to heal critical sized femur defects". *Bone* **39**(4), 922–931 (2006a).

Meinel, L., Fajardo, R., Hofmann, S., Langer, R., Chen, J., Snyder, B., Vunjak-Novakovic, G., and Kaplan, D. "Silk implants for the healing of critical size bone defects". *Bone* **37**(5), 688–698 (2005).

Meinel, L., Hofmann, S., Betz, O., Fajardo, R., Merkle, H.P., Langer, R., Evans, C.H., Vunjak-Novakovic, G., and Kaplan, D.L. "Osteogenesis by human mesenchymal stem cells cultured on silk biomaterials: Comparison of adenovirus mediated gene transfer and protein delivery of BMP-2". *Biomaterials* **27**(28), 4993–5002 (2006b).

Meinel, L., Karageorgiou, V., Fajardo, R., Snyder, B., Shinde-Patil, V., Zichner, L., Kaplan, D., Langer, R., and Vunjak-Novakovic, G. "Bone tissue engineering using human mesenchymal stem cells: Effects of scaffold material and medium flow". *Ann. Biomed. Eng.* **32**(1), 112–122 (2004b).

Meinel, L., Karageorgiou, V., Hofmann, S., Fajardo, R., Snyder, B., Li, C.M., Zichner, L., Langer, R., Vunjak-Novakovic, G., and Kaplan, D.L. "Engineering bone-like tissue in vitro using human bone marrow stem cells and silk scaffolds". *J. Biomed. Mater. Res. A* **71A**(1), 25–34 (2004c).

Miller, L.D., Putthanarat, S., Eby, R.K., and Adams, W.W. "Investigation of the nanofibrillar morphology in silk fibers by small angle X-ray scattering and atomic force microscopy". *Int. J. Biol. Macromol.* **24**(2–3), 159–165 (1999).

Min, B.M., Jeong, L., Lee, K.Y., and Park, W.H. "Regenerated silk fibroin nanofibers: Water vapor-induced structural changes and their effects on the behavior of normal human cells". *Macromol. Biosci.* **6**(4), 285–292 (2006).

Min, B.M., Lee, G., Kim, S.H., Nam, Y.S., Lee, T.S., and Park, W.H. "Electrospinning of silk fibroin nanofibers and its effect on the adhesion and spreading of normal human keratinocytes and fibroblasts in vitro". *Biomaterials* **25**(7–8), 1289–1297 (2004).

Minoura, N., Tsukada, M., and Nagura, M. "Physicochemical properties of silk fibroin membrane as a biomaterial". *Biomaterials* **11**(6), 430–434 (1990).

Mita, K., Ichimura, S., and James, T.C. "Highly repetitive structure and its organization of the silk fibroin gene". *J. Mol. Evol.* **38**(6), 583–592 (1994).

Mo, C.L., Wu, P.Y., Chen, X., and Shao, Z.Z. "Near-infrared characterization on the secondary structure of regenerated Bombyx mori silk fibroin" *Appl. Spectrosc.* **60**(12), 1438–1441 (2006).

Moreau, J.E., Chen, J.S., Horan, R.L., Kaplan, D.L., and Altman, G.H. "Sequential growth factor application in bone marrow stromal cell ligament engineering". *Tissue Eng.* **11**(11–12), 1887–1897 (2005).

Motriuk-Smith, D., Smith, A., Hayashi, C.Y., and Lewis, R.V. Analysis of the conserved N-terminal domains in major ampullate spider silk proteins. *Biomacromolecules* **6**(6), 3152–3159 (2005).

Motta, A., Fambri, L., and Migliaresi, C. "Regenerated silk fibroin films: Thermal and dynamic mechanical analysis". *Macromol. Chem. Phys.* 203, (10–11), 1658–1665 (2002).

Motta, A., Migliaresi, C., Faccioni, F., Torricelli, P., Fini, M., and Giardino, R. "Fibroin hydrogels for biomedical applications: Preparation, characterization and in vitro cell culture studies". *J. Biomat. Sci. Polym. Ed.* **15**(7), 851–864 (2004).

Nudelman, F., Gotliv, B.A., Addadi, L., and Weiner, S. "Mollusk shell formation: Mapping the distribution of organic matrix components underlying a single aragonitic tablet in nacre". *J. Struct. Biol.* **153**(2), 176–187 (2006).

O'Brien, J.P., Fahnestock, S.R., Termonia, Y., and Gardner, K.C.H. "Nylons from nature: Synthetic analogs to spider silk". *Advan. Mater.* **10**(15), 1185–1195 (1998).

Ohgo, K., Zhao, C.H., Kobayashi, M., and Asakura, T. "Preparation of non-woven nanofibers of Bombyx mori silk, Samia cynthia ricini silk and recombinant hybrid silk with electrospinning method". *Polymer* **44**(3), 841–846 (2003).

Oroudjev, E., Soares, J., Arcdiacono, S., Thompson, J.B., Fossey, S.A., and Hansma, H.G. "Segmented nanofibers of spider dragline silk: atomic force microscopy and single-molecule force spectroscopy". *Proc. Natl. Acad. Sci. U.S.A.* **99**(14), 6460–6465 (2002).

Park, W.H., Jeong, L., Yoo, D.I., and Hudson, S. "Effect of chitosan on morphology and conformation of electrospun silk fibroin nanofibers". *Polymer* **45**(21), 7151–7157 (2004).

Park, K.E., Jung, S.Y., Lee, S.J., Min, B.M., and Park, W.H. "Biomimetic nanofibrous scaffolds: Preparation and characterization of chitin/silk fibroin blend nanofibers". *Int. J. Biol. Macromol.* **38**(3–5), 165–173 (2006).

Park, R.J., and Meldrum, F.C. "Synthesis of single crystals of calcite with complex morphologies"*Adv. Mater.* **14**(16), 1167–1169 (2002).

Peng, X.N., Shao, Z.Z., Chen, X., Knight, D.P., Wu, P.Y., and Vollrath, F. "Further investigation on potassium-induced conformation transition of Nephila spidroin film with two-dimensional infrared correlation spectroscopy". *Biomacromolecules* **6**(1), 302–308 (2005).

Phillips, D.M., Drummy, L.F., Conrady, D.G., Fox, D.M., Naik, R.R., Stone, M.O., Trulove, P.C., De Long, H.C., and Mantz, R.A. "Dissolution and regeneration of Bombyx mori Silk fibroin using ionic liquids". *J. Am. Chem. Soc.* **126**(44), 14350–14351 (2004).

Phillips, D.M., Drummy, L.F., Naik, R.R., De Long, H.C., Fox, D.M., Trulove, P.C., and Mantz, R.A. "Regenerated silk fiber wet spinning from an ionic liquid solution". *J. Mater. Chem.* **15**(39), 4206–4208 (2005).

Phillips, D.M., Drummy, L.F., Naik, R.R., Trulove, P.C., De Long, H.C., and Mantz, R.A. "Silk regeneration with ionic liquids". Abstracts of Papers of the American Chemical Society (2006), p. 231.

Pins, G.D., Christiansen, D.L., Patel, R., and Silver, F.H. "Self-assembly of collagen fibers. Influence of fibrillar alignment and decorin on mechanical properties". *Biophys. J.* **73**(4), 2164–2172 (1997).

Porter, D., Vollrath, F., and Shao, Z. "Predicting the mechanical properties of spider silk as a model nanostructured polymer". *Eur. Phys. J. E* **16**(2), 199–206 (2005).

Poza, P., Perez-Rigueiro, J., Elices, M., and Llorca, J. "Fractographic analysis of silkworm and spider silk". *Eng. Fract. Mech.* **69**(9), 1035–1048 (2002).

Putthanarat, S., Eby, R.K., Naik, R.R., Juhl, S.B., Walker, M.A., Peterman, E., Ristich, S., Magoshi, J., Tanaka, T., Stone, M.O., Farmer, B.L., Brewer, C., and Ott, D. "Nonlinear optical transmission of silk/green fluorescent protein (GFP) films". *Polymer* **45**(25), 8451–8457 (2004).

Rathore, O., and Sogah, D.Y. "Self-assembly of beta-sheets into nanostructures by poly(alanine) segments incorporated in multiblock copolymers inspired by spider silk". *J. Am. Chem. Soc.* **123**(22), 5231–5239 (2001).

Raz, S., Weiner, S., and Addadi, L. "Formation of high-magnesian calcites via an amorphous precursor phase: Possible biological implications". *Advan. Mater.* **12**(1), 38–42 (2000).

Riekel, C., Muller, M., and Vollrath, F. "In situ X-ray diffraction during forced silking of spider silk". *Macromolecules* **32**(13), 4464–4466 (1999).

Riekel, C., and Vollrath, F. "Spider silk fibre extrusion: Combined wide- and small-angle X-ray microdiffraction experiments". *Int. J. Biol. Macromol.* **29**(3), 203–210 (2001).

Sapede, D., Seydel, T., Forsyth, V.T., Koza, M.A., Schweins, R., Vollrath, F., and Riekel, C. "Nanofibrillar structure and molecular mobility in spider dragline silk". *Macromolecules* **38**(20), 8447–8453 (2005).

Savage, K.N., Guerette, P.A., and Gosline, J.M. "Supercontraction stress in spider webs". *Biomacromolecules* **5**(3), 675–679 (2004).

Sehnal, F., and Zurovec, M. "Construction of silk fiber core in Lepidoptera". *Biomacromolecules* **5**(3), 666–674 (2004).

Seidel, A., Liivak, O., Calve, S., Adaska, J., Ji, G.D., Yang, Z.T., Grubb, D., Zax, D.B., and Jelinski, L.W. "Regenerated spider silk: Processing, properties, and structure". *Macromolecules* **33**(3), 775–780 (2000).

Seidel, A., Liivak, O., and Jelinski, L.W. "Artificial spinning of spider silk". *Macromolecules* **31**(19), 6733–6736 (1998).

Shimura, K., Kikuchi, A., Ohtomo, K., Katagata, Y., and Hyodo, A. "Studies on silk fibroin of *Bombyx-Mori*. 1. Fractionation of fibroin prepared from posterior silk gland". *J. Biochem.* **80**(4), 693–702 (1976).
Sponner, A., Schlott, B., Vollrath, F., Unger, E., Grosse, F., and Weisshart, K. "Characterization of the protein components of Nephila clavipes dragline silk". *Biochemistry* **44**(12), 4727–4736 (2005a).
Sponner, A., Unger, E., Grosse, F., and Klaus, W. "Differential polymerization of the two main protein components of dragline silk during fibre spinning". *Nat. Mat.* **4**(10), 772–775 (2005b).
Shao, Z., Hu, X.W., Frische, S., and Vollrath, F. "Heterogeneous morphology of Nephila edulis spider silk and its significance for mechanical properties". *Polymer* **40**(16), 4709–4711 (1999a).
Shao, Z., Vollrath, F., Sirichaisit, J., and Young, R.J. "Analysis of spider silk in native and supercontracted states using Raman spectroscopy". *Polymer* **40**(10), 2493–2500 (1999b).
Shao, Z.Z., and Vollrath, F. "The effect of solvents on the contraction and mechanical properties of spider silk". *Polymer* **40**(7), 1799–1806 (1999).
Shao, Z.Z., and Vollrath, F. "Materials: Surprising strength of silkworm silk." *Nature* **418**(6899), 741 (2002).
Shao, Z.Z., Vollrath, F., Yang, Y., and Thogersen, H.C. "Structure and behavior of regenerated spider silk". *Macromolecules* **36**(4), 1157–1161 (2003).
Shao, Z.Z., Young, R.J., and Vollrath, F. "The effect of solvents on spider silk studied by mechanical testing and single-fibre Raman spectroscopy". *Int. J. Biol. Macromol.* **24**(2–3), 295–300 (1999).
Shen, Y., Johnson, M.A., and Martin, D.C. "Microstructural characterization of Bombyx mori silk fibers". *Macromolecules* **31**(25), 8857–8864 (1998).
Shulha, H., Foo, C.W.P., Kaplan, D.L., and Tsukruk, V.V. "Unfolding the multi-length scale domain structure of silk fibroin protein"*Polymer* **47**(16), 5821–5830 (2006).
Shen, F.H., Feng, Q.L., and Wang, C.M. "The modulation of collagen on crystal morphology of calcium carbonate". *J. Cryst. Growth* **242**(1–2), 239–244 (2002).
Slocik, J.M., and Naik, R.R. "Biological assembly of hybrid inorganic nanomaterials". *Curr. Nanosci.* **3**(2), 117–120 (2007).
Sofia, S., McCarthy, M.B., Gronowicz, G., and Kaplan, D.L. "Functionalized silk-based biomaterials for bone formation". *J. Biomed. Mater. Res.* **54**(1), 139–148 (2001).
Sugihara, A., Sugiura, K., Morita, H., Ninagawa, T., Tubouchi, K., Tobe, R., Izumiya, M., Horio, T., Abraham, N.G., and Ikehara, S. "Promotive effects of a silk film on epidermal recovery from full-thickness skin wounds". *Proc. Soc. Exp. Biol. Med.* **225**(1), 58–64 (2000).
Sukigara, S., Gandhi, M., Ayutsede, J., Micklus, M., and Ko, F. "Regeneration of Bombyx mori silk by electrospinning – part 1: Processing parameters and geometric properties". *Polymer* **44**(19), 5721–5727 (2003).
Sukigara, S., Gandhi, M., Ayutsede, J., Micklus, M., and Ko, F. "Regeneration of Bombyx mori silk by electrospinning. Part 2. Process optimization and empirical modeling using response surface methodology". *Polymer* **45**(11), 3701–3708 (2004).
Sudo, S., Fujikawa, T., Nagakura, T., Ohkubo, T., Sakaguchi, K., Tanaka, M., Nakashima, K., and Takahashi, T. "Structures of mollusc shell framework proteins". *Nature* **387**(6633), 563–564 (1997).
Sumper, M., and Brunner, E. "Learning from diatoms: Nature's tools for the production of nanostructured silica". *Adv. Funct. Mater.* **16**(1), 17–26 (2006).
Sugawara, T., Suwa, Y., Ohkawa, K., and Yamamoto, H. "Chiral biomineralization: Mirror-imaged helical growth of calcite with chiral phosphoserine copolypeptides". *Macromol. Rapid Commun.* **24**(14), 847–851 (2003).
Takeuchi, A., Ohtsuki, C., Miyazaki, T., Tanaka, H., Yamazaki, M., and Tanihara, M. "Deposition of bone-like apatite on silk fiber in a solution that mimics extracellular fluid". *J. Biomed. Mater. Res. A* **65A**(2), 283–289 (2003).

Takeuchi, A., Ohtsuki, C., Miyazaki, T., Kamitakahara, M., Ogata, S., Yamazaki, M., Furutani, Y., Kinoshita, H., and Tanihara, M. "Heterogeneous nucleation of hydroxyapatite on protein: Structural effect of silk sericin". *J. R. Soc. Interface* **2**(4), 373–378 (2005).

Tanaka, K., Inoue, S., and Mizuno, S. "Hydrophobic interaction of P25, containing Asn-linked oligosaccharide chains, with the H-L complex of silk fibroin produced by Bombyx mori". *Insect Biochem. Mol. Biol.* **29**(3), 269–276 (1999).

Tanaka, T., Hirose, M., Kotobuki, N., Ohgushi, H., Furuzono, T., and Sato, J. "Nano-scaled hydroxyapatite/silk fibroin sheets support osteogenic differentiation of rat bone marrow mesenchymal cells". *Mater. Sci. Eng. C Biomimetic Supramol. Sys.* **27**(4), 817–823 (2007).

Terada, S., Sasaki, M., Yanagihara, K., and Yamada, H. "Preparation of silk protein sericin as mitogenic factor for better mammalian cell culture". *J. Biosci. Bioeng.* **100**(6), 667–671 (2005).

Termonia, Y. "Molecular modeling of spider silk elasticity". *Macromolecules* **27**(25), 7378–7381 (1994).

Tsubouchi, K., Igarashi, Y., Takasu, Y., and Yamada, H. "Sericin enhances attachment of cultured human skin fibroblasts". *Biosci. Biotechnol. Biochem.* **69**(2), 403–405 (2005).

Um, I.C., Ki, C.S., Kweon, H.Y., Lee, K.G., Ihm, D.W., and Park, Y.H. "Wet spinning of silk polymer – II. Effect of drawing on the structural characteristics and properties of filament". *Int. J. Biol. Macromol.* **34**(1–2), 107–119 (2004a).

Um, I.C., Kweon, H.Y., Lee, K.G., Ihm, D.W., Lee, J.H., and Park, Y.H. "Wet spinning of silk polymer – I. Effect of coagulation conditions on the morphological feature of filament". *Int. J. Biol. Macromol.* **34**(1–2), 89–105 (2004b).

Unger, R.E., Peters, K., Wolf, M., Motta, A., Migliaresi, C., and Kirkpatrick, C.J. "Endothelialization of a non-woven silk fibroin net for use in tissue engineering: Growth and gene regulation of hunian endothelial cells". *Biomaterials* **25**(21), 5137–5146 (2004).

Urry, D.W., Ohnishi, T., Long, M.M., and Mitchell, L.W. "Studies on Conformation and Interactions of Elastin – Nuclear Magnetic-Resonance of Polyhexapeptide". *Int. J. Pept. Protein Res.* **7**(5), 367–378 (1975).

Vepari, C., and Kaplan, D.L. "Silk as a biomaterial". *Prog. Polym. Sci.* **32**(8–9), 991–1007 (2007).

Vepari, C.P., and Kaplan, D.L. "Covalently immobilized enzyme gradients within three-dimensional porous scaffolds". *Biotechnol. Bioeng.* **93**(6), 1130–1137 (2006).

Vollrath, F. and Porter, D."Spider silk as archetypal protein elastomer". *Soft Matter* **2**(5), 377–385 (2006).

Vollrath, F., Madsen, B., and Shao, Z.Z. "The effect of spinning conditions on the mechanics of a spider's dragline silk". *Proc. R. Soc. London Ser. B Biol. Sci.* **268**(1483), 2339–2346 (2001).

Verraest, D.L., Peters, J.A., vanBekkum, H., and vanRosmalen, G.M. "Carboxymethyl inulin: A new inhibitor for calcium carbonate precipitation". *J. Am. Oil Chem. Soc.* **73**(1), 55–62 (1996).

Vollrath, F., and Knight, D.P. "Structure and function of the silk production pathway in the Spider Nephila edulis". *Int. J. Biol. Macromol.* **24**(2–3), 243–249 (1999).

Vollrath, F., and Knight, D.P. "Liquid crystalline spinning of spider silk" *Nature* **410**(6828), 541–548 (2001).

Wang, Y.Z., Blasioli, D.J., Kim, H.J., Kim, H.S., and Kaplan, D.L. "Cartilage tissue engineering with silk scaffolds and human articular chondrocytes". *Biomaterials* **27**(25), 4434–4442 (2006).

Wang, H., Shao, H.L., and Hu, X.C. "Structure of silk fibroin fibers made by an electrospinning process from a silk fibroin aqueous solution". *J. Appl. Polym. Sci.* **101**(2), 961–968 (2006).

Wang, H., Zhang, Y.P., Shao, H.L., and Hu, X.C. "Electrospun ultra-fine silk fibroin fibers from aqueous solutions". *J. Mater. Sci.* **40**(20), 5359–5363 (2005).

Wang, L., Li, C.Z., and Senna, M. "High-affinity integration of hydroxyapatite nanoparticles with chemically modified silk fibroin". *J. Nanoparticle Res.* **9**(5), 919–929 (2007).

Wang, M., Jin, H.J., Kaplan, D.L., and Rutledge, G.C. "Mechanical properties of electrospun silk fibers". *Macromolecules* **37**(18), 6856–6864 (2004).

Wang, M., Yu, J.H., Kaplan, D.L., and Rutledge, G.C. "Production of submicron diameter silk fibers under benign processing conditions by two-fluid electrospinning". *Macromolecules* **39**(3), 1102–1107 (2006).

Wang, T.P., Antonietti, M., and Colfen, H. "Calcite mesocrystals: 'Morphing' crystals by a polyelectrolyte". *Chem. Eur. J.* **12**(22), 5722–5730 (2006).

Wang, T.X., Colfen, H., and Antonietti, M. "Nonclassical crystallization: Mesocrystals and morphology change of CaCO$_3$ crystals in the presence of a polyelectrolyte additive". *J. Am. Chem. Soc.* **127**(10), 3246–3247 (2005).

Wang, X., Wenk, E., Hu, X., Castro, G.R., Meinel, L., Wang, X., Li, C., Merkle, H., and Kaplan, D.L. "Silk coatings on PLGA and alginate microspheres for protein delivery". *Biomaterials* **28**(28), 4161–4169 (2007).

Wang, X.Q., Wenk, E., Matsumoto, A., Meinel, L., Li, C.M., and Kaplan, D.L. "Silk microspheres for encapsulation and controlled release". *J. Controlled Release* **117**(3), 360–370 (2007).

Wei, H., Ma, N., Shi, F., Wang, Z.Q., and Zhang, X. "Artificial nacre by alternating preparation of layer-by-layer polymer films and CaCO3 strata". *Chem. Mater.* **19**(8), 1974–1978 (2007).

Weiner, S., and Traub, W. "X-ray-diffraction study of the insoluble organic matrix of Mollusk shells". *FEBS Lett.* **111**(2), 311–316 (1980).

Work, R.W. "Dimensions, birefringences, and force-elongation behavior of major and minor ampullate silk fibers from Orb-Web-Spinning Spiders – Effects of wetting on these properties". *Text. Res. J.* **47**(10), 650–662 (1977).

Xu, M., and Lewis, R.V. "Structure of a protein superfiber – spider dragline silk". *Proc. Natl. Acad. Sci. U.S.A.* **87**(18), 7120–7124 (1990).

Xu, A.W., Ma, Y.R., and Colfen, H. "Biomimetic mineralization". *J. Mater. Chem.* **17**(5), 415–449 (2007).

Yamada, H., Nakao, H., Takasu, Y., and Tsubouchi, K. "Preparation of undegraded native molecular fibroin solution from silkworm cocoons". *Mat. Sci. Eng. C Biomimetic Supramol. Sys.* **14**(1–2), 41–46 (2001).

Yang, Y., Chen, X., Shao, Z.Z., Zhou, P., Porter, D., Knight, D.P., and Vollrath, F. "Toughness of spider silk at high and low temperatures". *Adv. Mater.* **17**(1), 84–89 (2005).

Yao, J.M., Masuda, H., Zhao, C.H., and Asakura, T. "Artificial spinning and characterization of silk fiber from Bombyx mori silk fibroin in hexafluoroacetone hydrate". *Macromolecules* **35**(1), 6–9 (2002).

Yao, J.M., Wei, K.M., Li, L., Kong, X.D., Zhu, Y.Q., and Lin, F. "Effect of initial Bombyx mori silk fibroin structure on the protein biomineralization". *Acta Chimica Sinica* **65**(7), 635–639 (2007).

Yeo, J.H., Lee, K.G., Kim, H.C., Oh, Y.L., Kim, A.J., and Kim, S.Y. "The effects of PVA/ Chitosan/Fibroin (PCF)-blended spongy sheets on wound healing in rats". *Biol. Pharm. Bull.* **23**(10), 1220–1223 (2000).

Yu, S.H., and Colfen, H. "Bio-inspired crystal morphogenesis by hydrophilic polymers". *J. Mater. Chem.* **14**(14), 2124–2147 (2004).

Zarkoob, S., Eby, R.K., Reneker, D.H., Hudson, S.D., Ertley, D., and Adams, W.W. "Structure and morphology of electrospun silk nanofibers" *Polymer* **45**(11), 3973–3977 (2004).

Zarkoob, S., Reneker, D.H., Eby, R.K., Hudson, S.D., Ertley, D., and Adams, W.W. Structure and morphology of nano electrospun silk fibers. Abstracts of Papers of the American Chemical Society (1998), 216, U122–U122.

Zhang, Y.Q. "Applications of natural silk protein sericin in biomaterials". *Biotechnol. Adv.* **20**(2), 91–100 (2002).

Zhang, Y.Q., Shen, W.D., Xiang, R.L., Zhuge, L.J., Gao, W.J., and Wang, W.B. "Formation of silk fibroin nanoparticles in water-miscible organic solvent and their characterization". *J. Nanopart. Res.* **9**(5), 885–900 (2007).

Zhao, C.H., Yao, J.M., Masuda, H., Kishore, R., and Asakura, T. "Structural characterization and artificial fiber formation of Bombyx mori silk fibroin in hexafluoro-iso-propanol solvent system". *Biopolymers* **69**(2), 253–259 (2003).

Zhaorigetu, S., Sasaki, M., Watanabe, H., and Kato, N. "Supplemental silk protein, sericin, suppresses colon tumorigenesis in 1,2-dimethylhydrazine-treated mice by reducing oxidative stress and cell proliferation". *Biosci. Biotechnol. Biochem.* **65**(10), 2181–2186 (2001).

Zhou, C.Z., Confalonieri, F., Medina, N., Zivanovic, Y., Esnault, C., Yang, T., Jacquet, M., Janin, J., Duguet, M., Perasso, R., and Li, Z.G. "Fine organization of Bombyx mori fibroin heavy chain gene" *Nucleic Acids Res.* **28**(12), 2413–2419 (2000).

Zhou, C.Z., Confalonieri, F., Jacquet, M., Perasso, R., Li, Z.G., and Janin, J. "Silk fibroin: Structural implications of a remarkable amino acid sequence". *Proteins Struc Funct. Genet.* **44**(2), 119–122 (2001).

Zhou, G.Q., Chen, X., and Shao, Z.Z. "The artificial spinning based on silk proteins". *Prog. Chem.* **18**(7–8), 933–938 (2006a).

Zhou, L., Chen, X., Dai, W.L., and Shao, Z.Z. "X-ray photoelectron spectroscopic and Raman analysis of silk fibroin-Cu(II) films". *Biopolymers* **82**(2), 144–151 (2006b).

Zhou, L., Chen, X., Shao, Z.Z., Huang, Y.F., and Knight, D.P. "Effect of metallic ions on silk formation the mulberry silkworm, Bombyx mori". *J. Phys. Chem. B*, **109**(35), 16937–16945 (2005a).

Zhou, L., Chen, X., Shao, Z.Z., Zhou, P., Knight, D.P., and Vollrath, F. "Copper in the silk formation process of Bombyx mori silkworm". *FEBS Lett.* **554**(3), 337–341 (2003).

Zhou, L., Terry, A.E., Huang, Y.F., Shao, Z.Z., and Chen, X. "Metal element contents in silk gland and silk fiber of Bombyx mori silkworm". *Acta Chimica Sinica* **63**(15), 1379–1382 (2005b).

Zhou, W., Huang, Y.F., Shao, Z.Z., and Chen, X. "Effect of Fe and Mn on the conformation transition of Bombyx mori silk fibroin". *Acta Chimica Sinica* **65**(19), 2197–2201 (2007).

Zong, X.H., Zhou, P., Shao, Z.Z., Chen, S.M., Chen, X., Hu, B.W., Deng, F., and Yao, W.H. "Effect of pH and copper(II) on the conformation transitions of silk fibroin based on EPR, NMR, and Raman spectroscopy". *Biochemistry* **43**(38), 11932–11941 (2004).

Zhu, J.X., Shao, H.L., and Hu, X.C. "Morphology and structure of electrospun mats from regenerated silk fibroin aqueous solutions with adjusting pH". *Int. J. Biol. Macromol.* **41**(4), 469–474 (2007).

Zuo, B.Q., Liu, L., and Wu, Z. "Effect on properties of regenerated silk fibroin fiber coagulated with aqueous Methanol/Ethanol". *J. Appl. Polym. Sci.* **106**(1), 53–59 (2007).

CHAPTER 6

Surface- and Solution-Based Assembly of Amyloid Fibrils for Biomedical and Nanotechnology Applications

Sally L. Gras

Contents		
	1. Introduction	161
	2. Polypeptide Self-Assembly	162
	2.1 Amyloid fibrils and other fibrous structures	162
	2.2 Amyloid fibril assembly, stability, and disassembly	165
	3. The Role of Surfaces During and After Assembly	167
	3.1 Surface-based assembly	168
	3.2 Surface-directed assembly	175
	3.3 Surface interactions following assembly	182
	4. Applications	189
	4.1 Electronics and photonics	189
	4.2 Platforms for enzyme immobilization and biosensors	193
	4.3 Biocompatible materials	196
	5. Conclusion	205
	References	206

1. INTRODUCTION

Nature uses self-assembling polypeptides to build a staggering variety of protein structures. These proteins act as structural supports, as efficient enzymes, and complex machines. They also take part in sophisticated signaling pathways where they direct and control cell metabolism and growth.

Department of Chemical and Biomolecular Engineering and The Bio21 Molecular Science and Biotechnology Institute, The University of Melbourne, Victoria 3010, Australia.

E-mail address: sgras@unimelb.edu.au

This theme of self-assembly, where multiple units come together and form an organized structure, is essential to life and assembly processes can be observed across the phylogenetic kingdom from single-cell organisms to much larger and more complex organisms including humans.

As scientists and engineers, natural self-assembly processes represent a tremendous resource, which we can use to create our own miniature materials and devices. Our endeavors are informed by hundreds of years of curiosity-driven research interested in the natural world. Our toolbox is further expanded by modern synthetic chemistry which extends beyond the realm of natural molecules. We can also create artificial environments to control and direct assembly and use computer-based tools and simulations to model and predict self-assembly pathways and their resulting protein structures. Many researchers believe we can use these modern tools to simplify, improve, and refine assembly processes. We have much to do in order to reach this ambitious goal but the next 10 years are likely to be filled with exciting discoveries and advances as self-assembling polypeptide materials move from the laboratory to the clinic or the manufacturing assembly line.

This chapter describes the self-assembly of non-native protein fibers known as amyloid fibrils and the development of these fibrils for potential applications in nanotechnology and biomedicine. It extends an earlier review by the author on a related topic (Gras, 2007). In Section 1, the self-assembly of polypeptides into amyloid fibrils and efforts to control assembly and any subsequent disassembly are discussed. In Section 2, this review focuses on the important role of surfaces and interfaces during and after polypeptide assembly. It examines how different surfaces can influence fibril assembly, how surfaces can be used to direct self-assembly in order to create highly ordered structures, and how different techniques can be used to create aligned and patterned materials on surfaces following self-assembly.

In Section 3, the ever growing range of potential applications for fibrous self-assembled polypeptide materials is discussed including the development of electronic, photonic, or magnetic components for devices, progress toward platforms for enzyme immobilization and sensors, and advances in biocompatible materials. Comparisons are made between amyloid fibrils and other designed self-assembling fibrous structures throughout this chapter in order to illustrate the diversity of this rapidly expanding field and highlight similarities between the different protein structures and their potential applications.

2. POLYPEPTIDE SELF-ASSEMBLY

2.1. Amyloid fibrils and other fibrous structures

Amyloid fibrils are non-native protein structures that demonstrate a host of desirable properties, making these fibers attractive components for nanotechnology and biomedicine (Gras, 2007; Hamada et al., 2004; Macphee and Woolfson, 2004; Stevens and George, 2005; Waterhouse and Gerrard, 2004). These fibrils have a comparable size to other nanostructures; they typically span 10–30 nm in width and reach up to several microns in length. Fibrils are also rich in β-sheet secondary structure and share a cross-beta core structure where β-strands are arranged in β-sheets that stack perpendicular to the fiber axis (Pauling and Corey, 1951).

The highly ordered core of amyloid fibrils is thought to be the basis of their great strength (Smith et al., 2006). Fibrils are also resistant to enzymatic digestion (Zurdo et al., 2001) and can be dehydrated without loss of structural integrity (Squires et al., 2006). The regular placement of polypeptides within the core also allows additional pendant groups to be displayed on the surface of the fibril and ordered on a nanoscale (Baldwin et al., 2006; Baxa et al., 2003; Gras et al., 2008; MacPhee and Dobson, 2000b). Further molecules may also be recruited to the fibril surface using standard chemistries such as antibody interactions, receptor–ligand interactions, biotin–streptavidin binding, and gold–thiol interactions.

Amyloid fibrils are commonly known for their association with a range of protein misfolding diseases (Westermark et al., 2005), where proteins loose their native shape and associate through a series of sequential steps eventually adopting a cross-beta structure (Sunde et al., 1997). These fibrils were initially thought to be the cause of disease but it is now generally thought that the protein structures found early on the fibrillation pathway may be responsible for toxicity, with fibrils possibly being inert structures (Bucciantini et al., 2002; Mucke et al. 2000).

Increasingly amyloid fibrils are being discovered in nature where they may demonstrate a positive role, as structures that help bacteria to colonize new surfaces (Chapman et al., 2002), as bacterial extensions that modify surface tension (Claessen et al., 2003; Talbot, 2003), and as structures that aid the polymerization of melanin in our own cells (Fowler et al., 2006). This insight together with the hypothesis that the generic nature of a polypeptide backbone allows all proteins to adopt a highly regular β-sheet structure and form amyloid fibrils (Dobson, 1999; Fandrich et al., 2003; Guijarro et al., 1998) suggests that designed amyloid fibrils could be used as highly functional materials for a wide range of applications.

Amyloid fibrils can be constructed from a range of raw materials including de novo-designed peptides, cheap proteins that are typically found in the laboratory, inexpensive waste proteins (Garvey et al., 2008, Pearce et al., 2007), or proteins that can be produced by recombinant expression in a host cell such as a bacterial or plant cell. Computer models and simulations including phenomenological models and atomistic simulations will certainly play an important role in the design or selection of protein sequences for amyloid fibril formation due to their ability to predict aggregation-prone regions (Caflisch, 2006). Molecular dynamics, coarse-grained models, and Monte Carlo simulations are also valuable for their ability to model the behavior of monomers and the assembled structures (Colombo et al., 2007). An understanding of the energetic parameters involved in self-assembly can also be used to produce structures with different morphologies such as tapes and ribbons, and this understanding can be applied to promote gel formation (Aggeli et al., 1997).

Other self-assembling polypeptide systems form nanofibrous structures that resemble the appearance of amyloid fibrils by microscopy. These include but are not limited to diphenylalanine nanofibers (Reches and Gazit, 2003), peptide amphiphiles (Hartgerink et al., 2001), and ionic complementary self-assembling peptides (Holmes et al. 2000). These systems can be rich in β-sheet secondary protein structure but often differ in their self-assembly pathways, the forces driving assembly, and the final arrangement of polypeptides within the resulting fiber. The term "fiber" is used within this chapter to refer to these structures and distinguish from amyloid "fibrils" that have a cross-beta structure.

The Phe–Phe motif used to form diphenylalanine peptide nanotubes was originally derived from the hydrophobic portion of the Aβ peptide which forms amyloid fibrils associated with Alzheimer disease (Reches and Gazit, 2003). This motif was found to be sufficient to drive assembly of peptide nanotubes, which are hollow and thought to be stabilized by ring stacking (Reches and Gazit, 2003). The peptide amphiphiles developed by Stupp and other researches consist of a hydrophobic alkyl tail and a β-sheet-forming hydrophilic peptide which assembles into a cylindrical fiber with the hydrophilic peptide facing the solvent (Hartgerink et al., 2001). Ionic complementary self-assembling peptides also have an amphiphilic nature. They consist of alternating hydrophilic and hydrophobic residues, where the hydrophilic group has alternating positive and negative charges, providing ionic complementary (Hong et al., 2003). These motifs were originally isolated from a yeast Z-DNA-binding protein (Zhang et al., 1993). Each of these systems has great potential as structures for nanotechnology and biomedicine.

This chapter focuses on amyloid fibrils formed by polypeptides and contrasts developments in our understanding of fibril assembly with developments reported for the other selected self-assembling systems described above. Amyloid fibrils can be formed from protein–polymer hybrids (Channon and Macphee, 2008), and there is further scope to create fibrillar structures that incorporate lipids (Griffin et al., 2008) or carbohydrates (Alexandrescu, 2005). The range of possible self-assembling polypeptide sequences and structures is much wider than can be included within this chapter and the reader is directed to other studies describing structures built from cyclic peptides (Ghadiri et al., 1993), structures from helical proteins (Woolfson and Ryadnov, 2006) and the broad principles for the design of other self-assembling polypeptide systems (Bromley et al., 2008).

2.2. Amyloid fibril assembly, stability, and disassembly

If amyloid fibrils and other self-assembling polypeptide systems are to be developed as new materials and components for nanotechnology and biomedicine, it is important that we understand the pathways and kinetics of their assembly. It will also be desirable to be able to control this assembly and any subsequent disassembly.

Amyloid fibrils are highly stable once fully formed and can withstand extreme conditions including pressures as high as 1.3 GPa (Meersman and Dobson, 2006). However, there is increasing evidence that amyloid fibrils are dynamic structures that are in exchange with monomeric peptides in solution (Carulla et al., 2005). Assembly is also reversible under some conditions such as changes in pH, the addition of solvents, or use of antibodies or denaturants (MacPhee and Dobson, 2000a; Meersman and Dobson, 2006). While stability will be important for many functions and could be enhanced by cross-linking, disassembly may also prove useful for materials designed for temporary use or materials designed to release associated drugs, biomolecules, or other cargo into the surrounding environment.

A number of techniques can be used to monitor the growth of amyloid fibrils and provide information on the kinetics of fibril assembly or disassembly. These techniques include light scattering or dye binding assays where Thioflavin T binds to the emerging fibril structure resulting in an increase in fluorescence (Krebs et al., 2005). Fourier transform infrared spectroscopy and circular dichroism can be used to monitor a change in secondary structure as the polypeptide adopts a β-sheet-rich confirmation (Nilsson, 2004) and a quartz crystal oscillator used to follow an increase in fibril mass as a function of time (Knowles et al., 2007).

Complementary microscopy techniques can be used to follow the morphology and growth of fibrils either on a surface or in aliquots taken from the assembly solution including total internal reflection fluorescence microscopy (TIRFM) (Ban et al., 2004), transmission electron microscopy (TEM), or atomic force microscopy (AFM).

Fibril formation is a nucleated process that is characterized by an initial lag phase (Devlin et al., 2006; Hortschansky et al., 2005). Nucleation is followed by a fibril growth phase known as elongation. The assembly kinetics then plateau as the majority of polypeptide is incorporated into fibrillar structures. The final portion of polypeptide incorporated into fibrils is characteristic to a particular fibril system and can vary from low conversions to greater than 99% of all protein in the sample (Gras et al., 2008).

Seeding is one way of potentially controlling what is typically an uncontrolled fibril assembly process, where polypeptide building blocks are combined in solution under favorable conditions. Seeds are short fibril fragments that can nucleate fibril growth, eliminate the lag phase, and accelerate assembly. To be successful, seeding requires a high sequence identity between the sequence used to construct the seeds and the sequence of polypeptide which is being seeded (Krebs et al., 2004; Wright et al., 2005). This specificity has the added advantage of promoting the growth of a single fibril type from a solution containing multiple proteins, potentially allowing fibril growth from unpurified protein mixtures.

The conditions used to initiate fibril growth can also be used to control assembly. Parameters such as solution temperature, pH, and the ionic concentration are all known to influence the efficiency of the assembly process and can be used to exert control. For example, unfavorable low temperatures could be used to store polypeptide building blocks in order to prevent assembly. The temperature can then be raised to induce assembly. The designed "ccbeta switch system" behaves in this manner; changing from a coiled coil structure to an amyloid fibril when the temperature is increased (Kammerer and Steinmetz, 2006). The challenge with this approach is to design polypeptides so that they effectively respond to stimuli that are compatible with proposed applications. Agelli and colleagues (2003) have effectively illustrated this approach, rationally introducing Glu and Orn side chains so that peptide assembly is responsive to pH.

Other stimuli successfully used for temporal control of fibril assembly include chemical or enzymatic switches and light, often involving an O—N intramolecular acyl migration which increases peptide solubility (Sohma et al., 2004). Dos Santos used O—N intramolecular acyl migration to switch the Aβ peptide from a stable monomer to an assembly competent state (Dos Santos et al., 2005). Switch elements consisting of an ester and a flexible C—C bond were placed along the backbone increasing monomer stability. These residues were then removed using proteases, triggering O—N intramolecular acyl migration and fibril assembly. Other stimuli used by Dos Santos to induce O—N intramolecular acyl migration and assembly included pH and light. Similarly, Taniguchi and colleagues explored a phototriggered click reaction followed by O—N intramolecular acyl migration to generate an analog of Aβ1-42 (Taniguchi et al., 2006). Whereas Bosques used a photoliable linker to initiate fibril assembly (Bosques and Imperiali, 2003).

Pseudoproline units offer another approach for controlling fibril assembly (Tuchscherer et al., 2007). Serine, threonine, or cystiene residues may be replaced with pseudoproline. The pseudoproline ring introduces a cis–amide bond in the backbone preventing aggregation and increasing peptide solubility. The ring can be cleaved during the final deprotection stage of fluoren-9-ylmethyloxycarbonyl (FMOC) synthesis or by treatment with acid when assembly is desired. Tuchscherer and colleagues also present evidence that the pseudoproline approach may be more effective than incorporating O-acyl groups at preventing the stable monomer from forming fibrils (Tuchscherer et al., 2007).

The chemistry used to control fibril assembly may also be applied to influence morphology, structure, disassembly, and the spatial control of fibrils. Switch elements responding to different signals may be placed in distinct sections of a fibril-forming peptide potentially offering a programmable way to change fibril structure and morphology (Tuchscherer et al., 2007). O—N intramolecular acyl migration can also be used to trigger the transition from a β-sheet confirmation to an α-helical structure disrupting preformed amyloid fibrils (Mimna et al., 2007). A strategy of UV-cleavable linkers applied to disassemble peptide amphiphiles by Lowik and colleagues (2008) may also be applied to fibrils. Lowick suggests these linkers may prove useful for controlling the placement of fibers on a surface, where a UV source and mask may be used to spatially control patterned areas of assembled and disassembled fibers.

3. THE ROLE OF SURFACES DURING AND AFTER ASSEMBLY

Surfaces and interfaces play an important role in the formation of fibrous structures from polypeptides. While the majority of assembly processes are conducted in solution within a bulk liquid phase, this liquid will be bounded by a single interface or combination of interfaces including solid–liquid interfaces, liquid–liquid interfaces, or liquid–gas interfaces each which can influence the assembly process, as illustrated in Figure 1.

An understanding of the interaction between polypeptides and surfaces is imperative if self-assembly is to be achieved in a reproducible manner. This is especially important if the scale of assembly is to be increased from small-scale laboratory experiments to larger reaction vessels. Insights obtained into surface-based assembly can also be used to design surface-based "reactor platforms" which encourage surface interactions in order to achieve

Figure 1 Interfaces that can be present during the self-assembly of polypeptides in the bulk liquid phase. The left image depicts a common experimental procedure where structures are assembled in a solution contained by a small tube, exposing polypeptides to both solid–liquid and liquid–gas interfaces. A third possible interface is a liquid–liquid interface depicted in the image on the right that contains a solution of two immiscible liquids.

controlled assembly (Nielsen et al., 2001). In Section 3.1 below, the assembly of amyloid fibrils in the presence of surfaces is discussed, with emphasis on the influence of solid nonbiological surfaces. The placement of nanofibers will be critical to many potential applications for fibrils and the use of surfaces to direct fibril assembly and produce patterned materials is subsequently reviewed in Section 3.2.

Surfaces and interfaces remain an important consideration following self-assembly. Fibrous proteins may be specifically designed to interact with surfaces as part of devices either by adsorption or by covalent attachment to the surface. They can also be used to coat surfaces, changing their properties. Our ability to precisely place these fibrils on a surface will determine the range of potential uses for these materials. Interfacial contact can also occur during storage, for example, fibrils in liquid solutions that are stored in plastic containers will encounter both a liquid–polymeric interface and a liquid–air interface. Studies examining the interaction of preassembled fibrils with solid nonbiological surfaces are therefore reviewed in Section 3.3.

3.1. Surface-based assembly

A surface or interface can influence the assembly of fibrils by altering both the process and kinetics of fibril nucleation or elongation. This behavior is not surprising as surface properties are known to influence the absorption, confirmation, and destabilization of globular proteins or smaller peptides (Rocha et al., 2005). Surface properties influence fibril assembly in a similar way by altering the absorption, unfolding, and aggregation of monomers.

Key surface properties that can influence fibril assembly include hydrophobicity or hydrophilicity, the presence of negative or positive charges, and surface roughness. Two methods which can be used to characterize surface properties and rationalize interactions between proteins and surfaces are surface wettability and the root mean squared (RMS) roughness. Surface wettability is measured via the contact angle formed when a droplet of water is placed on a surface. A small contact angle indicates a highly wetting surface. The RMS can be measured using an AFM and reflects the variation in surface height about an average surface height. A high RMS indicates a rough surface. Techniques such as streaming potential and AFM can also be used to measure surface charge (Franks and Meagher, 2003) and further characterize a surface.

A number of studies have examined fibril formation in the presence of different solid nonbiological surfaces, as summarized in Table 1. While many of these studies have focused on the formation of fibrils, the wettability and RMS of some surfaces have been characterized. Typical surface contact angles are also presented in Table 1 to aid comparison between these surface-based experiments.

Table 1 Examples of surface-induced nucleation and growth of fibrils

Surface	Properties	Polypeptide	Fibril formation	Reference
Regenerated cellulose (RE)	Contact angle 27° ± 2, RMS ~ 87 nm	Human insulin	Nucleation faster than bulk	Nayak et al. (2008)
Poly(ethersulfone) (PES)	Contact angle 55° ± 2, RMS ~ 36 nm.	Human insulin	Nucleation faster than bulk	Nayak et al. (2008)
Poly(vinylidine difluoride (PVDF)	Contact angle 76° ± 2, RMS ~ 195 nm	Human insulin	Nucleation faster than bulk. Elongation not different to the bulk	Nayak et al. (2008)
Polyethylene (PE)	Contact angle 94° ± 4, RMS ~ 45 nm	Human insulin	Nucleation faster than bulk. Elongation not different to the bulk	Nayak et al. (2008)
Polytetrafluoroethylene (PTFE)	Contact angle 120° ± 3, RMS ~ 54 nm	Human insulin	Nucleation faster than bulk. Elongation not different to the bulk	Nayak et al. (2008)
N-isopropylacrylamide-co-N-tert-butylacrylamide (NIPAM/BAM) copolymer particles (polymer ratio 85:15)	Contact angle ~53° for equivalent flat film	Human β_2-microglobulin	Nucleation faster than bulk. Elongation not different to the bulk	Allen et al. (2003); Linse et al. (2007)

Table 1 (Continued)

Surface	Properties	Polypeptide	Fibril formation	Reference
NIPAM/BAM copolymer particles (polymer ratio 50:50)	Contact angle ~57° for equivalent flat film	Human β$_2$-microglobulin	Nucleation faster than bulk. Elongation not different to the bulk	Allen et al. (2003); Linse et al. (2007)
Hydrophilic polymer-coated quantum dots, cerium oxide particles, and multiwalled carbon nanotubes	—	Human β$_2$-microglobulin	Nucleation faster than bulk. Elongation not different to the bulk	Linse et al. (2007)
Polystyrene (PS)	Contact angle ~90°[a]	Bovine insulin	Nucleation faster than bulk. Elongation slower than the bulk	Smith et al. (2007); Narrainen et al. (2006)
Mica unmodified	Hydrophilic and negatively charged Contact angle <10°[b]	Light chain variable domain from light chain amyloidosis (SAM)	Growth observed but not divided into nucleation and elongation phases	Zhu et al. (2002); Yang et al. (2007)
Mica modified	Hydrophilic and positively charged	Light chain variable domain from light chain amyloidosis (SAM)	No growth observed	Zhu et al. (2002)

Mica modified	Hydrophobic	Light chain variable domain from light chain amyloidosis (SAM)	No growth observed	Zhu et al. (2002)
Highly oriented pyrolitic graphite (HOPG), ZYB grade	Hydrophobic Contact angle ~71°c	Aβ1-42	Growth observed but not divided into nucleation and elongation phases	Kowalewski and Holtzman, (1999); Yang et al. (2007)
Mica unmodified	Hydrophilic and negatively charged Contact angle <10°b	Aβ1-42	Growth observed but not divided into nucleation and elongation phases	Kowalewski and Holtzman, (1999); Yang et al. (2007)
Highly oriented pyrolitic graphite (HOPG), ZYH grade	Hydrophobic basal planes and hydrophilic step edges Contact angle ~71°c	Aβ1-40, Aβ1-42, and Aβ1-28.	Nucleation observed on hydrophobic surface and growth on hydrophilic surface for both Aβ1-40 and Aβ1-42. No growth observed for Aβ1-28	Losic et al. (2006); Yang et al. (2007)

Table 1 (*Continued*)

Surface	Properties	Polypeptide	Fibril formation	Reference
Mica unmodified	Hydrophilic and negatively charged Contact angle <10°[b]	Aβ1-40	No growth observed	Losic et al. (2006); Yang et al. (2007)
Mica unmodified	Hydrophilic and negatively charged Contact angle <10°[b]	Recombinant β$_2$-microglobulin	No growth observed	Relini et al. (2006); Yang et al. (2007)
Mica modified	Positively charged	Recombinant β$_2$-microglobulin	Fibril growth observed	Relini et al. (2006)

[a] Typical contact angle for a polystyrene film.
[b] Typical contact angle for freshly cleaved mica.
[c] Typical contact angle for HOPG.

It is clear that a solid surface can nucleate fibril formation under conditions where fibrils are not observed in a bulk liquid phase (Zhu et al., 2002). The surface may act to increase the local concentration of protein and alter the conformation of those proteins leading to an increased propensity for association (Linse et al., 2007; Nayak et al., 2008). However, not all surfaces are equal in their ability to nucleate fibrils, and differences are reported in the literature. One of the most systematic and quantitative studies in fibril nucleation was performed using a range of synthetic surfaces suspended in a bulk solution during the formation of insulin fibrils (Nayak et al., 2008). Nayak and colleagues found that fibril nucleation was accelerated on all surfaces relative to nucleation in a bulk solution and was fastest on rough hydrophobic surfaces. A near-linear inverse correlation was observed between both hydrophobicity or roughness and the lag phase and fibril growth. Similar results were reported by Smith who observed faster nucleation of insulin on a hydrophobic polystyrene surface compared to nucleation in the bulk phase (Smith et al., 2007).

A range of nanoparticles (copolymer particles, quantum dots, cerium oxide particles, and carbon nanotubes) with high surface area have also been shown to accelerate fibril growth relative to fibril growth in solutions lacking nanoparticles (Linse et al., 2007). Specifically, an increase in the surface area of polymeric nanoparticles decreased the lag phase of fibril growth. However, in contrast to Nayak's findings, Linse observed that small increases in copolymer hydrophobicity ($\sim 4°$, Table 1) delayed nucleation. Surface plasmon resonance was also used to show that protein binding to hydrophobic particles was weaker than protein binding to hydrophilic particles. Zhu similarly found that hydrophobic surfaces failed to stimulate the nucleation and growth of SAM fibrils (Zhu et al., 2002).

The difference between the findings of Nayak and those of Linse and Zhu could be due to the curved particle surface in the nanoparticle study (Nayak et al., 2008) or differences in surface roughness. These findings also indicate that the protein under consideration, insulin in the case of Nayak's study and β_2-microglobulin or SAM within Linse and Zhu's studies, may experience different surface interactions. It is certainly likely that the protein hydrophobicity, charge, structure, and size will all influence surface interactions, suggesting that a wide ranging investigation on the effect of surface-induced nucleation for different proteins is required.

Fibrils have been observed to grow on both hydrophobic and hydrophilic surfaces (Kowalewski and Holtzman, 1999). However, the effect of surface hydrophobicity on fibril elongation and growth appears unclear. Smith found fibril growth on polystyrene was slower than in the bulk phase (Smith et al., 2007) and suggested that growth on a polystyrene surface results in different associations between the proteins compared to growth in the bulk phase. However, the growth kinetics observed by Linse in the presence of various nanoparticles (Linse et al., 2007) and by Nayak in the presence of various polymeric surfaces did not appear significantly altered by the presence of these surfaces in these later studies (Nayak et al., 2008). Growth rates were also similar across different surface types. Interestingly, when both hydrophilic and hydrophobic surfaces are presented in the solution at the same time, as can occur for particular grades of highly ordered pyrolytic graphite (HOPG), Aβ1-40 fibrils grow on the hydrophilic surface (Losic et al., 2006). Losic suggests that nucleation may initiate on the hydrophobic surface.

Electrostatic charges on surfaces can also influence fibril nucleation and growth. The light chain variable domain from light chain amyloidosis (SAM) was found to form fibrils on negatively charged hydrophilic mica but not on positively charged mica (Zhu et al., 2002). Conversely, $β_2$-microglobulin forms fibrils on positively charged mica but not on negatively charged mica (Relini et al., 2006). Relini and colleagues observed a decrease in the charge of $β_2$-microglobulin during incubation and suggest that the positively charged surface helps dipole crowding and alignment when the protein comes in contact with the surface. These examples highlight the importance of electrostatics in determining which surfaces will support fibril growth. If we are to predict such interactions, we must also consider the charges on a protein and the distribution of these charges throughout a protein, as such charges will be placed in different parts of the protein and will change as a function of pH due to the ionization of protein side chains. The distribution and exposure of hydrophobic regions that are likely to mediate interactions with a hydrophobic surface will also vary within a protein.

There is certainly evidence that protein properties, including the propensity for fibril formation, influence fibril nucleation and growth on a surface. For example, the peptides Aβ1-40, Aβ1-42, and Aβ1-28 behave differently reflecting their different propensities to form fibrils (Losic et al., 2006). Aβ1-42 forms fibrils most readily, and Aβ1-42 fibrils are observed on both mica and HOPG (Kowalewski and Holtzman, 1999; Losic et al., 2006). In contrast, Aβ1-40 forms fibrils on HPOG but not mica, and Aβ1-28, the least prone to aggregation, forms fibrils on neither surface (Losic et al., 2006).

One technique which can be used to systematically examine the effect of surface properties on amyloid fibril growth is high-resolution AFM or kymography. This technique has recently been used to scan along the length of the fibril to determine the stepwise rate of fibril elongation from each of the fibril ends (Kellermayer et al., 2008), and it could be used to examine the influence of surface properties on individual fibrils. Ultimately, it will also be desirable to combine observations of fibril morphology on a surface with information about the kinetics of protein unfolding, rearrangement, and fibril assembly. This information would provide insight into the mechanism of fibril formation. It will also help us to direct fibril formation on surfaces where it may be desired and to prevent unwanted aggregation such as on surgical instruments and dialysis membranes (Nayak et al., 2008).

Fibril formation is known to occur at interfaces other than solid–liquid interfaces, such as lipid–water interfaces (Adams et al., 2002; Sharp et al., 2002). At lipid–water interfaces, fibril formation is strongly influenced by surface charge, and aggregation differs to processes in the bulk solution. Biological molecules such as collagen provide further types of interfaces (Relini et al., 2006) which can be used to induce fibril formation. These interactions offer further opportunities for directing the formation of fibrils but are beyond the scope of this chapter. The reader is directed to a recent review (Stefani, 2007) which discusses the role of membranes and biomacromolecules in stimulating fibril formation in greater detail.

3.2. Surface-directed assembly

Surfaces can do more than stimulate the growth of amyloid fibrils; the presence of a surface during assembly can also influence fibril morphology. Fibril diameter has been observed to be smaller on surfaces than in the bulk for some polypeptide systems (Zhu et al., 2002). This may be due to the constraints imposed by the surface or could result from a different mechanism of fibril growth. The shape of fibrils formed on surfaces can also differ to the shape of fibrils formed in the bulk solution.

One of the most interesting properties of highly orientated surfaces is their ability to direct fibril orientation, and this order can be used to create networks and patterns of fibrils on the surface. This trait could be used to determine where fibrillar components assemble within devices or to create patterns for nanoelectronic circuits. A number of biomaterials also involve β-sheet lamination on an inorganic surface (Brown et al., 2002). The surface growth of fibrils, known as epitaxial growth or surface templated assembly, relies on the underlying crystallographic symmetry of the surface material and is analogous to the growth of one crystal phase on top of a host crystal (Kowalewski and Holtzman, 1999).

Kowalewski and colleagues found that Aβ1-42 fibrils formed in three major directions on HOPG as shown in Figure 2 (Kowalewski and Holtzman, 1999). Similar growth has been observed by a range of researchers for Aβ fibrils on HPOG (Kellermayer et al., 2008; Losic et al., 2006). These patterns have also been observed for de novo-designed β-sheet-rich nanofibers constructed from alternating polar and nonpolar sequences (Brown et al., 2002) that resemble amyloid fibrils, as shown in Figure 3. This figure illustrates how epitaxial growth is not limited to individual fibers, with thick layers of more than 10 fibers assembling in the same direction.

The orientated growth of Aβ fibrils and peptide nanofibers on HPOG is due to the underlying six-fold symmetry of the graphite surface. This directs fiber formation in the three observed directions, each orientated at 120°, as illustrated in the schematic in Figure 3. Kowalewski suggested such growth is driven by hydrophobic forces (Kowalewski and Holtzman, 1999), where

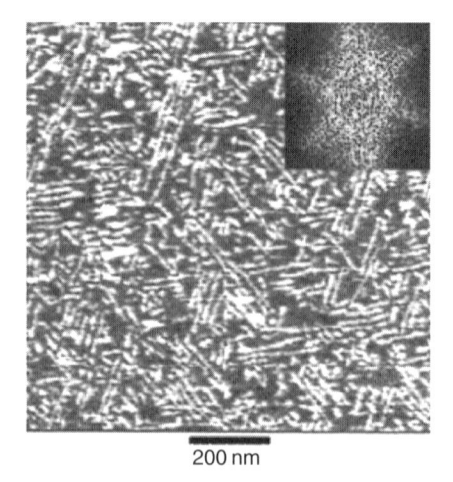

Figure 2 The orientated growth of Aβ1-42 fibrils from an aqueous solution onto a highly orientated pyrolytic graphite surface. The fibrils are orientated in three directions, each angled at 120°, reflecting the crystalline structure of the underlying material. The image was collected by atomic force microscopy. The inset is a two-dimensional Fourier transform showing the six-fold symmetry of the graphite material (Kowalewski, 1999) (copyright © The National Academy of Sciences of the United States of America, all rights reserved, 1999).

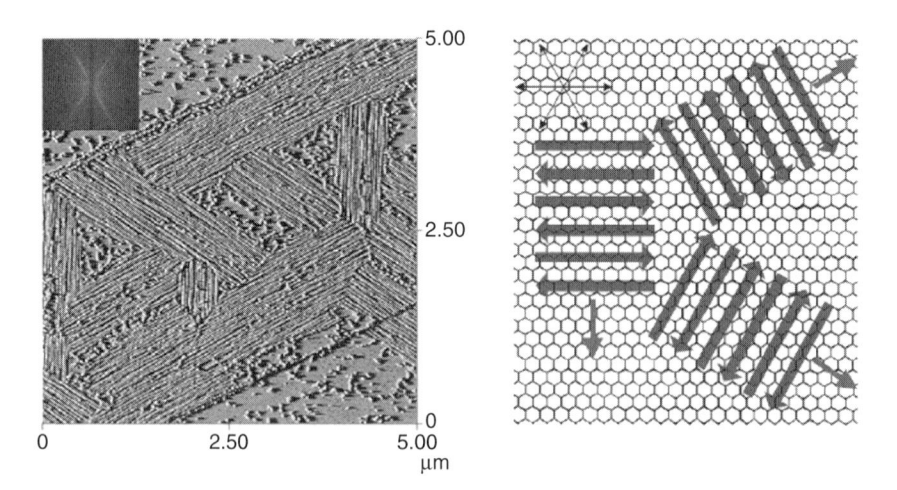

Figure 3 The orientated growth of peptide fibers formed from an alternating sequence of polar and nonpolar amino acids on a highly orientated pyrolytic graphite surface (left). The image was collected by atomic force microscopy. The inset is a two-dimensional Fourier transform showing the six-fold symmetry of the graphite material. This symmetry is also shown in the schematic (right) which shows how the surface symmetry directs the growth of the fibers, shown as six stranded assemblies, that grow in the direction of the arrow. Reprinted with permission from Brown et al., (2002) (copyright 2002 American Chemical Society).

hydrophobic interactions are maximized by contact between the carbons in graphite and the hydrophobic residues within the peptide chain.

Recently, Losic and colleagues (2006) used steps in HPOG, known as step edges, to direct the growth of Aβ fibrils. Losic specifically choose lower grade graphite with a mosaic angle of 3.5 ± 1.5°. This graphite is less ordered than the graphite used by Kowaleski, with a smaller grain size (~30–40 nm rather than 1 μm). The lower grade graphite still contains the regular hydrophobic basal planes containing fused hydrophobic rings, but it also contains more frequent changes in surface height of 0.3–3.6 nm called step edges. These edges are hydrophilic and display a number of functional groups (phenolic, carboxylic, and ketonic groups) (Losic et al., 2006). Losic found that Aβ1-40 and Aβ1-42 fibrils assembled preferentially along the hydrophilic step edges of HPOG rather than on the hydrophobic basal planes as occurs for more ordered graphite. The resulting long straight fibrils are shown in Figure 4.

It is worth noting that not all protein fibers will assemble along the edge of HPOG steps like Aβ fibrils. Yang and colleagues (2007) found that fibers formed from the EAK16-II polypeptide sequence did not form along HPOG edges. Instead these fibers draped over the step edges and their assembly was not influenced by the different chemistry along the edge of the graphite step. The polypeptides that make up these fibers consist of alternating polar and nonpolar residues, resulting in a hydrophilic and hydrophobic face, and this structure is likely to determine the behavior of these fibers on a HOPG surface.

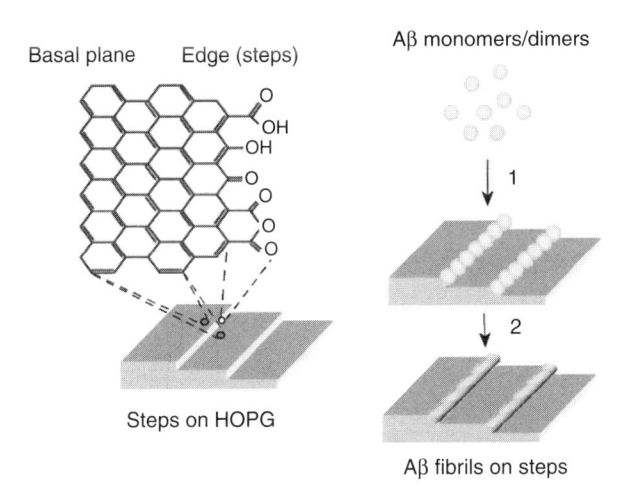

Figure 4 Schematic illustrating the orientated growth of Aβ1-40 fibrils along the hydrophilic step edges of highly orientated pyrolytic graphite (HOPG) in preference to assembly on the hydrophobic basal plane surface. Copyright (Losic, Martin et al., 2006). Reprinted with permission of John Wiley & Sons, inc.

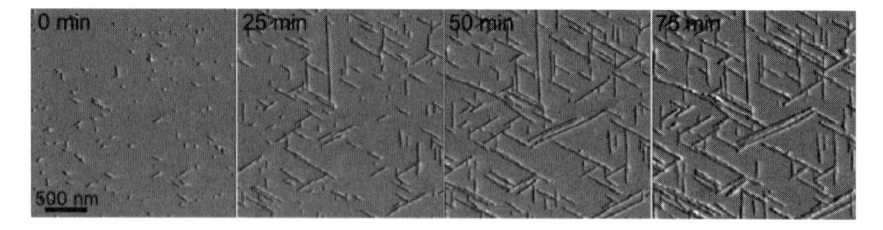

Figure 5 The orientated growth of Aβ25-35 fibrils from an aqueous solution on a mica surface. The fibrils grow along three main directions, each orientated at 120°. The images were collected by *in situ* time lapse Atomic Force Microscopy (Karsai, Grama et al. 2007). Copyright IOP Publishing Ltd. Reproduced with permission.

The hexagonal crystalline lattice of mica can also be used to direct fibril growth. Karsai and colleagues (2007, 2008) have assembled fibrils from the Aβ25-35 peptide and an N27C cystiene variant on mica. The resulting fibrillar networks are trigonally oriented at 120°, as shown in Figure 5. In contrast, no order is observed for the assembly of the same fibrils formed on a glass surface. Electrostatic attraction appears to govern this type of fibril binding. Specifically, amino groups of lysine residues or the N terminus of the peptide are thought to interact with the potassium binding pockets in mica. The growth of Aβ25-35 is sensitive to potassium ions or amine acetylation, consistent with this proposed mechanism (Karsai et al., 2007). The growth of fibrils from the N27C cystiene variant is also sensitive to sodium chloride (Karsai et al., 2008), and this sensitivity provides a further mechanism for the control of fibril growth.

Surface-based assembly can also be combined with different drying methods to create different surface effects and ordered arrays. Whitehouse and colleagues (2005) studied the ordered assembly of β-sheet tapes on mica and used a blotting method to dry the surface, where excess solution was removed from the side of the surface by wicking with filter paper. The shear forces created by this blotting created highly ordered arrays of fibrils suggesting that this method could be used to generate large aligned monolayers. It should be noted that methods that involve drying such as the blotting technique involve interactions with the liquid–air interface in addition to the solid–liquid interface. The blotting method also has some similarities to molecular combing method outlined in Section 3.3 below. A significant difference, however, is that the blotting method described by Whitehouse involves assembly and alignment rather than the alignment of preformed fibers.

Surface assembly can also be used to orientate fibrils relative to the surface, generating fibrils that are horizontal or perpendicular to a surface. Zhang observed this behavior for the artificial GAV9 peptide which contains the consensus sequences from α-synuclein, Aβ, and the prion protein (Zhang et al., 2006). This peptide assembles horizontally on HOPG surfaces and assembles perpendicular to the surface of mica.

The orientated growth of fibrils is dependent on the properties of both the surface and the polypeptide, similar to nonorientated fibril growth discussed in Section 3.1 above. For example, not all Aβ peptides assemble along the step edges on HPOG (Kellermayer et al., 2008; Losic et al., 2006). Hoyer also found that α-synuclein assembles on mica but not on HOPG (Hoyer et al., 2004), which the authors suggest is due to differences in electrostatic attraction between the two surfaces. Consequently, any insights that can be gained for surface nucleation and growth on nonorientated surfaces will help to advance our understanding for surface-directed assembly. It will also inform attempts to construct artificial surfaces with features designed to direct and exploit such fibrillar assembly.

External microscopy tools can be used to further control the surface-based assembly of fibrils on a surface. Kowalewski used an AFM tip to move the initial aggregates that first form during fibril assembly and change their position on a surface (Kowalewski and Holtzman, 1999). An AFM tip has also been used to induce fibril disassembly in restricted areas (Kellermayer et al., 2008), and both methods offer a way to create patterns of fibrils on a surface.

The growth of fibril ends will be important for creating arrays. In some cases fibril growth is unidirectional; in other cases growth appears to be bidirectional, although the speed of assembly can significantly differ between fibril ends (Inoue et al., 2001; Kellermayer et al., 2008; Scheibel et al., 2001). An interesting feature of fibril growth is that it aborts when the fibril reaches another fibril obscuring its path. When confronted with an obstacle, the fibrils do not change direction or grow over the obstacle; however, an AFM tip can be used to change the direction of fibril growth (Zhang et al., 2006). This behavior could only be induced by Zhang on mica and not on HPOG, possibly due to the different orientation of the fibrils formed by the GAV9 peptide on these surfaces.

One possible drawback of surface-directed assembly is that this phenomenon has been observed to decrease for some fibrils, such as Aβ25-35, as the fibrils reach the later stages of assembly (Karsai et al., 2007). This temporal change could possibly be avoided by using higher concentrations of Aβ25-35 peptide during assembly. Chemical cross-linking could also be used to capture the fibrils on the surface and stabilize the fibril network.

Information on the stability of fibrils will be critical for assessing the potential of fibril-coated surfaces for biotechnology applications. Fibrils such as insulin remain strongly bound to the surface after formation and remain on the surface after rinsing with water (Zhu et al., 2002). This is a desirable feature for the construction of new materials, although there may be some instances where surface transfer is required.

The surface stability of other self-assembling fiber systems such as EAK16-II have been extensively characterized (Yang et al., 2007). These fibers are stable on HOPG under both acidic and basic conditions, most likely due to the hydrophobic nature of the fiber–surface interaction. In contrast, on mica where electrostatic interactions are likely to dominate these EAK16-II, fibers are stable under acidic conditions but unstable under basic conditions. This study indicates that wide ranging tests are required to fully understand the conditions under which patterned fibers will remain stable on surfaces.

It will also be desirable to characterize the surface properties of fibril-modified surfaces. The self-assembling amphiphilic EAK16-II fibers provide a good example of this type of characterization (Yang et al., 2007). Adsorption of fibers on a surface significantly changes the surface properties, including increasing the contact angle of water on mica and decreasing the contact angle of water on HPOG. This change in contact angle will subsequently change interactions between the surface with water, solvents, or biomolecules.

Larger micron-scale fiber patterns can be created on a surface by employing physical techniques in addition to surface templating. Several examples can be drawn from studies that employ nonamyloidogenic self-assembling protein systems. Hung recently described how the assembly of fibrous peptide amphiphiles can be directed using spatial confinement in microcapillaries (Hung and Stupp, 2007). This method is a sonication-assisted solution embossing technique, where solvent evaporation drives assembly within a narrow channel. Sonication imparts energy to the system and improves alignment. Straight and curved patterns of aligned fibrils could be achieved using straight or curved channels cut by an electron beam as shown in Figure 6. The authors suggest that these techniques will be applicable to a wide range of other self-assembling systems.

In some cases, assembly can also be focused in the vertical direction away from the surface. This type of assembly can produce materials with very different properties from the same initial building blocks. Diphenylalanine-based peptides dissolved in 1,1,1,3,3,3-hexafluoro-2-propanol (HFP) solvent can be assembled vertically on a siliconized glass surface, as illustrated in Figure 7. The authors suggest that rapid solvent evaporation concentrates the peptide, generating a supersaturated solution that nucleates quickly on the surface (Reches and Gazit, 2006). The stacking of aromatic groups then leads to growth in the vertical direction. Electrostatic charges appear important to this growth and uncharged or negatively charged peptide analogs do not assemble vertically. Positively charged peptides do form highly orientated structures but only on negatively charged surfaces and not on positively charged surfaces, indicating that charge repulsion between the peptides and the surface stabilizes these structures. The specific interactions that mediate this assembly process suggest that it will be restricted to self-assembling systems containing aromatic motifs.

The studies highlighted in this section illustrate how surfaces can be used to direct the assembly of amyloid fibrils and other self-assembling

Surface- and Solution-Based Assembly of Amyloid Fibrils 181

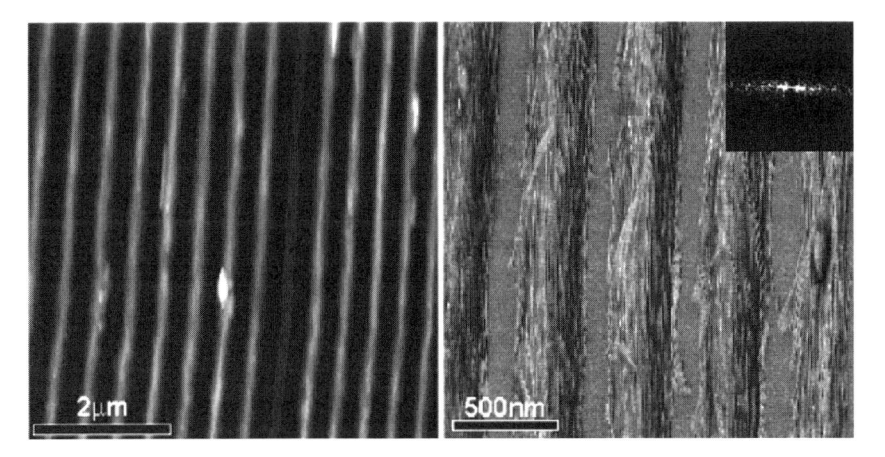

Figure 6 Peptide amphiphile nanofibers that have been aligned during assembly using a sonication-assisted solution embossing technique. The images are atomic force microscopy height (left) and phase (right) images. The inset is a fast Fourier transform of the phase image showing the periodicity of the pattern. Reprinted with permission from Hung and Stupp (2007) (copyright 2007 American Chemical Society).

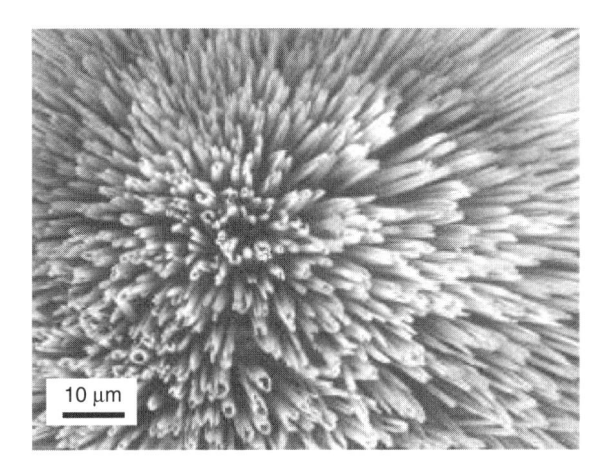

Figure 7 Scanning electron microscopy image of vertically aligned diphenylalanine-based peptide nanotubes assembled on a glass surface. The scale bar is 10 μm in length. Reprinted by permission from Macmillan Publishers Ltd., *Nature Nanotechnology* (Reches and Gazit, 2006), copyright 2006 (http://www.nature.com/nnano/index.html).

systems. For further details on the ordered assembly of amphiphilic peptides at an interface and methods that can be used to characterize these assemblies, the reader is directed to a recent review by Rapaport (2006).

3.3. Surface interactions following assembly

The simplest way to coat a surface with fibrils following fibril assembly is to deposit a droplet of fibrils suspended in solution onto a surface. The individual fibrils will then sediment to the bottom of the droplet, make contact with the surface and adsorb. A large volume of solution can also be brought in contact with the surface. The strength of fibril binding will depend on the electrostatic, hydrophobic, or Van der Waals interactions between the fibrils and the surface. These interactions cannot necessarily be predicted from behavior observed during fibril assembly, as different regions of the polypeptide chain are likely to be involved in each case. Indeed, Losic observed that preformed Aβ fibrils preferred to adhere to hydrophobic surfaces, whereas Aβ peptides preferentially assembled at a hydrophilic step edge on HOPG (see Section 3.2 above; Losic et al., 2006). Similarly, fibrils constructed from different polypeptide sequences will display different behaviors at a surface, as the residues on the outside of the fibrils will differ between fibrils, and it is these residues that are likely to mediate fibril–surface interactions.

Following fibril deposition, the surface can be rinsed to remove any unbound fibrils and dried. Studies on model protein systems (Squires et al., 2006) suggest that drying does not change structural integrity of fibrils. A degree of natural alignment is observed for some fibrils deposited on a surface, as shown in Figure 8, although it is not yet understood how buffers and surface properties influence this alignment (Scheibel et al., 2003). The concentration of fibrils in solution and interactions between fibrils will certainly contribute to fibril alignment.

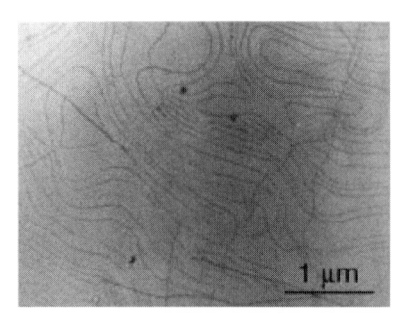

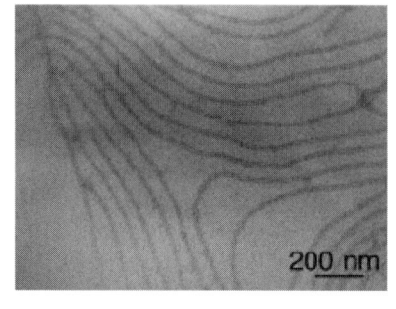

Figure 8 Transmission electron microscopy image of fibrils formed from the Sup35 NM protein. The fibrils demonstrate natural alignment and are shown at low and high magnification on the left and right respectively (Scheibel et al., 2003). The scale bars are 1 μm and 200 nm in length, respectively (copyright © The National Academy of Sciences of the United States of America, all rights reserved, 2003).

For the majority of fibril systems, simple deposition results in layers of fibrils that are largely unordered. The thickness of the fibril layer can also vary. This section explores a variety of methods that can be used to induce fibrillar alignment on a surface or create surface patterns. It also details methods that can be used to induce covalent and noncovalent attachment of fibrils to the surface.

Electrostatic interactions between fibrils and a surface can be exploited to create micron-sized patterns of fibrils. Mesquida and colleagues (2005) used this approach to selectively deposit fibrils onto negatively charged stripes of mica that were exposed between alternating stripes of deposited cationic poly-L-lysine, as shown in Figure 9. The striped pattern was formed by directing poly-L-lysine through the channels of a polydimethysiloxane (PDMS) stamp that was placed in contact with the mica surface.

The fibril deposition observed by Mesquida is thought to be governed largely by electrostatics, and fibril binding could be switched from the negatively charged mica surface to the positively charged poly-L-lysine by raising the solution pH. This is presumably due to the ionization of exposed amino acid side chains on the fibril. This type of fibril behavior also suggests that salt may be used to screen electrostatic charges and prevent adhesion. Although this method does not create order within the patterns and the fibrils are orientated randomly, the method does effectively achieve micron-sized arrays of fibrils.

In a second study, Mesquida again exploited electrostatic interactions between fibrils and surfaces to create surface patterns. This second study employed AFM charge writing to position fibrils in lines several microns

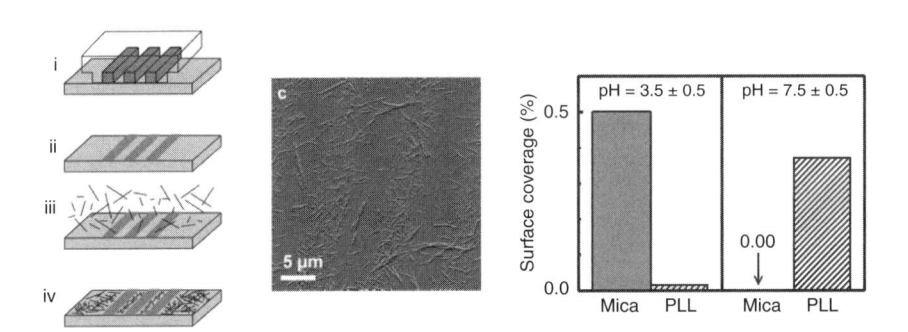

Figure 9 Electrostatic patterning of pre-formed fibrils on a surface. Positively charged poly-L-lysine is patterned in stripes across a negatively charged mica surface using a PDMS stamp (left). Atomic Force Microscopy image showing how fibrils selectively deposit on the exposed mica surface at low pH (centre). The scale bar is 5 mm in length. Fibrils deposition can be switched from mica to poly-L-lysine as the solution pH is increased (right) (Mesquida, Ammann et al., 2005). Copyright Wiley-VCH Verlag GmbH & Co. KGaA. Reproduced with permission.

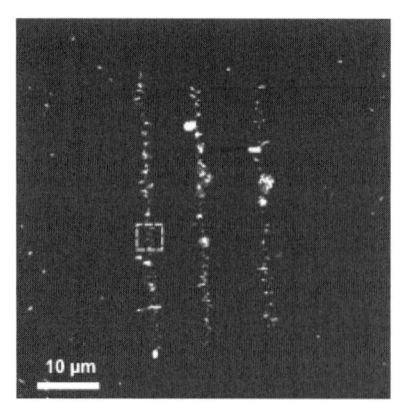

Figure 10 Atomic force microscopy image of a fibril microarray. The pattern was created by charge-writing using atomic force microscopy. The scale bar is 10 µm in length.
Reprinted with permission from Mesquida et al. (2006) (copyright 2006 American Chemical Society).

in length on a polymer surface, as shown in Figure 10 (Mesquida et al., 2006). The fibril cargo was suspended in water droplets within a water-in-perflurocarbon oil emulsion, generated by sonicating an aqueous solution of fibrils mixed with perflurocarbon. Three negatively charged lines were drawn on a polymethyl–methacrylate polymer surface using an AFM tip to which a small voltage was applied. The emulsion was then washed over this charged surface and fibrils deposited along the charge lines with an accuracy of 1–2 µm. Although the fibrillar structures did not form a continuous line and fibrils were not orientated along this line, fibril adsorption was specific to the charged areas. Once bound, the fibrils were also stable and were not removed from the surface by rinsing with water.

Molecular combing is another method that has been successfully employed to align individual amyloid fibrils on a surface (Herland et al., 2007). Molecular combing has been used extensively to align DNA molecules (Bensimon et al., 1994) and involves the movement of liquid along a surface. The receding air–liquid interface causes the alignment of partially adsorbed molecules largely due to viscous drag and surface tension. Herland and colleagues used this method to align insulin fibrils on glass. The authors stress the importance of achieving moderately strong interactions between the fibrils and the surface as this interaction facilitates alignment; if the interaction is too strong multiple parts of the fibril will adhere to the surface preventing alignment, if the interaction is too weak fibrils will not adsorb to the surface during combing preventing alignment. To overcome this problem, Herland incubated the fibril droplet on the surface for 1 min prior to combing to ensure sufficient adsorption.

Herland found treated glass to be the most successful surface for molecular combing (Herland et al., 2007). This glass was rendered hydrophobic using a PDMS stamp or silanized with dichlorodimethylsilane (Herland et al., 2008). Following treatment, insulin fibrils that had been wrapped in the water-soluble polyelectrolyte poly(thiophene acetic acid) (PTAA) could be combed on the surface. In contrast, these fibrils did not adhere or align well on unmodified hydrophilic glass. Molecular combing could also be successfully applied to insulin fibrils coated with the water-insoluble conjugated polymer APFO-12 (Herland et al., 2008) demonstrating the wide applicability of this technique.

Another variation in molecular combing involves aligning fibrils directly on the surface of a PDMS stamp. These fibrils can then be print transferred onto a clean glass surface, as illustrated in Figure 11. Fibril layers printed in this way can withstand brief washing in water, oil, or ethanol solutions. Aligned fibrils may have scientific interest in addition to their potential as materials and Herland also used the aligned fibrils to help determine the orientation of the polymer chains, which are perpendicular to the fibril axis (Herland et al., 2008).

Magnetism is yet another successful method for the alignment of individual peptide fibers. Magnetite (Fe_3O_4) particles approximately 10 nm in size were noncovalently attached to diphenylanalanine nanotubes during fiber

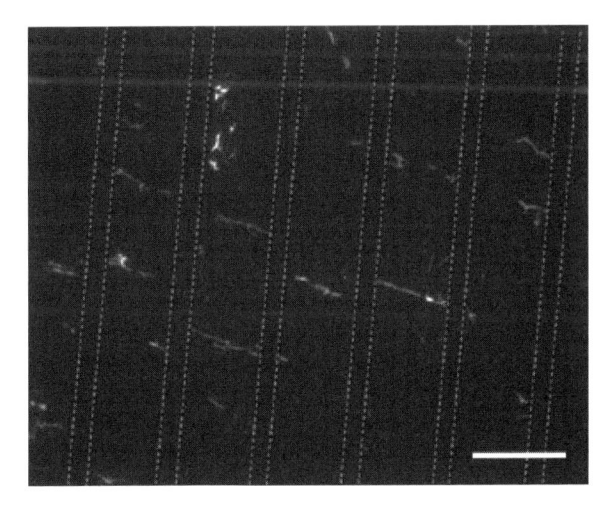

Figure 11 Atomic Force Microscopy image of aligned PTAA coated insulin fibrils that have been transferred to a glass surface after molecular combing directly onto the surface of a PDMS stamp. The lines have been added to illustrate gaps within the PDMS stamp. The scale bar is 2 mm in length (Herland, Björk et al., 2007). Copyright Wiley-VCH Verlag GmbH & Co. KGaA. Reproduced with permission.

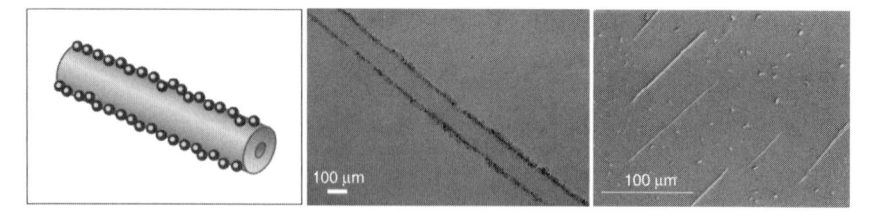

Figure 12 Magnetically induced horizontal alignment of diphenylalanine-based peptide nanotubes. A noncovalent layer of magnetic nanoparticles coats the nanotubes during assembly (shown in the schematic left and transmission electron microscope image, middle). The application of an electric field induces alignment of the nanotubes suspended in ferrofluid (scanning electron microscopy image, right). The scale bars are 100 nm and 100 μm in length. Reprinted with permission from Macmillan Publishers Ltd., *Nature Nanotechnology* (Reches and Gazit, 2006), copyright 2006 (http://www.nature.com/nnano/index.html).

assembly, as shown in Figure 12. This attachment is thought to be mediated by hydrophobic interactions (Reches and Gazit, 2006). An external magnetic field of 0.5 T can then be used to induce alignment of fibers that have been suspended in a ferrofluid. In this case, a constant magnetic field was applied until the solution dried.

Larger patterned surfaces can also be achieved using inkjet printing. This technique has been used to create micron-sized letters of the alphabet containing diphenylalanine peptide nanotubes [tertbutoxycarbonyl–Phe–Phe–OH (Boc–Phe–Phe–OH)] on either transparent foil or indium–tin oxide (Adler-Abramovich and Gazit, 2008) as shown in Figure 13. This later

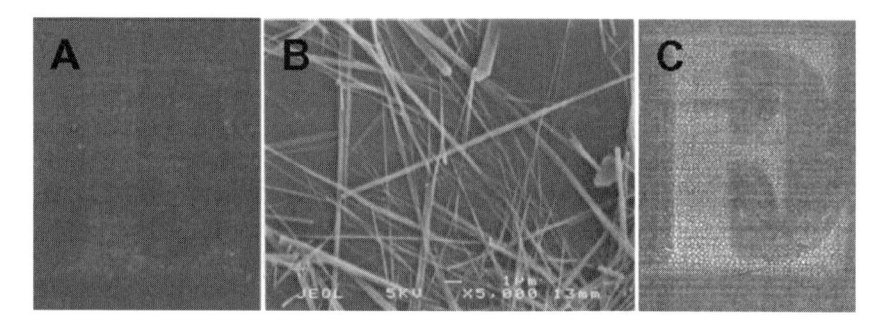

Figure 13 Printed letters of the alphabet containing di-phenylalanine peptide nanotubes. A: single letter printed on transparent foil, B: detail of fibers within the letter shown in A. C: multiple printing of the same letter. All images were collected using Scanning Electron Microscopy. The scale bar in image B is 1 mm in length (Adler-Abramovich and Gazit, 2008). Copyright Blackwell publishing. Reproduced with permission.

surface may be useful for creating conducting materials. The morphology of the patterned fibers was dependent on the solvent used for printing. When the peptide was suspended in a mixture of hexafluorpropane and water, the printed peptides form fibers but when the peptide was suspended in hexafluorpropane and ethanol the letters contained only spheres. Although the fibers are not aligned within the printed features, the density of fibrils could be changed by increasing the peptide concentration or by printing multiple times on the same surface, and inkjet printing represents an effective technique for building larger structures from self-assembling peptides.

Covalent binding offers another way to immobilize amyloid fibrils on a surface and control fibril arrangement. Ha and Park (2005) covalently attached insulin fibrils to a N-hydroxysuccinimide-activated glass surface. The insulin fibrils were partially formed and they act as seeds which can further template the assembly of fibrils. By washing monomeric insulin over the surface, Ha and Park could form longer fibrils on the activated glass surface. This type of seeded assembly is different to both the surface-mediated assembly and surface-directed assembly described above (Sections 3.1 and 3.2, respectively). However, the surface may still influence fibril growth to the same extent as fibrils were observed to be thinner than those observed to form in solution. This method of covalent attachment has been used to screen factors that inhibit fibril formation, and it was an advance on previous studies which physically absorbed fibrils onto either glass or plastic in order to further template fibril growth (Esler et al., 1997, 1999).

Streptavidin–biotin interactions are another means of coupling fibrils to a surface, as shown in Figure 14. The binding between streptavidin–biotin is a noncovalent interaction that is remarkably strong and stable. Inoue and

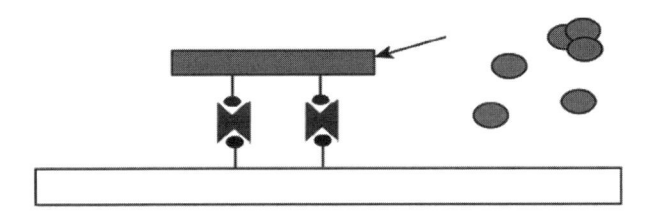

Figure 14 Covalent attachment of fibrils to a surface. Sup35 fibrils were created from a mixture of Sup35 protein labelled with a Cy5-fluorescent tag, Sup35 labelled with biotin and unlabelled Sup35 protein. Labels were attached via a cystiene added to the C terminus of the Sup35 protein. The resulting fluorescent fibrils (shown in grey) were attached to a glass surface (shown in white) via biotin displayed on their surface, which linked to a further strepavidin-biotin anchor on the surface. Cy3-fluorescently labelled monomers (shown as circular species) were then be added to the solution to promote fibril extension (as indicated by the arrow). These experiments have been used to observe single fibril growth using total internal reflection fluorescence microscopy (Inoue and Kishimoto 2001).

colleagues (2001) formed fibrils from the Sup35 protein that had been labeled at the C terminal Cys residue with either a fluorphore or with biotin PEAC–maleimide. The resulting fibrils are fluorescent and display biotin on the fibril surface. These fibrils were then immobilized via a streptavidin linker to a biotinylated casein surface and monomer labeled with a second flurophore was added to observe fibril growth. One advantage of this technique is that unfixed fibrils can easily be washed away. Collins has also immobilized Sup35 fibrils via a biotin–streptavidin linker to a glass slide in order to observe individual fibril growth by TIRFM (Collins et al., 2004).

Dinca and colleagues (2008) have combined fibril self-assembly, lithographic features, and covalent and noncovalent binding to create three-dimensional patterned fibrillar structures that could be used in electronics or tissue engineering. Images of these structures, which are described as fibril bridges, are shown in Figure 15.

The positioning of fibrils between the lithographic columns was made specific through the use of both noncovalent and covalent chemistry shown in Figure 16. Dinca first deposited a layer of photobiotin on the column surface, this layer was then exposed to ultraviolet light in order to activate

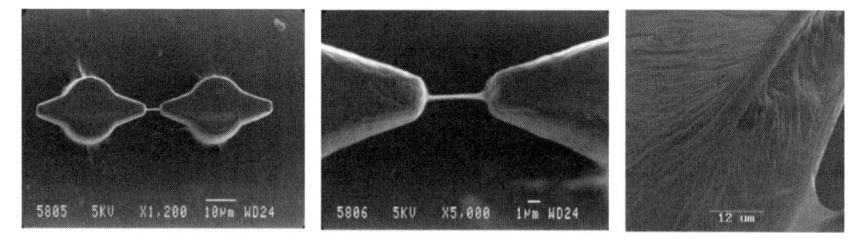

Figure 15 Scanning electron microscopy images of lithographic columns and a fibril bridge that has formed between two of the columns (left). The fibril bridge and the point of fibril attachment to the column are shown at higher magnification (center and right respectively). The scale bars are 10 (left), 1 (center), and 12 μm (right) in length. Reprinted in part with permission from Dinca et al. (2008) (copyright 2008 American Chemical Society).

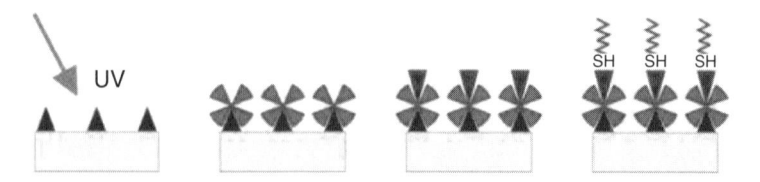

Figure 16 Schematic illustrating the chemistry used to specifically immobilize peptides to lithographic features. Photobiotin (triangles) was activated with UV light to initiate binding. Biotin was next coupled to the surface (crosses), followed by the idoacetamide-functionalized biotin (triangles) and the final cysteine containing peptides. Reprinted with permission from Dinca et al. (2008) (copyright 2008 American Chemical Society).

protein binding to the column surface. Avidin was noncovalently bound to this layer, followed by idoacetamide-functionalized biotin. Fibrils displaying cysteine residues could then be covalently immobilized to the outer biotin layer via a SH–iodoacetamide reaction.

A suspension of preformed fibrils formed from a peptide containing cystiene was washed over the modified lithographic features and the fibril bridge initiated by drying. The fibril bridge contains an ensemble of fibrils that appear moderately aligned in Figure 15. No bridge was observed on columns without the required chemistry. This process of drying fibrils between pointed features has some analogies to the process used to form dried fibril stalks that are prepared for fiber X-ray diffraction. The formation of these site-specific fibril bridges could be widely applicable for the creation of nanoelectronic circuits and devices.

The studies outlined in this section show a number of different techniques that can be used to align fibrils and other peptide nanofibers in order to create micron-sized patterns from these self-assembling materials. This section also explored some of the covalent and noncovalent strategies for attaching fibrils to surfaces, and how these methods may be exploited to create three-dimensional patterns of fibrils that have an exciting range of potential applications.

4. APPLICATIONS

In this section, the potential application for amyloid fibrils and other self-assembling fibrous protein structures are outlined. These include potential uses in electronics and photonics presented in Section 4.1, uses as platforms for the immobilization of enzymes and biosensors presented in Section 4.2, and uses as biocompatible materials presented in Section 4.3. Each of these applications makes use of the ability of polypeptides to self-assemble and form nanostructured materials, a process that can occur under aqueous conditions. These applications also seek to exploit the favorable properties of fibrils such as strength and durability, the ability to arrange ligands on a nanoscale, and their potential biocompatibility arising from the natural materials used for assembly.

4.1. Electronics and photonics

The high order and aspect ratio of protein nanofibers and nanotubes makes these structures attractive candidates to construct nanowires. Consequently, amyloid fibrils and diphenylanalnine protein nanotubes are among a growing list of biomolecular templates being investigated for their ability to organize inorganic materials (Gazit, 2007). Both amyloid fibrils and peptide nanotubes can withstand the harsh thermal and chemical conditions required for metal deposition and can be used to create wires up to several microns in length. Amyloid fibrils can also be constructed from a diverse range of peptide building blocks.

Wires have been made from amyloid fibrils using a modified version of amyloidogenic NM Sup 35 protein that displays cystiene residues along the fiber surface (Scheibel et al., 2003). Colloidal gold was covalently linked to these cystienes and the wires placed on gold electrodes where they were further coated with silver and gold layers by reductive deposition. Wires with diameters between 80 and 200 nm were produced, as shown in Figure 17. These wires demonstrate low resistance and conductive behavior ($R = 86\ \Omega$). In contrast, the uncoated fibers act as insulators and display no conductivity ($R > 10^{14}\ \Omega$). At higher voltages, the nanofibril wires also vaporize, effectively acting as fuses, illustrating the potential of these fibrils as wires for electrical circuits.

A different approach was taken by Reches and Gazit who exploited the hollow interior of diphenylalanine nanotubes to cast silver nanowires (Carny et al., 2006; Reches and Gazit, 2003). The peptide nanotubes were placed in a boiling ionic silver solution. Silver ions that had assembled inside the tube core were then reduced with citrate. The outer protein layer was next digested with the enzyme proteinase K, releasing the wire from its casing as shown in Figure 17.

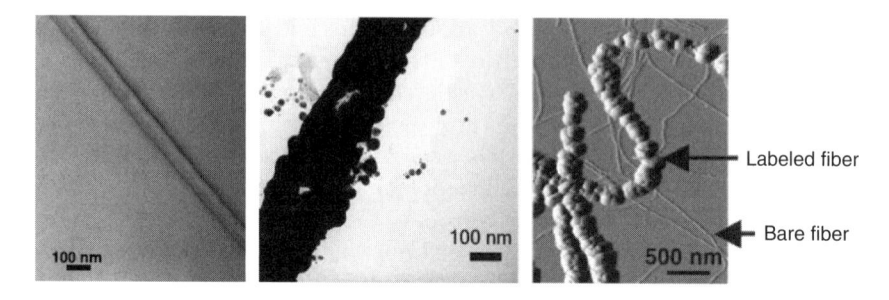

Figure 17 A range of nanowires and nanotubes. Silver nanowires are cast inside diphenylananine nanotubes and the outer peptide layer digested with the enzyme Proteinase K (left), coaxial cables created from diphenylananine nanotubes containing silver and coated with gold (center) both imaged by transmission electron microscopy. Nanowires can also be constructed from the amyloidgenic protein NM Sup35 where layers of gold and silver coat the fiber as shown in the atomic force microscopy image (right). The scale bars are 100 nm (left and center) and 500 nm (right) in length. Left from Reches and Gazit (2003), reprinted with permission from AAAS. Centre reprinted in part with permission from Carny et al. (2006) (copyright 2006 American Chemical Society). Right from Scheibel et al. (2003) (copyright © The National Academy of Sciences of the United States of America, all rights reserved, 2003).

Coaxial fibers were also formed from diphenylalanine fibers (Carny et al., 2006). Coaxial fibers contain two metal layers separated by an insulating layer, in this case the peptide nanotube. Silver was again cast within the peptide nanotube but peptide linkers containing diphenylalanine and cystiene were also recruited to the outside of the tube where they interacted noncovalently with the surface. The cystiene residues were then used to attract gold nanoparticles before further reductive deposition of gold. This led to a gold-covered surface as shown in Figure 17. The conductive properties of these wires have not yet been tested, nevertheless Carny and colleagues suggest that the coaxial geometry will shield electromagnetic interference.

Remarkably, some peptide nanofibers can act as nanowires without the addition of a conductive metal layer. Fibers constructed from a polypeptide containing the sequence Val-Gly-Gly-Leu-Gly from the hydrophobic domain of elastin display an intrinsic blue-green fluorescence and have been found to be conductive (del Mercato et al., 2007). When cast onto gold electrodes by simple drop deposition and dried, these fibers conduct and display a high current. This conductivity is likely to be due to the water molecules that reside along the length of the fibril, as charge transport is highly dependent on humidity and does not occur in a vacuum. The authors suggest that charge transport and luminescent properties in these fibers are also related. Importantly, this study suggests that designed sequences can potentially be used as label-free nanowires.

Another approach toward fibrillar nanowires has been taken by Baldwin and colleagues (2006), who assembled a porphyrin binding protein onto the surface of an amyloid fibril. This binding protein could incorporate heme to form a functional b-type cytochrome. These fibrils could be developed to create wires for electron transfer, similar to structures observed in nature that consist of chains of heme molecules.

Conducting fiber cores have also been created from peptide amphiphiles that incorporate the hydrophobic conductive polymer poly(3,4-ethylenedioxythiophene) (PEDOT) into the fiber core (Tovar et al., 2007). The addition of ammonium persulfate to these structures triggers polymerization of the polymer resulting in a film of fibers that display high conductivity. The authors suggest that this approach is general and could be applied to sequester other conducting polymer precursors. The peptide amphiphiles containing these polymers may also find use as biosensors (see Section 4.2 below), if biological interactions with the fibers can be used to change the signal transferred along the conducting polymer. Peptide amphiphiles have also been created from peptides containing a diacetylene group within the alkyl tail that can be polymerized following assembly (Hsu et al., 2008), providing further scope for linking electronic properties to bioactivity.

Other inorganic materials have been assembled on a wide range of self-assembling protein structures. Many of these studies have sought to mimic nature, which can create a vast range of templated inorganic structures.

Copper has been immobilized on glycylglycine bolaamphiphile peptide nanotubes that display histidine residues, and paramagnetic gadolinium, a magnetic resonance image contrast agent, has been immobilized on nanofibers produced from peptide amphiphiles (see Gazit, 2007 and references therein).

Peptide amphiphiles have also been used to induce the nonorientated growth of cadmium sulfide nanocrystals and orientated the growth of hydroxyapatite along the fiber axis (Hartgerink et al., 2001; Sone and Stupp, 2004). The authors suggest these fibers could act as semiconductors (cadmium sulfate), could promote the mineralization of other components (cadmium sulfate), or could be used as biomaterials (hydroxyapatite; see section below). Silicon nanotubes have also been created by the condensation of tetraethoxysilane (TEOS) on β-sheet fibrils (Meegan et al., 2004) and peptide–amphiphile nanofiber templates (Yuwono and Hartgerink, 2007). This TEOS layer can then be calcified by heating, and the interior organic peptide template degraded, resulting in hollow silicon nanotubes.

Optoelectronics is another area where fibrils could be used. Herland and colleagues (2005) formed fibrils from a solution containing bovine insulin protein and one of two types of semi conducting conjugated oligoelectrolytes. Although it is not clear how the oligoelectrolytes are included within the fibril core, the ratios of each component, or the structure of the fiber core, the oligoelectrolytes appear to be arranged symmetrically and the fibrils form electrically active luminescent bundles. These wires demonstrate fluorescent quenching when a small voltage (0.9 V) is applied and quenching is reversed when the surrounding electrolytic solution is reduced by a second voltage (−0.3 V), suggesting that electric transfer occurs along the length of the fibril.

Preformed amyloid fibrils have also been coated with both water-soluble and water-insoluble conjugated polymers (Herland et al., 2007, 2008). The water-soluble conjugated polymer PTAA was found to align along the length of the fibril. The water-insoluble conjugated polymer, APFO-12, is an uncharged polar alternating polyfluorene derivative that was also found to form a highly ordered structure along the length of the fibril. Materials similar to APFO-12 have been used in solar cells and for light emitting diodes, where their order is essential to properties such as electrical transport. The authors suggest that such coated fibrils may prove useful as nanowires or in ordered films.

The polycylic fluorphore fluorene has also been assembled on the outside of fibrils constructed from the amyloidogenic TTR105-115 peptide (Channon et al., 2008). The fluroene unit has the ability to drive self-assembly. However, in this case the fluorene unit was attached to the TTR105-115 sequence which was used to drive fibril formation so that the fluorene unit was displayed on the fibril surface. Channon observed the transfer of excitons from high energy absorbance sites to low-energy trapping sites along these fibrils, suggesting the fibrils could be used to construct light harvesting materials with properties analogous to those observed for photosynthesis in nature.

4.2. Platforms for enzyme immobilization and biosensors

An exciting potential use for protein fibers is as platforms for the immobilization of biologically active components, including enzymes or motor proteins (Gras, 2007; Karsai et al., 2008). These components can easily interface with protein nanofibers and could be linked to the protein unit prior to or following fiber assembly. In the case of amyloid fibrils, there are multiple sites where biomolecules could be attached. These include the polypeptide termini or amino acids that are typically exposed at multiple locations along the length of the fibril. The nanoscale order of amyloid fibrils can be used to control the position of biologically active elements, and immobilization can offer advantages such as enhanced enzymatic properties and stability. Moreover, multiple complementary components can be displayed on a single fibril, and this approach could be used to achieve sequential or cooperative biological processes.

Biosensors and devices may also include fiber components. Biosensors essentially contain two main components. These are the sensing element, which senses the molecule of interest, and the transducer, which generates a signal. Amyloid fibrils and other protein nanofibers could be used in biosensors in two ways. The first is as a scaffold to immobilize the sensing element on a nanoscale as described above. The second use is as a coating on the transducer where the presence of the protein fiber could enhance device performance. This enhancement could be carried out either with or without the aid of a further conductive coating (see Section 4.1 above).

The amyloidogenic Sup35 protein has been successfully used to immobilize three enzymes: barnase, carbonic anhydrase, and glutathione *S*-transferase (Baxa et al., 2002). Each of these enzymes was linked to the Sup35 sequence which drives assembly creating three different fusion proteins which successfully formed fibrils displaying functional enzymes on the fibril surface.

Artificial catalysts have also been incorporated into amphiphilic structures (Guler and Stupp, 2007). These catalysts were imidazolyl-functionalized peptides, which demonstrate a greater rate of 2,4-dinitrophenyl acetate hydrolysis when immobilized on the peptide amphiphile than the rate observed when the same enzyme is present in solution. Although the density of the enzymes on the fiber surface has not been established, the authors attribute the increase in enzymatic activity to the likely concentration of enzyme along the fiber surface, and this study illustrates one of the advantages of enzyme immobilization.

The hydrophobins are a case where protein nanofibers can play a dual role in creating a biosensor. They can aid in the immobilization of bioactive components within a biosensor and also add further functionality to the transducing element of a biosensor device. Hydrophobins are self-assembling β-sheet structures observed on the hyphae of filamentous fungi. They are surface active and aid the adhesion of hyphae to hydrophobic surfaces (Corvis et al., 2005). These properties can be used to create hydrophobin layers on glass electrodes. These layers can then facilitate the adsorption of two model enzymes: glucose oxidase (GOX) and hydrogen peroxidase (HRP) to the electrode surface. The hydrophobin layer also enhances the electrochemical properties of the electrodes.

Enzyme immobilization on a hydrophobin layer does not appear to significantly decrease enzyme properties, and these enzymes display similar substrate affinity to free enzymes. The specific activity of these enzymes may be lower than enzymes that are free in solution, but the immobilized enzymes also have the advantage of increased stability from 1 to 3 months (HRP and GOX, respectively).

The increased sensitivity of the electrodes coated with the hydrophobin layer is also appealing. The hydrophobin-modified electrodes allow H_2O_2, the product of glucose oxidation, to be detected at a concentration of 0.2 mM. This level is significantly lower than the detection limit of the bare electrode, which is 0.9 mM. The authors attribute the increase in electrode performance to the local accumulation of H_2O_2 at the electrode surface or a possible decrease in current in the presence of the hydrophobin layer.

Diphenylalanine nanotubes are a second example where self-assembling protein structures can have a dual role in biosensing. Yemini and colleagues (2005a) used diphenylalanine nanotubes to immobilize enzymes and to enhance the transducer component of the biosensor. In this case, the enzymes were linked to the preformed nanofibers by nonspecific chemical cross-linking using gluteraldehyde. The nanotubes were then covalently attached to a gold electrode via a thiol linkage. A further polyethyleneimine matrix was then deposited across the surface to capture and increase the concentration of reactive intermediates, enhancing sensor performance. This setup is shown in Figure 18.

The diphenylalanine nanotube sensors were based on the observation that peptide nanotubes improve the electrochemical properties of graphite and gold electrodes when deposited directly onto the electrode surface (Yemini et al., 2005b). The high surface area of the nanotubes and the potential alignment of aromatic residues are thought to contribute to the observed increase in conductivity. This property makes nanotube-coated electrodes and hydrophobin-coated electrodes suitable for use as amperometric biosensors that produce a current in response to an electrical potential across two electrodes.

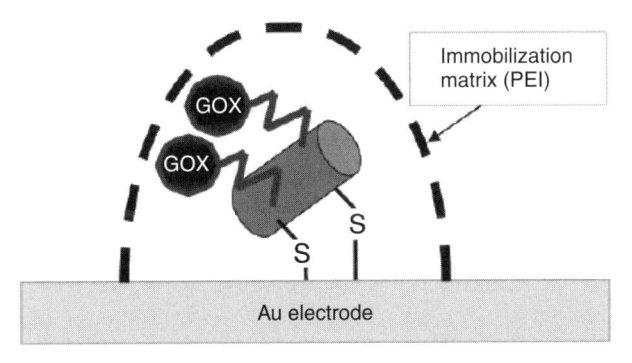

Figure 18 Schematic of a glucose biosensor assembled from diphenylalanine peptide nanotubes. The enzyme GOX has been cross-linked to these nanotubes, which are further linked to the gold (Au) electrode and immobilized in a polyethyleneimine (PEI) matrix. The nanofibers act in two ways; they immobilize the sensing enzyme and enhance the transducer. Reprinted in part with permission from Yemini et al. (2005a) (copyright 2005 American Chemical Society).

Diphenylalanine nanotube sensors and hydrophobin sensors could be widely used for diabetic monitoring, food, and environmental sensing. For example, GOX and ethanol dehydrogenase were successfully immobilized on the surface of the diphenylalanine nanotube sensors (Yemini et al., 2005a). The presence of glucose resulted in the production of hydrogen peroxide by GOX and an increase in current across the electrodes. The presence of ethanol similarly resulted in the production of Nicotinamide adenine dinucleotide (NADH) by ethanol dehydrogenase and an increase in current across the electrodes. In both cases, the current generated by the nanotube sensors was greater than for uncoated sensors suggesting the potential utility of these devices.

Another biosensor for the sensitive detection of viruses is based on antibody functionalized peptide nanotubes (MacCuspie et al., 2008). Viral antibodies were immobilized on the end of peptide nanotubes formed from the bolaamphiphile peptide monomer, bis(N-a-amidoglycylglycine)-1,7-heptane dicarboxylate while flurophores were displayed on the side of the nanotubes. When the virus is mixed with these nanotubes, they aggregate leading to an increase in light scattering and fluorescence. This assay is highly sensitive and trace amounts (10^3 pfu/ml) of pathogens including herpes simplex virus type 2 (HSV-2), vaccinia, adenovirus, and influenza type B could be detected. This sensitivity is due to the high number of flurophores along the nanotube length which increase the sensitivity to 10^5 times greater than traditional antibody–dye conjugates. This illustrates the broad potential of this nanofiber assay which could also utilize other self-assembling peptide systems.

The examples highlighted in this section illustrate the broad applicability of fibrils and other self-assembling systems as components in biosensors and devices. These sensors may be applied to detect a host of biomolecular reactions or biomolecules. The model enzymes attached to these scaffolds also present proof of principle for the concept of enzyme immobilization on fibrils and other fibers.

4.3. Biocompatible materials

Self-assembling peptide systems may be developed as advanced materials for biomedical applications. These nanofibers can be used to study how cells respond to nanostructures; an important area which we are only beginning to understand. Protein fibers can also be designed to support cells and provide an environment that mimics natural tissue, providing the physical and chemical cues to promote particular cell behavior such as cell adhesion, cell migration, cell differentiation, or tissue regeneration. Structures that are shown to be biocompatible may also find roles as materials for in vitro and in vivo use. This could include applications as diverse as soft and hard tissue engineering, composites materials, coatings to increase device biocompatibility, and vehicles for drug delivery.

Functionalization is a central to many studies exploring the potential of polypeptide fibers as biocompatible materials, and fibers are often decorated with short bioactive tags that originate from the extracellular matrix (ECM). These tags can be displayed on protein fibers at much higher densities than that which occurs in nature and can also be displayed in unique and complimentary combinations.

There are two main ways to incorporate bioactive tags; the first is to include the sequence prior to assembly (see Sections 4.1 and 4.2 above for other examples). The second is to link the sequence following assembly either by covalent capture or by chemical cross-linking. Fibers that display these tags not only have typical dimensions of ECM proteins but can interact with cells in a similar way to the ECM.

A range of functionalized and unfunctionalized self-assembling fibrous structures have been tested for their biocompatibility and ability to provide cells with a favorable micro- and nanoenvironments for soft tissue engineering. In this section, studies that focus on amyloid fibrils, on peptide amphiphiles, on ionic complementary peptides, and on dipeptide structures are reviewed. Hard tissue engineering, composites, and coating are also explored followed by macroscopic structures and networks that can be created from fibrous protein structures.

4.3.1. Amyloid fibrils

Fibrils have broad potential for many of the bioapplications listed above. While some amyloid fibrils have negative associations with protein misfolding diseases, other fibrils demonstrate positive functions such as the fibrils that thought to occur in mammalian cells (Fowler et al., 2006); reviewed in

Section 2.1 above). These functional fibrils suggest that designed fibrils can be developed as compatible materials. Toxicology will be a potential issue for all structures that self-assemble on a nanoscale. However, the natural building blocks used to construct protein fibers may increase the biocompatibility of these structures compared to other man-made materials. Polypeptide self-assembly also represents a route for generating ECM-like structures that are not only simpler than their natural equivalents but also easier to prepare.

Recently, TTR1 fibrils have been decorated with the classic RGD tripeptide motif isolated from fibronectin, which encourages cell adhesion via integrin cell surface receptors (Gras et al., 2008). This bioactive tag was added to the TTR1 sequence which drives fibril assembly and the tag shown to be exposed on the fibril surface and accessible to cells following assembly. The RGD-modified fibrils were bioactive in a cell dissociation assay which measures the ability of fibrils to competitively bind to cells and induce cell detachment from a surface, as illustrated in Figure 19. In

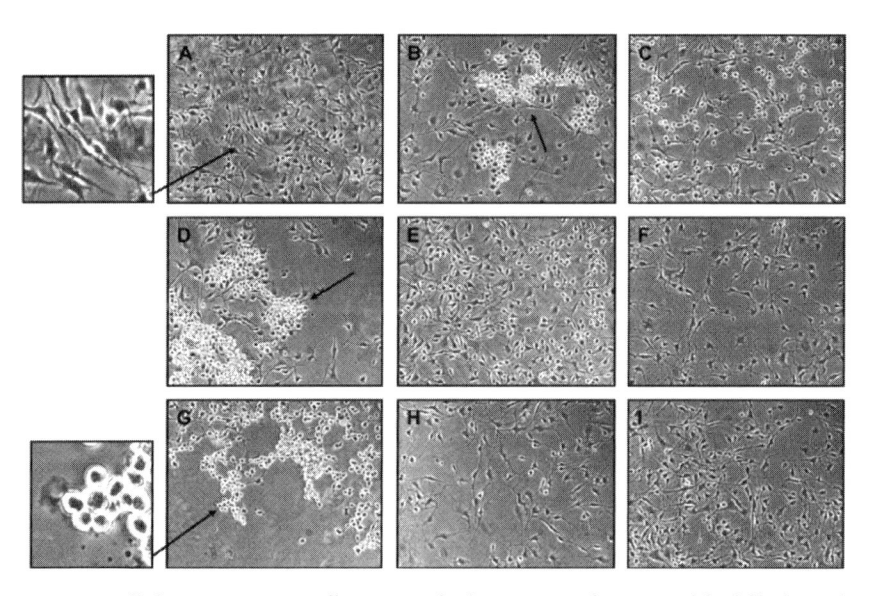

Figure 19 Cell dissociation assay illustrating the bioactivity of RGD-modified fibrils. In this assay, a layer of 3T3 mouse fibroblast cells on tissue culture plastic (a) is exposed to a range of peptides and fibrils (b–i). Samples that are bioactive compete for cell binding and induce the dissociation of cells from the surface. The images show (b) after incubation with positive control RGDS peptide, (c) untreated cells following incubation, (d) after incubation with TTR–RGD1 peptide, (e) after incubation with TTR–RAD1 peptide, (f) after incubation with TTR1 peptide, (g) after incubation with TTR1–RGD fibrils, (h) after incubation with TTR1–RAD fibrils, and (i) after incubation with TTR fibrils. The arrows and enlarged image for (g) indicate clumps of dissociated cells. Reprinted from Gras et al. (2008) (copyright 2008, with permission from Elsevier).

contrast, an RAD control motif did not encourage cell interactions, nor did the base fibril structure. All three fibrils systems could also support non-specific cell attachment in a cell adhesion assay where cells were deposited on a surface coated with fibrils.

The density of the bioactive tags added to fibrous structures will determine the strength of binding, and the density of RGD tags on TTR1 fibrils was estimated to be as high as 2,000 RGD/µm equating to a distance of 0.5 nm between RGD tags (Gras et al., 2008). This density could be controlled by forming mixed fibrils from a solution containing TTR1–RGD doped with the plain TTR1 peptide. Using this method, RGD density could be reduced to as low as 12.8 RGD/µm (with 1% doping) equating to 85 nm between ligands. This range of RGD densities is biologically relevant as integrin receptors are typically spaced ~50 nm apart (Puleo and Bizios, 1991). Significantly, the cross-beta core of the fibril structure was also conserved suggesting that this method can be used to generate other combinations of fibril-forming peptides with a diverse range of bioactive tags.

The protein elastin presents another opportunity to create amyloid-like fibrils from natural proteins for the purpose of developing biomaterials. Elastin is found in tissue where it imparts elastic recoil, and fibrils formed from this protein may demonstrate some of the elastic properties of the constituent elastic proteins (Bochicchio et al., 2007). Elastin typically contains the sequence poly(ZaaGlyGlyYaaGly) (where Zaa, Yaa = Val or Leu) (Tamburro et al., 2005), and short stretches of the protein retain the ability to form structures similar to the original protein. Simple proline to glycine mutations in the hydrophobic domains of elastin can induce the formation of amyloid-like fibrils (Miao et al., 2003), suggesting that fibrillar materials can be easily generated from these sequences.

4.3.2. Peptide amphiphiles

Other polypeptide systems such as peptide amphiphiles show great promise as biocompatible materials for applications including tissue engineering. Peptide amphiphiles incorporating phosphoserine that are stabilized by four cysteine residues have been shown to induce mineralization, generating hydroxyapatite crystals aligned along the peptide amphiphile that resemble the natural structures found in bone (Hartgerink et al., 2001).

The ligand RGD has been incorporated into a range of peptide amphiphilic structures. This ligand was included in the self-assembling sequence used by Hartgerink (Hartgerink et al., 2001) and has been added to peptide amphiphiles following assembly using the activity of the enzyme tissue transglutaminase (Collier and Messersmith, 2003). RGD has also been shown to help induce the differentiation of mesenchymal stem cells and bone formation observed in the center of gels formed by peptide amphiphiles (Hosseinkhani et al., 2006c).

Complementary bioactive epitopes have since been displayed on peptide amphiphiles and used to support cells. These epitopes include RGD from fibronectin, or IKVAV and YIGSR from laminin which are thought to interact with neurons encouraging outgrowth and adhesion, respectively (Niece et al., 2003). The density of IKVAV can be as high as 7.1×10^{14} epitopes per cm^2, which represents an increase of 10^3 compared to the density of IKVAV naturally displayed in laminin (Silva et al., 2004). Murine neural progenitor cells were also encapsulated in these fibers, resulting in cell growth and rapid differentiation.

Synergistic sites have also been explored for display. Mardilovich displayed both RGD and the synergy site PHSRN (proline–histidine–serine–arginine) from fibronectin on the surface of a peptide amphiphile (Mardilovich et al., 2006). Importantly, the spacing and the ratio of hydrophobic to hydrophilic residues in the sequence between RGD and PHSRN was maintained in the new construct and cells adhered, spread and moved over the structures following assembly. The cells also deposited extracellular proteins indicating that the new construct performed as well as the original fibronectin protein.

Peptide branching has been used as a way to increase the density of ligand display and improve interactions with cells (Guler et al., 2006; Harrington et al., 2006). Harrington used a lysine dendron to improve the accessibility and number of ligands on the fiber surface, while Guler used orthogonal protecting group chemistry to introduce a branch at amine or lysine groups. Guler also introduced cyclic chemistry to improve the affinity between RGD and cell surface integrin receptors.

Branched peptide amphiphiles demonstrate superior properties when compared to their linear counterparts and enhance cell growth when coated on poly(glycolic acid) (PGA) tissue engineering scaffolds (Harrington et al., 2006). This finding suggests such materials may have potential in bladder tissue regeneration. Cells have also been shown to migrate through gels displaying very high densities of branched and linear RGD ligands (Storrie et al., 2007).

Blood vessels are another tissue engineering target where fibrous structures can be used to assist regeneration (Rajangam et al., 2008). Peptide amphiphiles containing a heparin binding consensus sequence have been used to bind to heparin and promote angiogenesis, these structures may also protect heparin from proteolytic cleavage in a biological environment. Fibroblast growth factor was also released from these fibers, effectively enhancing the growth of vessels in an angiogenesis assay.

These same peptide amphiphiles have been shown to assist mice to recover from spinal cord injuries (Tysseling-Mattiace et al., 2008). Both motor and sensory axons were able to cross the site of injury when peptide amphiphiles were injected and the peptide amphiphiles were stable in this in vivo environment for 1–2 weeks. Although recovery was only partial, this represents a remarkable achievement.

Gels formed from peptide amphiphiles mixed with basic fibroblast growth factor (bFGF) have also been used to successfully promote blood vessel formation in mice (Hosseinkhani et al., 2006a). Tabata and colleagues followed the release of bFGF, which is a chemo attractant that acts to promote the growth of fibroblast and endothelial capillary cells. bFGF release was prolonged and lasted for 750 h. The peptide amphiphiles were also degraded, although the rate was slow.

The fate of peptide amphiphiles will be critical to their development as materials for in vivo use. Studies by Beniash and colleagues found that cells entrapped within a network of peptide amphiphiles internalize fibers by endocytosis and store these fibers in the lysosome. In this case the fibers displayed a KGE motif, with similar charge properties to RGD, and the cells could proliferate within these scaffolds for at least 3 weeks. Studies of glucose and lactate concentrations suggest the cell may metabolize the peptide amphiphiles (Beniash et al., 2005), indicating that clearance of these peptides from the site of application is unlikely to be an issue for these structures.

Peptide amphiphiles may also be used for drug delivery. For example, Marini used fluorescently labeled peptide amphiphiles that were functionalized with the cyclo-RGD ligand and found that these structures were internalized by cells (Marini et al., 2002).

4.3.3. Ionic complementary peptides

Ionic complementary peptides have been extensively examined for biocompatibility in a series of short-term in vitro and in vivo assays. The majority, but not all, of these studies have focused on two model systems: RAD16 and EAK16.

RAD16 scaffolds have been shown to support the growth and differentiation of a number of cell types including a range of mammalian cells (Zhang et al., 1995). Neural cells attach and differentiate on RAD16 gels and growth is enhanced in the presence of neural growth factor (Holmes et al., 2000). RAD16 gels also provide a way to entrap migrating glial cells and neurons from hippocampal brain tissue sections, creating an effective way to recover these cells from a tissue section (Semino et al., 2004). Condrocytes have also been entrapped within a KLD-12 gel where they grow for 4 weeks, maintaining their morphology and depositing ECM rich in proteoglycans and collagen (Kisiday et al., 2002). This behavior suggests KLD-12 gels could be used to repair cartridge.

One of the most significant studies employing RAD16 gels was able to achieve axonal regrowth after damage to the optical nerve of a hamster (Ellis-Behnke et al., 2006). The RAD16-I peptide scaffold was injected into the wound area where it permitted significant regrowth allowing partial recovery of the optical nerve and return of some optical function for the animal.

Not all approaches to tissue engineering have been successful with RAD16 gels, suggesting that gels need to be tailored for particular applications. Davis and colleagues used RAD16 gels to create an environment appropriate for the culture of endothelial cells within heart tissue in vivo and found that endothelial cells invaded the gel and isolated cardio myocytes injected into the gel could survive (Davis et al., 2005). Gels were also found to be biocompatible and did not elicit an immune response. In contrast, Dubois found the RAD16-I gels with and without additional adhesion signals including YIG, RGD, and IKV did not support the survival of cells (Dubois et al., 2008). Inflammation was also observed and staining revealed the presence of inflammatory macrophage cells.

The primary sequence certainly has a significant effect on the interaction between nanofibers and cells. Sieminski illustrated this well by showing that human umbilical vein endothelial cells behaved differently on gels made from RAD16-I, RAD16-II, KF8, and KDL12 (Sieminski et al., 2008). The cells attached readily and formed microcapillary-like structures on RAD16-I and RAD 16-II but not the other gels, where cells remain rounded. Differences in protein adsorption to the fibers could not explain the differences in cell behavior, indicating that the chemistry of the primary sequence used to form the gel can determine cell interactions.

Genove and colleagues (2005) found that functionalized peptide nanofibers could effectively support human aortic endothelial cells. The N terminus of the RAD16 sequence was extended with two adhesion sequences from the ECM protein laminin and one from collagen: YIGSR, RYVVLPR, and TAGSCLRKFSTM, respectively. The new structures promote cell growth and improved cell function including the deposition of basement membrane. Competition assays with soluble peptides indicate that the cells do indeed recognize the adhesion sequences displayed on these fibers.

Other functional sequences have also been added to RAD16-I peptides to encourage bone growth (Horii et al., 2007). These sequences include osteogenic growth peptide ALK (ALKRQGRTLYGF) that is released following bone injuries, osteopontin cell adhesion motif DGR (DGRGDSVAYG) that regulates a number of cell functions, and a sequence containing RGD that is designed to promote adhesion called PGR (PRGDSGYRGDS). These new scaffolds enhanced osteoblast differentiation. Gels incorporating at least 70% of the PGR peptide also promoted cell migration to a depth of 300 µm within the scaffold. RAD16-I gels combined with a PolyHIPE polymer displaying hydroxyapetite have also shown high levels of osteoblast growth and mineralization, again suggesting the use of this gel for bone tissue engineering (Bokhari et al., 2005).

Another use for fibers formed from EAK16 and RAD16 peptides is as vehicles for drug delivery. EAK16 gels are able to effectively stabilize model hydrophobic cargo such as the fluorphore pyrene (Keyes-Baig et al., 2004). The cargo can then be released into a simulated membrane made from phosophatidylcholine vesicles. Although this delivery mechanism is non-specific, this study suggests that ionic complementary peptides can be used to deliver compounds to the cell membrane.

The hydrophobic anticancer drug ellipticine is one of a number of other compounds that have been stabilized by association with fibrous proteins in aqueous solution (Fung et al., 2008 and references therein). Variants of EAK16 with different charge and hydrophobicity (EAK16II and EAK16-IV and EFK) were tested for their ability to stabilize the drug. The charge on the former two peptides enables these fibers to bind a protonated version of the drug, whereas the EFK can bind a neutral version of the drug using hydrophobic contacts. The bioactivity of these drugs was found to vary and this study illustrates that peptide properties will influence their behavior as drug carriers.

Nagai and colleagues (2006) have also examined the potential of gels formed from the peptide RAD16 to release drugs, and their study examined the diffusion of model compounds through the peptide amphiphile gel.

The properties of peptide fibers formed by ionic complementary peptides may need some tailoring to meet the different requirements for scaffolds and materials for drug delivery. However, these roles could be quite complementary. It is easy to envisage that mixtures of peptide fibers with different stabilities could be used to simultaneously stimulate and support the growth of soft tissues.

4.3.4. Dipeptide nanotubes

Other research has focused on the interaction of cells with peptide structures that assemble via aromatic interactions. Cells have been grown on the surface of gels formed by Fluorenylmethoxycarbonyl (FMOC) modified diphenylalanine nanotubes (Mahler et al., 2006). Although cells were only grown for short time frames (24 h), the cells were viable on these scaffolds.

Similar work was performed by Jayawarna and colleagues. Chondrocytes were grown on a number of different hydrogels formed by FMOC-modified dipeptides and shown to grow for up to 10 days on these scaffolds (Jayawarna et al., 2007). In their earlier paper, Jayawarna also grew chondrocytes on top of hydrogels or incorporated cells within hydrogels formed by FMOC-dipeptides up to 7 days (Jayawarna et al., 2006).

4.3.5. Hard tissue engineering, coatings, and other applications

Hard tissue engineering can include structures such as bone, and several examples of bone tissue engineering are presented in the sections above, other hard tissues that can potentially be repaired by self-assembling peptides include human dental enamel. Kirkhan and colleagues (2007) used self-assembling anionic peptides named P_{11}-4 to promote remineralization in lesions in enamel. The observed effect was two fold; remineralization occurred at conditions of neutral pH and demineralization was inhibited at conditions of low pH. X-ray diffraction suggests these materials are capable of triggering hydroxyapetite formation without additional functional groups, suggesting this approach is suitable for the repair of enamel.

Other applications use self-assembling fibers to coat materials. For example, preassembled peptide amphiphiles have been covalently immobilized on titanium implant surfaces via a silane layer (Sargeant et al., 2008). Primary bovine artery endothelial cells or mouse calvarial preosteoblastic cells spread on these coated surface and proliferated to a far greater extent than on samples where the peptide amphiphiles had been drop cast onto the metal surface. This study therefore suggests that covalent attachment is required in order to prevent fibers lifting from the coated surface and to encourage maximal cell growth.

A further use for self-assembling peptides is as lubricants to reduce joint friction between cartilage surfaces (Bell et al., 2006). The amphiphilc peptide P_{11}-9 which resembles hyaluronic acid (HA), the natural lubricant in joints, successfully reduced joint friction in simulated experiments. Bell and colleagues suggest that this peptide may be a superior synthetic treatment compared to injection to replace HA, which is not suitable in all osteoarthritic cases.

Other applications for self-assembling fibrous structures not explored in depth here include applications as natural bioadhesives (Mostaert and Jarvis, 2007) or as components in membranes or other devices for filtration and bioseparation.

4.3.6. Macroscopic structures and networks

Many biomaterial applications require large macroscopic structures that can be easily handled. The ability of peptide fibrils and nanofibers to form larger ordered macroscopic structures will therefore be important to their use as biomaterials. Many self-assembling peptide systems do form gels and peptides such as EAK16 form membrane-like structures (Zhang et al., 1993). Other options for making larger structures include linking fibers to existing materials (see above) and forming composite materials.

There are a number of ways β-sheet structures can be altered so that they do form gels. One method is to link the chemical termini of peptides already incorporated into β-sheet fibers to form stiffer gels (Jung et al., 2008). Jung and colleagues used native chemical ligation (NCL), which links an N terminal cystiene to a C terminal thioester, to increase the storage modulus of gels formed by peptide amphiphiles. The resulting gels successfully induced the proliferation of primary human umbilical vein endothelial cells and when functionalized showed even higher levels of cell adhesion. The authors suggest that this approach can be

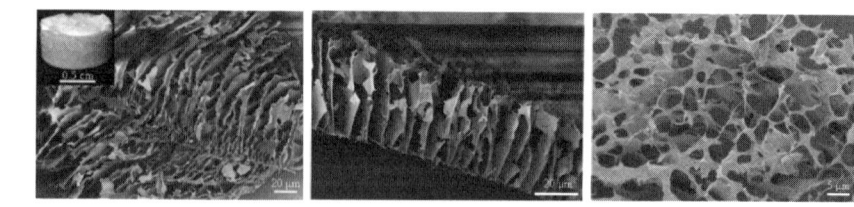

Figure 20 Aerogels created from freeze-dried protein networks. Scanning electron microscopy images of aerogels made from protein gels (40 mg/ml) in water (left and center) and protein gels (40 mg/ml) in 95% water and 5% 1,1,1,3,3,3-hexafluoroisopropanol (HFIP). The inset on the left image shows the aerogel shape. The scale bars are 20 μm in length (left and center), 5 μm length (right), and 0.5 cm in length (inset) (Scanlon, Aggeli et al., 2007). Figure reproduced by kind permission of the IET and Scanlon, Aggeli et al. 2007.

used to modulate gel stiffness independently from peptide/fibrillar concentration.

The formation of porous aerogels is another way to make biomaterials from self-assembling peptide networks that are structured on a micron scale (Scanlon et al., 2007). Scaffolds are created by freezing the fibrous sample in liquid nitrogen. The ice is then sublimed in a vacuum. The resulting scaffolds have a micron-sized layered structure, as illustrated in Figure 20. Scanlon and colleagues suggest that the size and shape of the pores could be determined by controlling the rate and direction of freezing and ice crystal nucleation. The presence of solvents can also be used to influence the structure, as shown in Figure 20 (right). Freeze drying proved more reliable than supercritical fluid drying, possibly due to the complications of peptide solubility in liquid CO_2, and the freeze drying method could be applied to create similar structures from other protein nanofiber networks. The authors suggest that these aerogels could also find use as sensors or packing materials.

Peptide amphiphiles have also been incorporated within the interconnected pores of collagen sponges that were reinforced with polyglycolic acid (PGA) for mechanical strength, as illustrated in Figure 21 (Hosseinkhani et al., 2006b). These structures were shown to enhance the differentiation of osteogenic cells both in vitro and in vivo. In a second study, the peptide amphiphiles were mixed with growth factor bFGF prior to incorporation within the collagen sponges (Hosseinkhai et al., 2006a). The presence of the peptide amphiphiles made a significant difference to scaffold performance; the release profile of bFGF was sustained and bone formation was induced in scaffolds that had been implanted within a rat. Both these studies illustrate how macroscopic structures can be developed that successfully incorporate protein nanofibers.

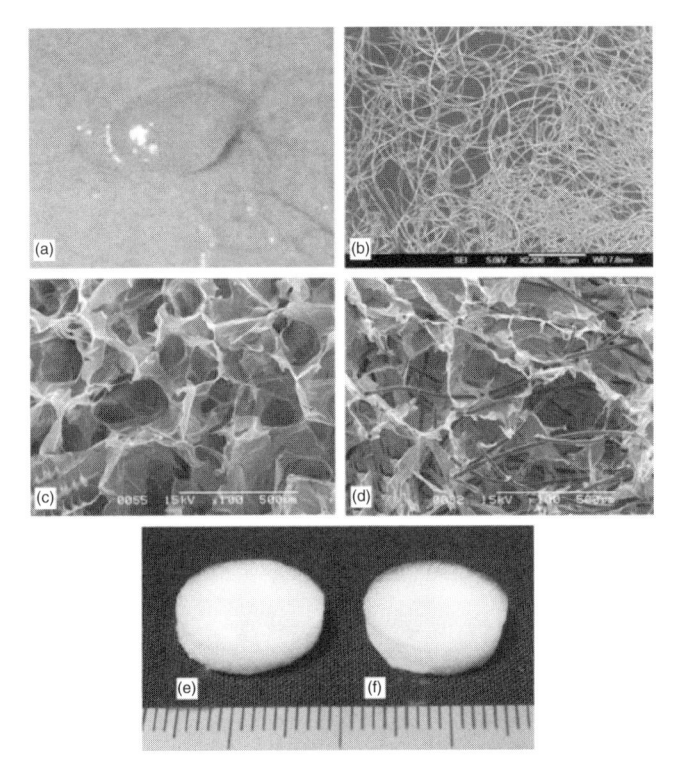

Figure 21 Light microscopy image of a gel formed by adding culture media to a solution of peptide amphihiles (a), scanning electron microscopy image of the peptide amphiphilic structures (b), the collagen sponge without PGA fibers (c) and with PGA fibers (d) and light microscopy images of the sponges without (e) and with (f) the PGA fibers. Reprinted from Hosseinkhani et al. (2006b) (copyright 2006, with permission from Elsevier).

5. CONCLUSION

The discovery of different peptide sequences that can readily self-assemble to form protein fibers is significant as it provides a way to organize peptides and create fibrous structures on a nanoscale. Among these different systems, amyloid fibrils hold particular promise, as all polypeptides are thought to be capable of forming these resilient structures. Structures formed from self-assembling peptides are not only smaller than many man-made materials, they are also constructed from natural building blocks and can be tailored to display a range of useful properties. We already have extensive knowledge about how self-assembling systems organize in solution and on solid surfaces but there is an opportunity to extend this knowledge and exploit these interactions to form new structures and assemblies.

Polypeptide fibers are being rapidly developed as wires and structures for applications as diverse as enzyme immobilization, biosensing, and biomaterials. Larger macroscopic structures are also being developed from protein fibers. As our ability to manipulate these structures grows so too will the range of applications to which these fibers can be applied.

REFERENCES

Adams, S., Higgins, A.M. et al. *Langmuir* **18**, 4854–4861 (2002).
Adler-Abramovich, L., and Gazit, E. *J. Pept. Sci.* **14**, 217–223 (2008).
Aggeli, A., Bell, M. et al. *Nature* **386**, 259–262 (1997).
Aggeli, A., Bell, M. et al. *J. Am. Chem. Soc.* **125**, 9619–9628 (2003).
Alexandrescu, A.T. *Protein Sci.* **14**, 1–12 (2005).
Allen, L.T., Fox, E.J. et al. *Proc. Natl. Acad. Sci. U.S.A.* **100**, 6331–6336 (2003).
Baldwin, A.J., Bader, R. et al. *J. Am. Chem. Soc.* **128**, 2162–2163 (2006).
Ban, T., Hoshino, M. et al. *J. Mol. Biol.* **344**, 757–767 (2004).
Baxa, U., Speransky, V. et al. *Proc. Natl. Acad. Sci. U.S.A.* **99**, 5253–5260 (2002).
Baxa, U., Taylor, K.L. et al. *J. Biol. Chem.* **278**, 43717–43727 (2003).
Bell, C.J., Carrick, L.M. et al. *J. Biomed. Mater. Res. A* **1**, 189–199 (2006).
Beniash, E., Hartgerink, J.D. et al. *Acta Biomater.* **1**, 387–397 (2005).
Bensimon, A., Simon, A. et al. *Science* **265**, 2096–2098 (1994).
Bochicchio, B., Pepe, A. et al. *Biomacromolecules* **8**, 3478–3486 (2007).
Bokhari, M.A., Akay, G. et al. *Biomaterials* **26**, 5198–5208 (2005).
Bosques, C.J., and Imperiali, B. *J. Am. Chem. Soc.* **125**, 7530–7531 (2003).
Bromley, E.H., Channon, K. et al. *ACS Chem. Biol.* **3**, 38–50 (2008).
Brown, C.L., Aksay, I.A. et al. *J. Am. Chem. Soc.* **124**, 6846–6848 (2002).
Bucciantini, M., Giannoni, E. et al. *Nature* **416**, 507–511 (2002).
Caflisch, A. *Curr. Opin. Chem. Biol.* **10**, 437–444 (2006).
Carny, O., Shalev, D.E. et al. *Nano Lett.* **6**, 1594–1597 (2006).
Carulla, N., Caddy, G.L. et al. *Nature* **436**, 554–558 (2005).
Channon, K., and Macphee, C.E. *Soft Matter* **4**, 647–652 (2008).
Channon, K.J., Devlin, G.L. et al. *J. Am. Chem. Soc.* **130**, 5487–5491 (2008).
Chapman, M.R., Robinson, L.S. et al. *Science* **295**, 851–855 (2002).
Claessen, D., Rink, R. et al. *Genes Dev.* **17**, 1714–1726 (2003).
Collier, J.H., and Messersmith, P.B. *Bioconjug Chem.* **14**, 748–755 (2003).
Collins, S.R., Douglass, A. et al. *PLoS Biol.* **2**, e321 (2004).
Colombo, G., Soto, P. et al. *Trends Biotechnol.* **25**, 211–218 (2007).
Corvis, Y., Walcarius, A. et al. *Anal Chem.* **77**, 1622–1630 (2005).
Davis, M.E., Motion, J.P. et al. *Circulation* **111**, 442–450 (2005).
del Mercato, L.L., Pompa, P.P. et al. *Proc. Natl. Acad. Sci. U.S.A.* **104**, 18019–18024 (2007).
Devlin, G.L., Knowles, T.P. et al. *J. Mol. Biol.* **360**, 497–509 (2006).
Dinca, V., Kasotakis, E. et al. *Nano Lett.* **8**, 538–543 (2008).
Dobson, C.M. *Trends Biochem. Sci.* **24**, 329–332 (1999).
Dos Santos, S., Chandravarkar, A. et al. *J. Am. Chem. Soc.* **127**, 11888–11889 (2005).
Dubois, G., Segers, V.F. et al. *J Biomed. Mater. Res. B Appl. Biomater.* **87B**, 1, 222–228 (2008).
Ellis-Behnke, R.G., Liang, Y.X. et al. *Proc. Natl. Acad. Sci. U.S.A.* **103**, 5054–5059 (2006).
Esler, W.P., Stimson, E.R. et al. *Nat. Biotechnol.* **15**, 258–263 (1997).
Esler, W.P., Stimson, E.R. et al. *Methods Enzymol.* **309**, 350–374 (1999).
Fandrich, M., Forge, V. et al. *Proc. Natl. Acad. Sci. U.S.A.* **100**, 15463–15468 (2003).
Fowler, D.M., Koulov, A.V. et al. *Public Library Sci. Biol.* **4**, 100–107 (2006).

Franks, G., and Meagher, L. *Colloids Surf. A* **214**, 99–110 (2003).
Fung, S.Y., Yang, H. et al. *PLoS One* **3**, e1956 (2008).
Gazit, E. *FEBS J.* **274**, 317–322 (2007).
Genove, E., Shen, C. et al. *Biomaterials* **26**, 3341–3351 (2005).
Ghadiri, M.R., Granja, J.R. et al. *Nature* **366**, 324–327 (1993).
Gras, S.L. *Aust. J. Chem.* **60**, 333–342 (2007).
Gras, S.L., Tickler, A.K. et al. *Biomaterials* **29**, 1553–1562 (2008).
Griffin, M.D., Mok, M.L. et al. *J. Mol. Biol.* **375**, 240–256 (2008).
Guijarro, J.I., Sunde, M. et al. *Proc. Natl. Acad. Sci. U.S.A.* **95**, 4224–4228 (1998).
Guler, M.O., Hsu, L. et al. *Biomacromolecules* **7**, 1855–1863 (2006).
Guler, M.O., and Stupp, S.I. *J. Am. Chem. Soc.* **129**, 12082–12083 (2007).
Ha, C., and Park, C.B. *Biotechnol. Bioeng.* **90**, 848–855 (2005).
Hamada, D., Yanagihara, I. et al. *Trends Biotechnol.* **22**, 93–97 (2004).
Harrington, D.A., Cheng, E.Y. et al. *J. Biomed. Mater. Res. A* **78**, 157–167 (2006).
Hartgerink, J.D., Beniash, E. et al. *Science* **294**, 1684–1688 (2001).
Herland, A., Bjork, P. et al. *Advan. Mater.* **17**, 1466–1471 (2005).
Herland, A., Björk, P. et al. *Small* **3**, 318–325 (2007).
Herland, A., Thomsson, D. et al. *J. Mater. Chem.* **18**, 126–132 (2008).
Holmes, T.C., de Lacalle, S. et al. *Proc. Natl. Acad. Sci. U.S.A.* **97**, 6728–6733 (2000).
Hong, Y., Legge, R.L. et al. *Biomacromolecules* **4**, 1433–1442 (2003).
Horii, A., Wang, X. et al. *PLoS One* **2**, e190 (2007).
Hortschansky, P., Schroeckh, V. et al. *Protein Sci.* **14**, 1753–1759 (2005).
Hosseinkhani, H., Hosseinkhani, M. et al. *Biomaterials* **27**, 5836–5844 (2006a).
Hosseinkhani, H., Hosseinkhani, M. et al. *Biomaterials* **27**, 5089–5098 (2006b).
Hosseinkhani, H., Hosseinkhani, M. et al. *Biomaterials* **27**, 4079–4086 (2006c).
Hosseinkhai, H., Hosseinkhani, M. et al. *Tissue Eng.* **13**, 11 (2007).
Hoyer, W., Cherny, D. et al. *J Mol Biol* **340**, 127–139 (2004).
Hsu, L., Cvetanovich, G.L. et al. *J. Am. Chem. Soc.* **130**, 3892–3899 (2008).
Hung, A.M., and Stupp, S.I. *Nano Lett.* **7**, 1165–1171 (2007).
Inoue, Y., Kishimoto, A. et al. *J. Biol. Chem.* **276**, 35227–35230 (2001).
Jayawarna, V., Ali, M. et al. *Advan. Mater.* **18**, 611–614 (2006).
Jayawarna, V., Smith, A. et al. *Biochem. Soc. Trans.* **35**, 535–537 (2007).
Jung, J.P., Jones, J.L. et al. *Biomaterials* **29**, 2143–2151 (2008).
Kammerer, R.A., and Steinmetz, M.O. *J. Struct. Biol.* **155**, 146–153 (2006).
Karsai, A., Grama, L. et al. *Nanotechnology* **34**, 345102/1–7 (2007).
Karsai, A., Murvai, U. et al. *Eur. Biophys. J.*, 1133–1137 (2008).
Kellermayer, M.S.Z., Karsai, A. et al. *Proc. Natl. Acad. Sci. U.S.A.* **105**, 141–144 (2008).
Keyes-Baig, C., Duhamel, J. et al. *J. Am. Chem. Soc.* **126**, 7522–7532 (2004).
Kirkham, J., Firth, A. et al. *J. Dental Res.* **86**, 426–430 (2007).
Kisiday, J., Jin, M. et al. *Proc. Natl. Acad. Sci. U.S.A.* **99**, 9996–10001 (2002).
Knowles, T.P., Shu, W. et al. *Proc. Natl. Acad. Sci. U.S.A.* **104**, 10016–10021 (2007).
Kowalewski, T., and Holtzman, D.M. *Proc. Natl. Acad. Sci. U.S.A.* **96**, 3688–3693 (1999).
Krebs, M.R., Bromley, E.H. et al. *J. Struct. Biol.* **149**, 30–37 (2005).
Krebs, M.R., Morozova-Roche, L.A. et al. *Protein Sci.* **13**, 1933–1938 (2004).
Linse, S., Cabaleiro-Lago, C. et al. *Proc. Natl. Acad. Sci. U.S.A.* **104**, 8691–8696 (2007).
Losic, D., Martin, L.L. et al. *Biopolymers* **84**, 519–526 (2006).
Lowik, D.W., Meijer, J.T. et al. *J. Pept. Sci.* **14**, 127–133 (2008).
MacCuspie, R.I., Banerjee, I.A. et al. *Soft Matter* **4**, 833–839 (2008).
MacPhee, C.E., and Dobson, C.M. *J. Mol. Biol.* **297**, 1203–1215 (2000a).
MacPhee, C.E., and Dobson, C.M. *J. Am. Chem. Soc.* **122**, 12707–12713 (2000b).
MacPhee, C.E., and Woolfson, D.N. *Curr. Opin. Solid State Mater. Sci.* **8**, 141–149 (2004).
Mahler, A., Reches, M. et al. *Advan. Mater.* **18**, 1365–1370 (2006).

Mardilovich, A., Craig, J.A. et al. *Langmuir* **22**, 3259–3264 (2006).
Marini, D.M., Hwang, W. et al. *Nanoletters* **2**, 295–299 (2002).
Meegan, J.E., Aggeli, A. et al. *Advan. Funct. Mater.* **14**, 31–37 (2004).
Meersman, F., and Dobson, C.M. *Biochim. Biophys. Acta* **1764**, 452–460 (2006).
Mesquida, P., Ammann, D.L. et al. *Advan. Mater.* **17**, 893–897 (2005).
Mesquida, P., Blanco, E.M. et al. *Langmuir* **22**, 9089–9091 (2006).
Miao, M., Bellingham, C.M. et al. *J. Biol. Chem.* **278**, 48553–48562 (2003).
Mimna, R., Camus, M.S. et al. *Angew. Chem. Int. Ed. Engl.* **46**, 2681–2684 (2007).
Mostaert, A.S., and Jarvis, S.P. *Nanotechnology* **18**, 044010 (2007).
Mucke, L., Masliah, E. et al. *J. Neurosci.* **20**, 4050–4058 (2000).
Nagai, Y., Unsworth, L.D. et al. *J. Control Release* **115**, 18–25 (2006).
Nayak, A., Dutta, A.K. et al. *Biochem. Biophys. Res. Commun.* **369**, 303–307 (2008).
Narrainen, A.P., Hutchings, L.R., *Soft Matter* **2**, 126–128 (2006).
Niece, K.L., Hartgerink, J.D. et al. *J. Am. Chem. Soc.* **125**, 7146–7147 (2003).
Nielsen, L., Khurana, R. et al. *Biochemistry* **40**, 6036–6046 (2001).
Nilsson, M.R. *Methods* **34**, 151–160 (2004).
Pauling, L., and Corey, R.B. *Proc. Natl. Acad. Sci. U.S.A.* **37**, 251–256 (1951).
Pearce, F.G., Mackintosh, S.H. et al. *J. Agric. Food Chem.* **55**, 318–322 (2007).
Puleo, D.A., and Bizios, R. *Bone* **12**, 271–276 (1991).
Rajangam, K., Arnold, M.S. et al. *Biomaterials* **29**, 3298–3305 (2008).
Rapaport, H. *Supramol. Chem.* **18**, 445–454 (2006).
Reches, M., and Gazit, E. *Science* **300**, 625–627 (2003).
Reches, M., and Gazit, E. *Nat. Nanotechnol.* **1**, 195–200 (2006).
Relini, A., Canale, C. et al. *J. Biol. Chem.* **281**, 16521–16529 (2006).
Rocha, S., Krastev, R. et al. *Chem. Phys. Chem.* **6**, 2527–2534 (2005).
Sargeant, T.D., Rao, M.S. et al. *Biomaterials* **29**, 1085–1098 (2008).
Scanlon, S., Aggeli, A. et al. *Micro. Nano Lett.* **2**, 24–29 (2007).
Scheibel, T., Kowal, A.S. et al. *Curr. Biol.* **11**, 366–369 (2001).
Scheibel, T., Parthasarathy, R. et al. *Proc. Natl. Acad. Sci. U.S.A.* **100**, 4527–4532 (2003).
Semino, C.E., Kasahara, J. et al. *Tissue Eng.* **10**, 643–655 (2004).
Sharp, J.S., Forrest, J.A. et al. *Biochemistry* **41**, 15810–15819 (2002).
Sieminski, A.L., Semino, C.E. et al. *J. Biomed. Mater. Res. A* **87A**, 2, 494–504 (2008).
Silva, G.A., Czeisler, C. et al. *Science* **303**, 1352–1355 (2004).
Smith, J.F., Knowles, T.P. et al. *Proc. Natl. Acad. Sci. U.S.A.* **103**, 15806–15811 (2006).
Smith, M.I., Sharp, J.S. et al. *Biophys. J.* **93**, 2143–2451 (2007).
Sohma, Y., Hayashi, Y. et al. *Biopolymers* **76**, 344–356 (2004).
Sone, E.D., and Stupp, S.I. *J. Am. Chem. Soc.* **126**, 12756–12757 (2004).
Squires, A.M., Devlin, G.L. et al. *J. Am. Chem. Soc.* **128**, 11738–11739 (2006).
Stefani, M. *Neuroscientist* **13**, 519–531 (2007).
Stevens, M.M., and George, J.H. *Science* **310**, 1135–1138 (2005).
Storrie, H., Guler, M.O. et al. *Biomaterials* **28**, 4608–4618 (2007).
Sunde, M., Serpell, L.C. et al. *J. Mol. Biol.* **273**, 729–739 (1997).
Talbot, N.J. *Curr. Biol.* **13**, R696–R698 (2003).
Tamburro, A.M., Pepe, A. et al. *J. Biol. Chem.* **280**, 2682–2690 (2005).
Taniguchi, A., Sohma, Y. et al. *J. Am. Chem. Soc.* **128**, 696–697 (2006).
Tovar, J.D., Rabatic, B.M. et al. *Small* **3**, 2024–2028 (2007).
Tuchscherer, G., Chandravarkar, A. et al. *Biopolymers* **88**, 239–252 (2007).
Tysseling-Mattiace, V.M., Sahni, V. et al. *J. Neurosci.* **28**, 3814–3823 (2008).
Waterhouse, S.H., and Gerrard, J.A. *Aust. J. Chem.* **57**, 519–523 (2004).
Westermark, P., Benson, M.D. et al. *Amyloid* **12**, 1–4 (2005).
Whitehouse, C., Fang, J. et al. *Angew. Chem. Int. Ed. Engl.* **44**, 1965–1968 (2005).
Woolfson, D.N., and Ryadnov, M.G. *Curr. Opin. Chem. Biol.* **10**, 559–567 (2006).

Wright, C.F., Teichmann, S.A. et al. *Nature* **438**, 878–881 (2005).
Yang, H., Fung, S.Y. et al. *PLoS One* **2**, e1325 (2007).
Yemini, M., Reches, M. et al. *Anal. Chem.* **77**, 5155–5159 (2005a).
Yemini, M., Reches, M. et al. *Nano Lett.* **5**, 183–186 (2005b).
Yuwono, V.M., and Hartgerink, J.D. *Langmuir* **23**, 5033–5038 (2007).
Zhang, F., Du, H.-N. et al. *Angew. Chem. Int. Ed.* **45**, 3611–3613 (2006).
Zhang, S., Holmes, T. et al. *Proc. Natl. Acad. Sci. U.S.A.* **90**, 3334–3338 (1993).
Zhang, S., Holmes, T.C. et al. *Biomaterials* **16**, 1385–1393 (1995).
Zhu, M., Souillac, P.O. et al. *J. Biol. Chem.* **277**, 50914–50922 (2002).
Zurdo, J., Guijarro, J.I. et al. *J. Am. Chem. Soc.* **123**, 8141–8142 (2001).

CHAPTER 7

Hybrid Systems Engineering: Polymer–Peptide Conjugates

Conan J. Fee

Contents		
	1. Introduction	211
	2. Peptide–Polymer Conjugation	213
	3. Peptide-Directed Formation of Gels	214
	4. Cyclopeptide Nanotubes	215
	5. Polyelectrolytic and Organometric Polymer–Peptide Conjugates	216
	6. β-Sheet Suprastructures	217
	7. Rod–Coil Conformations	218
	8. Physical Manipulation of Nanostructural Orientation	220
	9. Conclusions	220
	References	221

1. INTRODUCTION

Synthetic polymeric materials are normally designed with properties that are determined by the nature of the monomers making up their molecular structure. Improvements of function can be achieved by synthesizing block copolymers, which generally rely for their properties on the phase separation and differential solubility's of the blocks concerned (Park et al., 2003). For example, linear A-B block copolymers might incorporate hydrophilic A blocks with hydrophobic B blocks to achieve controllable degrees of crystallinity, water swelling, or rates of hydrolytic degradation. In designing such materials, gross material characteristics can be modified by altering the relative lengths of the blocks, creating branched structures or cross-linking. The introduction of a third block provides further control of properties through the increase in combinations available (Volker Abetz, 2000),

Department of Chemical & Process Engineering, University of Canterbury, Christchurch 8020, New Zealand.

E-mail address: conan.fee@canterbury.ac.nz

including the orders (Hückstädt et al., 2000) and ratios of the three blocks involved. Even so, the ability to tune the properties of synthetic, amorphous copolymers at the nanoscale level is limited.

By comparison, nature makes use of an astonishing array of biopolymers with properties that appear to be finely tuned to their biological functions, not only in terms of their chemical makeup but also in terms of their structural organization. Although a number of monomers are utilized in nature, including nucleotides and saccharides, the amino acid polymers, or peptides, are the most interesting for use as designed materials because of the range of properties available and the relative stability of amide bonds. The 20 "natural" amino acids found in nature can be combined in N^{20} ways in a linear chain, where N is the number of monomers in the chain. By adding branched structures and using "unnatural" or noncanonical amino acids, the combinations become virtually limitless. Moreover, peptides can be produced with precise sequences and uniform molecular weights by harnessing the power of the protein synthesis machinery of living cells through genetic manipulation. For example, Tirrell and his colleagues have introduced noncanonical amino acid analogs into designed polypeptides and proteins by this means, including fluorinated amino acids (Wang et al., 2003). With this addition to the chemists' synthesis toolkit, the boundary between "synthetic" and "bio" polymers is becoming blurred. Nevertheless, it should be understood that the term "biopolymer" herein refers to polymers that are made up of the building blocks used by nature (or closely related ones) however they are synthesized and be they inspired by naturally occurring biopolymers or designed de novo.

In contrast to synthetic polymers, which are usually biologically inert, biopolymers are generally bioactive. Bioactivity can convey both advantages and disadvantages. Proper biological signals in biopolymers can promote cell and protein adhesion or elicit biological responses and can allow enzymatic degradation. On the other hand, peptides can be allergenic, immunogenic, or even highly toxic, as the combinatorial cocktail of peptides present in spider venoms attests (Sollod et al., 2005). Biorecognition factors can be built into synthetic polymers, for example, the arginine–glycine–aspartic acid (RGD) sequence that promotes cell adhesion in biomaterials conjugated to the surface of a polycarbonate polyurethane material for human monocyte adhesion (Ernsting et al., 2007). However, the biological inertness of synthetic polymers may have advantages over biopolymers in avoiding toxic, immunogenic, or allergic responses in biomaterials and also can offer resistance to biodegradation.

An interesting approach to the design of biomaterials and biologically inspired materials is to combine the best features of synthetic and biological polymers into copolymers (Klok, 2005; Vandermeulen and Klok, 2004). Thus, peptide sequences may be used to create self-assembled structural motifs at the nanoscale, most commonly β-sheets or α-helices, while the amorphous (or in some cases liquid crystalline) synthetic polymers can be used to provide inert, hydrophilic, or hydrophobic regions and to limit gross aggregation of the peptides. Cells contacting artificial fibers with diameters measuring in the 10's of micrometers behave as though they are on a flat surface, taking on a flat, nonphysiological morphology. Much smaller diameter fibers can now be made, with diameters of 10's of nanometers, but it is difficult to produce these in a network of fibers with pore dimensions appropriate for cellular migration. Therefore, such networks must be formed around cells without damaging them, which is a challenging task. Instead, the use of supramolecular assembly of nanofibrillar and coiled-coil (CC) (nanotubular) structures can be advantageous, particularly if coupled with amphipilic polymer blocks which can gel under normal physiological conditions (Lutolf and Hubbell, 2005).

In this chapter, some methods for peptide–polymer conjugate synthesis are described and a number of examples are reviewed to illustrate the use of self-assembling peptide sequences to drive nanoscale organization. Conjugation of polymers with proteins or peptides for reasons other than self-assembly, for example, therapeutic protein or peptide PEGylation to prolong in vivo circulation half-lives, are not covered.

2. PEPTIDE–POLYMER CONJUGATION

Klok (2005) described a number of synthetic pathways for peptide–polymer conjugates, classed first by liquid- or solid-phase synthesis methods. Within these, synthesis was further classified as being either convergent or divergent. In convergent synthesis, a preformed polymer is attached to a preformed peptide, most frequently by site-specific conjugation, although nonspecific conjugation can be used. Nonspecific conjugation is usually achieved conveniently under mildly basic conditions by amine coupling between an N-hydroxy succinimidyl group on the polymer and the ε-amino group available on lysine (K) residues of the peptide. Lysine is one of the most abundant amino acids in proteins so in longer, biologically inspired peptide sequences, it is likely to occur several times in the sequence, thus leading to a number of positional isomers. For shorter sequences, a single lysine residue may be available for amine coupling. Alternatively, the N-terminal amine can be targeted somewhat selectively by reductive alkylation using an aldehyde-functionalized polymer. Because the pKa values of the N-terminus amine group ranges from 7.6 to 8.0 and those of lysine residues ranges from 10.0 to 10.2, the reaction can be made more selective by carrying it out under slightly acidic conditions (pH 5) (Klok, 2005).

The most common chemoselective conjugation is achieved through thiol coupling to free cysteine (C) groups via maelimide chemistry. In proteins, cysteine residues make up only 1.7% of amino acids, and because they tend to form disulphide bonds, it is rare to find a free cysteine group. In designed peptides, this limitation does not exist and a single cysteine can be engineered into the sequence to provide a specific conjugation site. Moreover, the cysteine can be located away from parts of the sequence that determine the structural organization so that the conjugated polymer does not sterically hinder self-assembly of biological function (Kopecky, 2006). Yang et al. (2008) used this technique to synthesize hydrogels by self-assembly.

Another biologically compatible, site-specific method for poly(ethylene glycol) (PEG) conjugation to proteins was described by Sato (2002), who reported the use of the transglutaminase enzyme to catalyze the acyl transfer between the γ-carboxyamide group of glutamine and alkylamines under mild conditions. This method could easily be adopted for peptide–PEG conjugation.

Solid-phase peptide synthesis, first introduced by Merrifield (1963), utilizes a system in which an amino acid is tethered to an insoluble particle by a cleavable link, followed by addition of amino acids one step at a time with protection and deprotection of reactive groups other than the conjugation site. It has the advantage of allowing high conversions at each step when adding each peptide to the sequence since large excesses of reagents can be used and then washed away at the end of each step. Felix and coworkers used this technique extensively during peptide synthesis to conjugate a PEG molecule at a precise location in the sequence (Felix, 1997). It can be used with either convergent or divergent conjugation strategies and has the advantage that it can be automated. However, the method is limited by economic considerations to sequences no longer than about 20–30 residues because the overall conversion is given by X^N, where X is the average fractional conversion at each step. Therefore, even with $X=0.98$, the cumulative effect leads to overall conversion rates of about 55% for 30 residues, which drops drastically to just 21% with a (still respectable) fractional conversion per step of $X=0.95$.

Examples of divergent addition strategies are given by Lutz and coworkers (Lutz and Hoth, 2006; Lutz et al., 2006) and Loschonsky et al. (2008), specifically via atom transfer radical polymerization (ATRP). This method may lead to a distribution in the synthetic polymer molecular weight, though this might not necessarily be any broader than might be present in a preformed polymer attached via a convergent method. Divergent addition can be combined with either solid phase synthesis or genetic engineering techniques to introduce amino acids with unusual side groups for targeted conjugation.

3. PEPTIDE-DIRECTED FORMATION OF GELS

Natural extracellular matrices in vivo consist of various protein fibrils and fibers, interwoven within a hydrated network of glycosaminoglycan chains and may require less than 1% solid materials to form mechanically robust structures (Lutolf and Hubbell, 2005). Yang et al. (2008) synthesized a hydrogel from hydrophilic synthetic N-(2-hydroxypropyl)methacrylamide (HPMA) copolymerized separately with oppositely charged pentaheptad CC peptides. Such peptides possess the tendency to form CC conformations if their primary heptad sequences "abcdefg" have hydrophobic residues at positions a and d. In this case, one pentaheptad sequence (CCK) had mostly lysine (K) residues at positions e and g, while the other (CCE) had glutamic acid (E) residues in the same positions. By mixing equimolar quantities of these two peptide–polymer hybrids, a hydrogel was formed, driven by the antiparallel alignment of the CCK and CCE peptides. When modified versions of the two peptides were used, designed to form random coils by replacing glutamic acid with tryptophan (W) and lysine with tyrosine (Y), no hydrogel formed. Also, the hydrogel formed with the CCK and CCE peptide-conjugated HPMA hybrids collapsed after addition of a chaotropic agent (guanidine hydrochloride) or pure CCK or CCE peptides. This evidence supports the conclusion that the CCK and CCE portions of the hybrid molecules self-assembled as antiparallel coil–coil heterodimers, acting as physical cross-linkers to form the hydrogel.

Tonegawa et al. (2004) created a cationic polylysine with a tetrapeptide end sequence (glycine–tyrosine–glycine–lysine), which is a motif common to the consensus sequences of mussel adhesive proteins. They then cross-linked this with the anionic polysaccharide, gellan, enzymatically. The polyionic complexation between the cationic peptide and the anionic polysaccharide formed a hybrid fiber at the aqueous solution interface that, when cross-linked, mimicked the byssus gel that marine mussels use to adhere to surfaces, despite the presence of water and salt.

4. CYCLOPEPTIDE NANOTUBES

Unlike cyclopeptides that consist of peptides with uniform chirality, cyclic peptide sequences of 8–12 residues with an alternating arrangement D- and L-α-amino acids adopt a flat (planar) ring conformation (Ghadiri et al. 1993; Hartgerink et al., 1996). These (D-alt-L)-peptide rings have a high tendency to assemble into tubular ring stacks, occurring by β-sheet-like antiparallel arrangements. Such an arrangement has the H-bonds following the tube axis, with the amino acid side chains oriented outward, so that the exterior is functionalizable and the interior is hollow. ten Cate et al. (2006) noted that cyclopeptide nanotubes grafted with polymers after assembly can suffer from cross-ring linking or disruption with high side-chain grafting densities so investigated the assembly of (D-alt-L)-cyclopeptide-polymer conjugates. Poly(n-butyl acrylate) chains were conjugated to opposite sides of a pre-formed cyclic (D-alt-L)-α-octapeptide, thereby forming a coil–ring–coil conjugate through a convergent approach. Atomic force microscopy of the

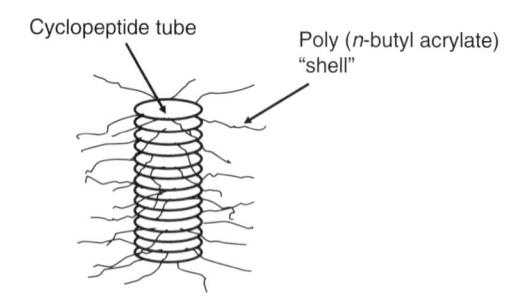

Figure 1 Nanotube-shell conformation of cyclopeptide–polymer coil copolymers (ten Cate et al., 2006).

resulting self-assembled nanotubes on a mica surface revealed aggregates with a height of 1.4 nm and a width of 5 nm and lengths of 200–300 nm. These dimensions were explained by the formation of a tubular peptide core built via stacking of the cyclopeptides, with a poly(*n*-butyl acrylate) shell wrapped around the tube (Figure 1). Loschonsky et al. (2008) observed similar behavior with a cyclic (D-alt-L) octapeptide conjugated to butylacrylate chains at three positions in the rings. The hybrid molecules displayed a solvent-induced self assembly, yielding rod-like structures having a core shell morphology with an internal beta-sheet peptide assembly surrounded by a soft poly(*n*-butyl acrylate) exterior (Loschonsky et al., 2008). They used a divergent strategy with in situ polymerization of the butylacrylate and were able to control the final molecular weight between 5,000 and 30,000 g/mol by controlling the monomer conversion extent.

Although such nanotubes form an interesting structure with potential for exterior functionalization, they are currently limited to passive transport/release of molecules in the interior. Future work may focus on using non-canonical amino acids to impart functionalizable interior surfaces to allow orthogonal functionalization of the interior and exterior surfaces (ten Cate et al., 2006).

5. POLYELECTROLYTIC AND ORGANOMETRIC POLYMER–PEPTIDE CONJUGATES

Chiral conjugate polymers such as polythiophenes with an optically active substituent in the 3 position have been studied for use in optoelectronic devices, sensors, and catalysis (Nilsson et al., 2004). These materials show activity based on their chiral behaviors, which can be influenced by solvent or temperature. Helical biomolecules can be used to induce chirality in optically inactive polythiophenes (Ewbank et al., 2001). Nilsson et al. (2004) extended this to use a positively charged random-coil peptide conjugated with a negatively charged, optically inactive polythiophene to create a hybrid supermolecule with a three-dimensional ordered structure determined by the biomolecule and electronic properties determined by the polythiophene. The polymer backbone conformation was altered in the presence of the peptide to give a main-chain chirality with a predominantly α-helical structure due to acid–base complexation between the two components.

The incorporation of metal centers in polymeric structures creates molecules with interesting functions that have uses as precursors to carbon nanotubes, wire-like structures, or iron-rich ceramic materials (Vandermeulen et al., 2006). The oligotetrapeptide sequence found in silk, glycine–alanine–glycine–alanine (GAGA), was shown to retain the ability to form antiparallel β-sheets when conjugated with polyferrocenylsilane in both block and graft conformations. The block copolymer formed a fibrous network consisting of a core containing self-assembled antiparallel β-sheets with a corona of organometric polyferrocenylsilane. This result was similar to that of Rathore and Sogah (2001), who showed the conservation of GAGA-directed β-sheet formation when conjugated with PEG, as a structure inspired by spider silk.

6. β-SHEET SUPRASTRUCTURES

β-sheet structures are of significant interest, partly because of their ability to form fibrillar strands similar to those that are associated with the lesions seen in Alzheimer's Disease and partly because they form the basis of nanoscale architectures such as fibers and nanotubes. Smeenk et al. (2005) created a repetitive sequence of alanine–glycine interspersed with a glutamic acid–glycine block, conjugated with PEG at both the C and the N terminal cysteine residues via thiol coupling. The formation of β-sheets was observed, while the PEG decoration prevented the further aggregation that normally forms needle-like crystals in its absence.

To extend the self-assembling properties of peptide–polymer conjugates into the organic domain, an organosoluble peptide–polymer conjugate comprising a sequence-defined peptide and a poly(*n*-butyl acrylate) was created with incorporation of structure-breaking switch defects in the peptide segment that allowed control of self-assembly (Hentschel and Borner, 2006). The defects consisted of substituting two amine bonds with esters, which were later restored by an O—N acyl transfer reaction with a pH change. Aggregation tendency was thus suppressed and then triggered by a pH change which restored the native peptide backbone, leading to formation of densely twisted tape-like microstructures. The helical superstructures underwent defined entanglement to form superhelices, leading to the formation of soft, continuous organogels. Again, the structural model proposed to explain the observed behavior was a core-shell tape in which the peptide segments form an antiparallel β-sheet surrounded by a polymer shell. Hentschel et al. (2007) used a similar switch defect approach to suppress aggregation for a range of peptide–polymer conjugates with constant peptide sequences but varying polymer molecular weights and found that once the switch defects had been repaired, the rates of self-assembly correlated with the molecular weights of the poly(n-butylacrylate) blocks. Oligopeptide contents of as little as 3.5 wt% could direct nanostructural arrangement.

Self-assembly in an organic phase, followed by in situ cross-linking to preserve the nanostructure, was achieved by a combining diacetylene macromonomers with a parallel β-sheet forming tetrapeptide. After self-assembly, in which the parallel arrangement of the β-sheets was critical to align the functional endgroups, the diacetylene macromonomers were cross-linked by UV irradiation (Jahnke et al., 2007). Hsu et al. (2008) followed a similar path to create a nanostructured, cross-linked polydiacetylene structure. Linear and branched conjugates formed gels at concentrations of 2.0 wt% in water in the presence of ammonium hydroxide vapor. Samples irradiated at 256 nm after gel formation resulted in a characteristic color change from colorless to an intense blue color. Irradiation of the diacetylene macromonomers in the absence of the conjugated peptide resulted in a purple color, indicating both some ordered (blue) and disordered (red) states in the polydiacetylene acids. After irradiation, the gels appeared more mechanically robust, a result of the formation of polydiacetylene backbones within the nanofibers. The β-sheet structure was able to induce a chiral structure in the backbone, reflecting a high degree of order.

7. ROD–COIL CONFORMATIONS

Rod–coil diblocks can be synthesized by conjugating an α-helical peptide sequence to a random coil synthetic polymer (Klok et al., 2000). The self-assembly is driven both by the microphase separation of the blocks and by a tendency for the rods to form anisotropic liquid–crystal domains, giving rise to a range of morphologies that are distinct from those found in coil–coil copolymers (Stupp et al., 1997). According to Klok et al. (2000), asymmetry in the stiffness of the rod and coil parts of the copolymer results in an increase in the heat of mixing (as indicated by the Flory–Huggins parameter) over that of coil–coil copolymers, yielding phase separation-driven self-assembly at lower degrees of polymerization than coil–coil copolymers, that is, at smaller length scales. These rod–coil copolymers may therefore show self-assembled domains with sizes of the order of a few nanometers, although coil–coil copolymers can take on similar dimensions at the lower end of their range (5–100 nm). Unlike the earlier work of Gallot et al. (1996), who worked with polypeptides, Klok et al. (2000) utilized this expected phase separation in studying the behaviors of oligopeptides with just 10–20 units of γ-benzyl-L-glutamate as the rod and 10 repeat units of styrene as the random coil. The resulting morphologies were distinct from the lamellar structures observed with high-molecular-weight components. At 120°C, (styrene)$_{10}$-b-(γ-benzyl-L-glutamate)$_{10}$ conjugates possess partly α-helical and partly β-sheet organization, whereas increasing the peptide length to 20 units increased the proportion of α-helices, resulting in a "double-hexagonal" organization (Figure 2). Increasing the temperature tended to disrupt the hexagonal structure, yielding only β-sheet assemblies.

The advantage of utilizing peptides as rod segments is that their conformation as either α-helices or random coils can be controlled by pH, solvent, ionic strength, or temperature, which may therefore allow reversible switching of the nanoscale organization since this is driven largely by hydrogen-bonding interactions between neighboring rods in the α-helical conformation.

Mori et al. (2005) used a convergent method to fabricate rod–coil diblock copolymers of oligo(p-phenylenevinylene) (OPV) as a hydrophobic rod and poly(ethylene oxide) (PEO) as a hydrophilic random coil. Their aggregation behavior in a tetrahydrofuran/H$_2$O solvent was explored and parallel alignment of the OPV blocks was observed. Copolymers with PEO weight fractions of 41 and 62% formed cylindrical aggregates (diameter 6–8 nm and lengths of several hundred nanometers), but with higher PEO content (79 wt%), the conjugates formed distorted spherical aggregates of average diameter 13 nm. The aggregates were able to solubilize homo-oligomer OPV within the OPV cores of the aggregates, and their morphologies changed from "ambiguous" nanosized structures to cylindrical structures with

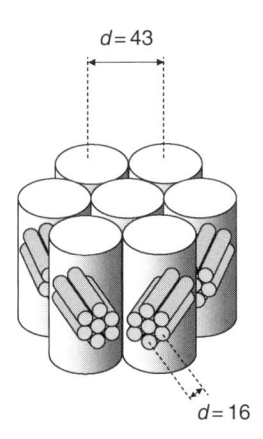

Figure 2 Klok et al.'s model of the self-assembly of rod–coil copolymer (styrene)$_{10}$-b-(γ-benzyl-L-glutamate)$_{20}$ at 120°C. Small cylinders represent the peptide rods. Styrene coils, not shown for clarity, are proposed to protrude randomly from both sides of the oligopeptide clusters (Klok et al., 2000).

diameters that depended on the mixing ratio of the OPV homo-oligomer. As suggested by the authors, solubilization inside the cores of aggregates may comprise a method of fine tuning the nanostructural organization.

8. PHYSICAL MANIPULATION OF NANOSTRUCTURAL ORIENTATION

In an attempt to mimic the process of biosilicification, Kessel and Borner recently developed an intriguing method for creating aligned polymer-silica fiber networks by 2D-plotting of PEO–peptide nanotape "ink" (Kessel and Borner, 2008). According to the authors, the PEO–peptide nanotapes mimic the functional structures of proteins like silaffin and silicatein that catalyze and guide the formation of biosilica from dilute silicic acid solution at neutral pH. The plotting process relied on local injection of the nanotape solution into a thin layer of a dilute solution of prehydrolyzed tetramethylorthosilicate (TMOS), a silicic acid equivalent. The "plotting" was achieved by moving a 250 µm inner diameter capillary connected to a syringe pump (flow rate 0.2 mL/min) with a computer-controlled (CNC) lathe, able to move at rates of between 0.5 and 2 m/min. The width of the nanotape fibers was directly related to the speed of plotting at a constant ink flow rate, giving fibers between 700 µm at the slowest plotting speed and 170 µm at the fastest plotting speed. Silicification was dependent on the presence of the β-sheet-aligned peptide blocks and increased with ageing of the fibers in the TMOS solution. The fibers were seen to align with the direction of plotting. Calcination at 450°C resulted in decomposition of the organic constituents, leaving the silica network intact and preserving the nanostructured morphology to give mesoporous silica fabrics with high surface areas and aligned cylindrical pores. The authors suggested that 3D plotting may provide more complex structures of importance for regenerative bone repair as active scaffolds for cell growth and directional cell organization in artificial tissues.

9. CONCLUSIONS

Peptides can be used to direct the nanoscale assembly of amphiphilic synthetic polymers. A common feature is that the self-assembly of the peptides proceeds as it would do in the absence of the polymer conjugates, with the peptide suprastructure forming a core, surrounded by the polymer random coil. The polymer shell acts to limit aggregation of the peptides beyond a certain size limit. A particularly striking example of this is the self-assembly of cyclopeptide–polymer composites, which form hollow nanotubes surrounded by a polymer shell, resulting in a uniform size distribution. The polymer shell could also be used to add functionality to the exterior of the nanotubes.

Potential applications of peptide–polymer conjugates include drug delivery materials, optoelectronics, biosensors, tissue scaffolds, tissue replacement materials, hydrogels, adhesives, biomimetic polymers, lithographic masks, and templates for metallic or silica nanostructures.

REFERENCES

ten Cate, M.G.J., Severin, N., and Borner, H.G. "Self-assembling peptide-polymer conjugates comprising (D-alt-L)-cyclopeptides as aggregator domains". *Macromolecules* **39**(23), 7831–7838 (2006).

Ernsting, M.J., Labow, R.S., and Santerre, J.P. "Human monocyte adhesion onto RGD and PHSRN peptides delivered to the surface of a polycarbonate polyurethane using bioactive fluorinated surface modifiers". *J. Biomed. Mater. Res. A* **83A**(3), 759–769 (2007).

Ewbank, P.C., Nuding, G., Suenaga, H., McCullough, R.D., and Shinkai, S. "Amine functionalized polythiophenes: Synthesis and formation of chiral, ordered structures on DNA substrates". *Tetrahedron Lett* **42**(2), 155–157 (2001).

Felix, A.M. Site-specific poly(ethylene glycol)ylation of peptides, *in* "Poly(ethylene glycol): Chemistry and Biological Applications" (J.M. Harris and S. Zalipsky, Eds.) ACS symposium Series **680**, 218–238 (1997). American Chemical Society, Washington DC.

Gallot, B. "Comb-like and block liquid crystalline polymers for biological applications". *Prog. Polym. Sci.* **21**(6), 1035–1088 (1996).

Ghadiri, M.R., Grnja, J.R., Milligan, R.A., McRee, D.E., and Khazanovich, N. "Self-assembling organic nanotubes based on a cyclic peptide architecture". *Nature* **366**, 324 (1993).

Hartgerink, J.D., Granja, J.R., Milligan, R.A., and Ghadiri, M.R. "Self-assembling peptide nanotubes". *J. Am. Chem. Soc.* **118**(1), 43–50 (1996).

Hentschel, J., and Borner, H.G. "Peptide-directed microstructure formation of polymers in organic media". *J. Am. Chem. Soc.* **128**(43), 14142–14149 (2006).

Hentschel, J., ten Cate, M.G.J., and Borner, H.G. "Peptide-guided organization of peptide-polymer conjugates: expanding the approach from oligo- to polymers". *Macromolecules* **40**(26), 9224–9232 (2007).

Hsu, L., Cvetanovich, G.L., and Stupp, S.I. "Peptide amphiphile nanofibers with conjugated polydiacetylene backbones in their core". *J. Am. Chem. Soc.* **130**(12), 3892–3899 (2008).

Hückstädt, H., Göpfert, A., and Abetz, V. "Influence of the block sequence on the morphological behavior of ABC triblock copolymers". *Polymer* **41**(26), 9089–9094 (2000).

Jahnke, E., Millerioux, A.-S., Severin, N., Rabe, J.P., and Frauenrath, H. "Functional, hierarchically tructured poly(diacetylene)s via supramolecular self-assembly". *Macromol. Biosci.* **7**, 136–143 (2007).

Kessel, S., and Borner, H.G. "Self-assembled PEO-peptide nanotapes as ink for plotting nonwoven silica nanocomposites and mesoporous silica fiber networks". *Macromol. Rapid Commun.* **29**, 316–320 (2008).

Klok, H.-A. "Biological-synthetic hybrid block copolymers: combining the best from two worlds". *J. Polym. Sci. A Polym. Chem.* **43**(1), 1–17 (2005).

Klok, H.-A., Langenwalter, J.F., and Lecommandoux, S. "Self-assembly of peptide-based diblock oligomers". *Macromolecules* **33**, 7819–7826 (2000).

Kopecky, E.M., Greinstetter, S., Pabinger, I., Buchacher, A., Römisch, J., and Jungbauer, A. "Effect of oriented or random PEGylation on bioactivity of a factor VIII inhibitor blocking peptide". *Biotechnol. Bioeng.* **93**(4), 647–655 (2006).

Loschonsky, S., Couet, J., and Biesalki, M. "Synthesis of peptide/polymer conjugates by solution ATRP of butylacrylate using an initiator-modified cyclic D-alt-L-peptide". *Macromol. Rapid Commun.* **29**, 309–315 (2008).

Lutolf, M.P., and Hubbell, J.A. "Synthetic biomaterials as instructive extracellular microenvironments for morphogenesis in tissue engineering". *Nat. Biotechnol.* **23**(1), 47–55 (2005).

Lutz, J.F., Borner, H.G., and Weichenhan, K. "Combining ATRP and 'click' chemistry: a promising platform toward functional biocompatible polymers and polymer bioconjugates". *Macromolecules* **39**(19), 6376–6383 (2006).

Lutz, J.F., and Hoth, A. "Preparation of Ideal PEG Analogues with a tunable thermosensitivity by controlled radical copolymerization of 2-(2-Methoxyethoxy)ethyl methacrylate and oligo(ethylene glycol) methacrylate". *Macromolecules* **39**(2), 893–896 (2006).

Merrifield, R.B. "Solid phase peptide synthesis I: the synthesis of a tetrapeptide". *J. Am. Chem. Soc.* **85**, 2149 (1963).

Mori, T., Watanabe, T., Minagawa, K., and Tanaka, M. "Self-assembly of oligo(p-phenylenenylene)-block-poly(ethylene oxide) in polar media and solubilization of an olig(p-phenylenevinylene) homooligomer inside the assembly". *J. Polym. Sci. Polym. Chem.* **43**, 1569–1578 (2005).

Nilsson, K.P., Rydberg, J., Baltzer, L., and Inganas, O. "Twisting macromolecular chains: self-assembly of a chiral supermolecule from nonchiral polythiophene polyanions and random-coil synthetic peptides". *Proc. Natl. Acad. Sci. U.S.A.* **101**(31), 11197–11202 (2004).

Park, C., Yoon, J., and Thomas E.L. "Enabling nantechnology with self assembled block copolymer patterns". *Polymer* **44**, 6725–6760 (2003).

Rathore, O., and Sogah, D.Y. "Nanostructure formation through β-sheet sheet self-assembly in silk-based materials". *Macromolecules* **34**(5), 1477–1486 (2001).

Sato, H. "Enzymatic procedure for site-specific pegylation of proteins". *Adv. Drug Deliv. Rev.* **54**(4), 487–504 (2002).

Smeenk, J.M., Otten, M.B.J., Thies, J., Tirrell, D.A., Stunnenberg, H.G., and van Hest, J.C.M. "Controlled assembly of macromolecular β-sheet fibrils". *Angew. Chem. Int. Ed.* **44**, 1968–1971 (2005).

Sollod, B.L., Wilson, D., Zhaxybayeva, O., Gogarten, J.P., Drinkwater, R., and King, G.F. "Were arachnids the first to use combinatorial peptide libraries?" *Peptides* **26**(1), 131–139 (2005).

Stupp, S.I., Lebonheur, V., Walker, K., Li, L.S., Huggins, K.E., Keser, M., and Amstutz, A. "Supramolecular materials: self-organized nanostructures". *Science* **276**(5311), 384–389 (1997).

Tonegawa, H., Kuboe, Y., Amaike, M., Nishida, A., Ohkawa, K., and Yamamoto, H. "Synthesis of enzymatically crosslinkable peptide-poly(L-lysine) conjugate and creation of bio-inspired hybrid fibres". *Macromol. Biosci.* **29**, 316–320 (2004).

Vandermeulen, G.W.M., Kim, K.T., Wang, Z., and Manners, I. "Metallopolymer-peptide conjugates: synthesis and self-assembly of polyferrocenylsilane graft and block copolymers containing a **beta**-sheet forming Gly-Ala-Gly-Ala tetrapeptide segment". *Biomacromolecules* **7**(4), 1005–1010 (2006).

Vandermeulen, G.W.M., and Klok, H.-A. "Peptide/protein hybrid materials: enhanced control of structure and improved performance through conjugation of biological and synthetic polymers". *Macromol. Biosci.* **4**(4), 383–398 (2004).

Volker Abetz, T.G. "Formation of superlattices via blending of block copolymers". *Macromol. Rapid Commun.* **21**(1), 16–34 (2000).

Wang, P., Tang, Y., and Tirrell, D.A. "Incorporation of trifluoroisoleucine into proteins in vivo". *J. Am. Chem. Soc.* **125**(23), 6900–6906 (2003).

Yang, J., Wu, K., Konák, C., and Kopecek, J. "Dynamic light scattering study of self-assembly of HPMA hybrid graft copolymers". *Biomacromolecules* **9**(2), 510–517 (2008).

SUBJECT INDEX

A
A-B block copolymers, 7, 211
$A\beta_{11-26}$, 109
Acetone, 137
ACL. *See* Anterior cruciate ligaments
Actin, 50, 61
  polymerization, 55
Activation rate, 58
Adenovirus, 195
Aerogels, 204f
AFM. *See* Atomic force microscopy
AG repeats, 109–110
Aggregates, 16, 22
Aggregation numbers, 16–17, 52, 71
Aggregation-prone regions, 163
Agrobacterium tumefaciens, 93
Alanine, 98, 122, 123, 217
Alcohol oxidase 1 (AOX1), 103
Alcohol systems, 139
Aliphatic groups, 46
Alkanes, 55
Alzheimer disease, 164, 217
Amino acids, 2, 47, 100, 148
  motifs, 124f
Ammonium hydroxide vapor, 218
Amorphous matrix, 129–130
Amphiphiles, 5, 32, 164
Ampullate gland, 121
  schematic representation
    of, 135f
Amyloid fibrils
  applications, 189–205
  assembly, 165–167
  biocompatible materials, 196–204
  biosensors and, 193–196
  disassembly, 165–167
  electronics in, 189–193
  in enzyme immobilization, 193–196
  in photonics, 189–193
  polypeptide self-assembly, 162–167
  stability, 165–167
Animal cells, 80
Anionic peptides, 35
Anterior cruciate ligaments
  (ACL), 143
Antibody interactions, 163
AOX1. *See* Alcohol oxidase 1
APFO-12, 185, 192
Aqueous gels, 33
Aqueous solutions, 68
  macroscopic properties of, 19t
Aragonite polymorphism, 146
Arg. *See* Arginine
Arginine (Arg), 17–18, 29,
    38, 83, 124
Aromatic groups, 180
Artificial spinning, 133–142
Aspergillus nidulans, 91
Assembly active, 51
Assembly inactive, 51
Atom transfer radical polymerization
    (ATRP), 214
Atomic force microscopy (AFM), 165,
    168, 179
  of fibril microarray, 184f
  high-resolution, 174
Atomistic simulations, 68
Atoms, 3
Atrina, 145
ATRP. *See* Atom transfer radical
    polymerization
Attraction energy, 15
AUG codon, 82
Autoinduction, 88
Autosteric binding, 52
Avidin, 189

223

B

Bacillus subtilis, 88
Bacitracin, 96
Bacteria, 80
  disadvantages of, 89
  high-density cultures, 88
  as host organism, 87–90
  protein production in, 88
Barnase, 193
Barrel channel β-sheets, 6
Basic fibroblast growth factor (bFGF), 200, 204
Benzene, 3
Best-fit energy values, 20, 22
bFGF. *See* Basic fibroblast growth factor
Bioactivity, 212
Bioapplication, fibroin, 142–144
Biocompatible materials, 196–204
Biodegradability, 143
Biofactories, 81
Bioinspired protein-like self-assembly, 12
Biomimetic molecules, 62
Biomimetic supramolecular polymers, 56f
Biomineralization, 144–147
Biopolymers, 212
Biosensors, 193–196
  glucose, 195f
Biotin-streptavidin, 163
Birefringent solutions, 24–25
Bjerrum length, 69
BL21, 104
Bolaamphiphile, 192
Boltzmann weight, 63
Boltzmann's constant, 47
Bombyx mori, 119, 122, 127f, 128, 132
Butyl acrylate, 215

C

$Ca(NO_3)$, 138
CAC. *See* Critical aggregate concentration
$CaCO_3$, 146, 147
Cadmium sulfate, 192
Calcium phosphate, 146
CaMV 35S, 94, 98
Canaliculi, 132
Carbon, 3
Carbonic anhydrase, 193
Cathelicidins, 105–107
Cationic peptides, 35
  indolicidin derivative, 107
CC. *See* Coiled coil
ccbeta switch system, 166
CDAP. *See* 1-cyano-4-dimethylaminopyridium tetrafluoroborate
CDS. *See* Peptide-coding sequence
Cecropins, 103–105
Cell dissociation, 197f
Cellulose-binding domain, 107
Chiral rods, hierarchical self-assembling, 13–25
Chirality, 6
  fibril formation and, 23–24
  of monomers, 15
Chitin-binding domain, 104
Chondrocytes, 202
Circular dichroism spectroscopy, 65f
Citrate, 190
Cloning target coding regions, 82f
Clostridium, 107
CM4, 104
CNBr. *See* Cyanogen bromide
Coarse grained models, 163
Coarse graining reversed, 66–70, 71
Coat proteins, 47
  assembled state of, 70f
Coatings, 202–203
Coaxial fibers, 191
CodenPlus, 83
Codon optimization, 83
Coiled coil (CC), 213
Collagen, 80, 201
Colloidal gold, 190
Complementary peptides, 35–36
Conformational frustration, 62, 63
Conformational switching, 71
$CONH_2$, 18, 20
COO^-, 30
COOD, 29
Cooperativity, nucleation and, 61–65
Copolymers, 212
Copper, 192

Subject Index

Coulomb repulsion, 68–69
  free energy of, 69
Couper, Archibald Scott, 3
CR1, 131f
Critical aggregate concentration (CAC), 111
Critical slowing down, 60
Crop Tech Corp., 95
C-terminal, 84, 125, 187f, 203
Curing, 135f
1-cyano-4-dimethylaminopyridium tetrafluoroborate (CDAP), 111
Cyanogen bromide (CNBr), 84, 101, 102f, 112
Cyclopeptide nanotubes, 215
Cyclopeptide-polymer coil copolymers, 216
Cysteine, 35, 187f, 213
  residue, 188

D
Darwin, Charles, 4
Debye length, 69
Dermcidin, 105
DHBCs. *See* Double-hydrophilic block copolymers
Dichlorodimethylsilane, 185
Dimensionless relaxation rate, 60f
2,4-dinitrophenyl acetate, 193
Dipeptide nanotubes, 202
Diphenylalanine, 164, 180, 186f
  peptide nanotubes, 181f, 186, 190, 194–195
Discotic compounds, 65
Distortion energy, 15
DLVO theory, 31, 34
Double-hydrophilic block copolymers (DHBCs), 145
Dragline silk, 126, 134
Drying methods, 178, 182
Dry-wet spinning, 136

E
EAK II, 27
EAK16, 26, 200, 201–202
EAK16-II, 177
  fibers, 179–180

ECM. *See* Extracellular matrix
EFK, 202
Elastin, 98–100, 197
Elastin-like protein polymers (ELP), 99
Electron micrographs
  of fibers, 21f
  of ribbons, 21f
Electronics, 189–193
Electrospinning, 140–142
Electrostatic Coulombic effects, 3
Electrostatics, 174, 180, 183
Elongation, 165
  fibril, 168
Elongation rate, 58
ELP. *See* Elastin-like protein polymers
End association, 57
End interchange, 57
Engineering materials, 2–4
  future challenges, 7–8
  ordering in, 3
Enthalpy, 49, 53
Entrepreneurialism, 2
Entropy, 72
Enzymatic dephosphorylation, 34
Enzymatic digestion, 163
Enzyme immobilization, 193–196
Enzyme proteinase K, 190
Epitaxial growth, 175
Escherichia coli, 83, 88, 99, 101, 104, 106, 107
Ethanol dehydrogenase, 195
EtOH, 138
Expression, 81–87
  microbial peptide, 108
  peptides, 85t–86t, 102–103
  transient, 93
External triggers, 25–37
  ionic strength and, 28–34
  pH, 25–27
Extracellular matrix (ECM), 196

F
F-actin, 50
Fibers
  coaxial, 191
  EAK16-II, 179–180
  electron micrographs, 21f

nematic fluids and, 24–25
nematic gels and, 24–25
peptide, 176f
Fibril(s), 15
AFM of, 184f
amyloid, 162–167, 189–205
β-sheet, 192
chirality, 23–24
coupling, 187
covalent attachment, 187f
deposition, 182, 183
electron micrographs of, 21f
elongation, 168
formation of, 16f, 175
forming peptides, 20–23
growth, 166, 169t–172t, 173, 178
incubation, 184
insulin, 173, 185f, 187
lithographic columns and, 188
nematic fluids and, 24–25
nematic gels and, 24–25
nucleation, 168
P_{11}-2 and, 23
pre-formed, 183f
SAM, 173
stability of, 179
sup35, 187f, 188
TEM, 33
Fibrillation pathway, 163
Fibroin, 97, 120–123, 128, 132, 136, 139, 148. *See also* Silk; Spinning
artificial spinning of, 133
bioapplication of, 142–144
electrospun, 140f
regenerated, 136, 137t
schematic representation of, 123f
ultrathin, 132
Fibronectin, 197
Filamentous fungi, 91–92, 194
advantages of, 91
disadvantages of, 92
Fischer, Emil, 4
Flagellin, 50
Flocculate, 29
insoluble, 31
Flory-Huggins parameter, 218

Fluorenylmethoxycarbonyl (FMOC), 166, 202
Fluorophore, 188
FMOC. *See* Fluorenylmethoxycarbonyl
4 kDa Scorpion Defensin, 103
Fourier transforms, 165, 181f
two-dimensional, 176f
Fraction helical bonds, 64f
Fraction polymerized material, 59f
Free energy
of binding, 47
of Coulumb repulsion, 69
penalty, 63
FTIR spectra, 18, 28, 30, 37f
Function, 4
Fungi, 80
filamentous, 91–92, 194
Fusion proteins, 100

G

G. *See* Glycine
G-actin, 50
Gadolinium, 192
β-galactosidase, 101
GAV9, 178, 179
Gelatin, 140
Gelation, 24–25
Gels, 214–215
Generally recognized as safe (GRAS), 90, 91
GFP. *See* Green fluorescent protein
Gln. *See* Glutamine
Globule formation, 135f
Glu. *See* Glutamate
Glucose, 6
biosensor, 195f
Glucose oxidase (GOX), 194
Glutamate (Glu), 17–18, 38, 110, 124, 166
self-assembly and, 25
Glutamic acid, 26, 29
Glutamine (Gln), 17–18, 20, 23, 124
Glutathione S-transferase (GST), 99, 104, 193
Glycine (G), 98, 123, 128, 130, 217
Glycosylation, 89, 95
Glycylglycine, 192
Gold, 190, 194

Subject Index

Gold-thiol interactions, 163
GOX. *See* Glucose oxidase
Graphite, 194
GRAS. *See* Generally recognized as safe
Green fluorescent protein (GFP), 95
GST. *See* Glutathione *S*-transferase
Guanidine, 215

H

HA. *See* Hyaluronic acid
HAP. *See* Hydroxyapatite
Hard tissue engineering, 202–203
HBsAg. *See* Hepatitis B surface antigen
hCAP-18, 105
HDEL, 94, 95
Helical configuration, 14
Helical tape, 14f
Helix-coil transition, 62
Heparin, 199
Hepatitis B, 93
Hepatitis B surface antigen (HBsAg), 96
Herpes simplex virus type 2 (HSV-2), 195
Heteroaggregates, 35–36
Hevea brasiliensis, 91
1,1,1,3,3,3-hexafluoro-2-propanol (HFP), 180
Hexafluoroisopropanol (HFIP), 137, 139
H-fibroin, 121
HFIP. *See* Hexafluoroisopropanol
HFP. *See* 1,1,1,3,3,3-hexafluoro-2-propanol
Hierarchical self-assembling chiral rods, 13–25
  experimental evidence of, 17–25
  model of, 14f
  theoretical rational of, 13–17
Hierarchical self-assembly
  macroscopic level, 46f
  mesoscopic level, 46f
  microscopic level, 46f
Highly oriented pyrolytic graphite (HOPG), 174, 175, 176f, 178, 180
  hydrophilic step edge, 182
  stability of, 179
  step edges of, 177f, 179
  steps in, 177
  ZYB grade, 171t
  ZYH grade, 171t

High-performance liquid chromatography (HPLC), 101, 104
His. *See* Histidine
Histidine (His), 84
  residues, 192
Histonin, 107–108
HOPG. *See* Highly oriented pyrolytic graphite
Host organisms, 80
  bacteria, 87–90
  for recombinant expression, 87–97
HPLC. *See* High-performance liquid chromatography
HRP. *See* Hydrogen peroxidase
HSV-2. *See* Herpes simplex virus type 2
Human β_2-microglobulin, 169t–170t, 173
Hyalophora ceropia, 92
Hyaluronic acid (HA), 203
Hydrogels, 5, 6
Hydrogen, 3, 123
  bonding, 71
Hydrogen peroxidase (HRP), 194
Hydrophilic A blocks, 211
Hydrophilic B blocks, 211
Hydrophilic regions, 213
Hydrophobic interactions, 68
Hydrophobic regions, 213
Hydrophobic rods, 219
Hydrophobicity, 173
Hydrophobins, 194, 195
Hydroxyapatite (HAP), 140, 144, 147, 198
N-hydroxy succinimidyl, 213
N-hydroxysuccinimide-activated glass surface, 187
Hypersensitivity, 144

I

IgM, 92
IKV, 201
IKVAV, 199
Imidazolyl, 193
Inclusion bodies, 89
Industrial Revolution, 2
Influenza, 195
Insulin fibrils, 173, 185f
  formation of, 187

Interfacial activity, 7
Interleukin-18, 96
Internal drawdown, 134, 135
Intramolecular ordering transition, 56f
Iodoacetamide, 188f
Ionic complementary peptides, 200–202
Ionic strength, 28–34
Ions, 6
IPTG. *See* Isopropyl-B-D-thiogalactopyransoide
Ising model, 62
Isodesmic assembly, 48, 50f, 61, 63, 64, 71
Isodichroic points, 20, 22
Isopropyl-B-D-thiogalactopyransoide (IPTG), 88
Isotropic fluid, 19t

J
JAVA Codon Adaptation Tool, 83
Jelly-roll structure, 47

K
KDEL, 94, 95
KDL12, 201
Kekulé, Friedrich August, 3
Keratin, 2
Ketosteroid isomerase (KSI), 101, 102f, 112
Kevlar, 125
KEX2, 103
KF8, 201
KFE12, 33
KGE motif, 200
Kinetics
  of nucleated assembly, 55–61
  polymerization, 59
  sigmoidal, 57, 59
KLD-12, 200
Kluyveroyces lactis, 90
KR-20, 105, 106
KSI. *See* Ketosteroid isomerase
Kymography, 174

L
Ladder model, 48
Laminin, 201
Landau model, 58, 59f
Law of mass action, 47

Leu, 83
L-fibroin, 121
LiBr, 138
Light chain amyloidosis, 174
Light microscopy, 205f
Linear assemblies, 48
Linear chain, 212
Linear Taylor expansion, 49
Lipid-water interfaces, 175
Liquid-gas interfaces, 167
Liquid-liquid interfaces, 167
Literature, self-assembling peptides, 4–7
Lithographic columns, 188
LL-23, 105, 106
LL-29, 105, 106
LL-37, 105
Lysine, 26, 98, 213
Lysis, 83

M
M. *See* Met
Macroscopic Landau free-energy density, 58, 59f
Macroscopic odd-even events, 55
Macroscopic phase transitions, 61
Magnetite, 185
MagnICON, 93
Mass action, law of, 47
MAX1, 34
Mean field theory, 54
Mechanical gene activation (MeGA), 95
MeGA. *See* Mechanical gene activation
MeOH, 138
Met (M), 83
Metal coordination, 3
Metalnikowin-2A, 103
Methanol, 137
Methionine, 84, 92, 102f
Mica, 38, 179
  hexagonal crystalline lattice of, 178
  modified, 170t–172t
  negatively charged, 174
  positively charged, 174
  surface, 178f
  unmodified, 170t–172t
Micelles, 46
  formation, 135f

Subject Index

Micro phase separation, 45
Microbial peptide expression, 108
Microscopic origin, 51
Molecular dynamics, 163
Molecular parameters, 19t
Molecular weight distribution, 48
Molecules, 3
Monomers, 13
   chirality of, 15
Monte Carlo simulations, 163–164
Multiple antimicrobial peptide expression, 102–103
Multistage open association, 48
Musca domestica, 104

N
N27C cysteine, 178
NaCL, 30, 32
NADH. *See* Nicotinamide adenine dinucleotide
Nano phase separation, 45
Nanofibers, 175
   peptide amphiphile, 181f
Nanofibrils, 132, 136
Nanomaterials, 12
Nanostructural orientation, 220
Nanotubes, 190
   dipeptide, 202
   diphenylalaline-based peptide, 181f, 186, 190, 194–195
Nanowires, 190
Native chemical ligation (NCL), 203
Natural selection, 4
Natural silk, 125–127
   fabrication of, 134–136
Nature
   repetitive and self-assembling peptides in, 97–100
   repetitive and self-assembling proteins in, 97–100
   self-assembly in, 80
n-butanol, 65f
NCL. *See* Native chemical ligation
Nematic fluid, 19t, 31
   fibers and, 24–25
   fibrils and, 24–25
Nematic gels, 19t

   fibers and, 24–25
   fibrils and, 24–25
Nephila clavipes, 98, 134, 140
Nephila edulis, 120f, 127
Newtonian fluids, 31
Nicotinamide adenine dinucleotide (NADH), 195
NIPAM/BAM. *See N*-isopropylacrylamide-co-*N*-tert-butrylacrylamide
N-isopropylacrylamide-co-*N*-tert-butrylacrylamide (NIPAM/BAM), 168t
Nitric acid, 144
Nonlinear equations, 57
North Western University, 5
Novagen, 101
NT1, 96
N-terminus, 35, 84, 125, 178, 212
Nucleated assembly, 48f, 50f, 51–55, 66, 70, 71
   cooperativity and, 61–65
   kinetics of, 55–61
   non-self-catalyzed, 52
   self-catalyzed, 53
   solvents in, 55
Nucleated equilibrium polymerization, 70
Nucleated supramolecular polymerization, 70
Nucleation, 165
   fibril, 168
   surface-induced, 169t–172t

O
O-acyl, 166
OAR1, 131f
OAR2, 131f
Odd-even events, 55
Oligo(phenylene vinyelene), 54f, 55, 62, 65
Oligoelectrolytes, 192
Oligomeric prenuclei, 56f
Oligopeptides, 51, 219
   chains, 66, 67
O-N acyl transfer reaction, 217
On Growth and Form (Thompson), 4

Subject Index

O-N intramolecular acyl migration, 166, 167
OPH. *See* Organophosphohydrolase
Ordering, 3
  intramolecular, 56f
Organometric peptide-polymer conjugation, 216–217
Organophosphohydrolase (OPH), 89
Orn, 31, 166
  self-assembly and, 25
Osteogenic cells, 204

P

P. *See* Proline
P$_{11}$-1, 18
  ε_{trans}, 23
  properties of, 19t
  self-assembling properties of, 18f
P$_{11}$-2, 111–112
  dimeric, 39f
  ε_{trans}, 23
  fibril formation and, 23
  phase behavior of, 21f
  properties of, 19t
  ribbons, 23
  self-assembling behavior of, 22f
P$_{11}$-4, 112, 202
  initial random coil state, 37f
  phase behavior, 27f
  self-assembly of, 26f
  in water, 27f
P$_{11}$-5
  initial random coil state, 37f
  molecular structure of, 36f
    molecular structure of, 36f
  phase behavior, 28f
  self-assembly of, 26f
P$_{11}$-9, 28, 203
  self-assembly of, 29f
  TEM, 30f
P$_{11}$-12, 31
  self-assembly of, 32f
  TEM, 33f
P$_{11}$-X family, 13
P25, 121
PA-1, 34
pacC, 91

palA, 91
palB, 91
palC, 91
palF, 91
palH, 91
palI, 91
Palmitoyl, 5
Palmitoyl-pentapeptides, 5
PDMS. *See* Polydimethylsiloxane
PECs. *See* Polyelectrolyte ß-sheet complexes
PEDOT. *See* Poly(4,5-ethylene-dioxythiophene)
PEGylation, 213
PEO. *See* Poly(ethylene oxide); Polyethyleneoxide
Pepfactants, 7
Peptide(s)
  amphiphiles, 181f, 191, 198–200, 204, 205f
  anionic, 35
  branching, 199
  C, 35
  cationic, 35
  complementary, 35–36
  defined, 2
  expression, 85t–86t, 102–103
  fibers, 176f
  fibril-forming, 20–23
  fractions of, 22f
  in gel formation, 214–215
  hydrophilic, 164
  hydrophobic, 164
  nanotubes, 181f, 186, 190, 194–195
  in nature, 97–100
  primary structure, 5–6
  recombinant bioactive, 102–108
  recombinant expression of, 108–112
  recombinant production, 100–108
  self-assembling, 4–7
  self-assembly on surfaces, 37–39
  β-sheet-forming, 111–112
  solid phase synthesis, 214
  stag, 106
  surfactants, 6
  tape-forming, 17–20
  vesicle-forming, 110–111

Subject Index

Peptide-coding sequence (CDS), 102f
Peptide-polymer conjugation, 213–214
　organometric, 216–217
　polyelectric, 216–217
pET31b, 101, 112
Petroleum, 3
PGA. *See* Polyglycolic acid
pH, 25–27, 174
　in water, 27f
Phase behavior
　P_{11}-2, 21f
　P_{11}-4, 27f
　P_{11}-5, 28f
Phase diagram, 16f
Phase transitions
　macroscopic, 61
　phenomenological theory of, 57–58
　thermodynamic, 57
Phenomenological relaxation rates, 58
Phenomenological theory, 57–58
Phenylalanine, 20
Phe-Phe motif, 164
Photobiotin, 188f
Photocage, 35
Photonics, 189–193
PHSRN, 199
Pichia pastoris, 90, 102
Pi-pi interactions, 3
Piscidins, 106–107
Plant cells, 80
　culture, 96–97
Plastics, 3
Poly(4,5-ethylene-dioxythiophene) (PEDOT), 191
Poly(ethersulfone), 169t
Poly(ethylene glycol), 214
Poly(ethylene oxide) (PEO), 219
Poly(glycolic acid), 199
Poly(thiophene acetic acid) (PTAA), 185
Poly(vinylidine diflouride), 169t
Poly-aramide, 125
Polydiacetylene, 218
Polydimethylsiloxane (PDMS), 182, 183
　stamp, 185
Polyelectric peptide-polymer conjugation, 216–217
Polyelectrolyte ß-sheet complexes (PECs), 35
Polyelectrolytes, 145
Polyethyleneimine, 194
Polyethyleneoxide (PEO), 136
Polyglycolic acid (PGA), 204, 205f
Poly-L-lysine, 183
Polymerization
　actin, 55
　atom transfer radical, 214
　kinetics, 59
　nucleated equilibrium, 70
　nucleated supramolecular, 70
　reversible, 51, 70
　temperature, 54f
Polymerized state, 49f
Polymers
　biomimetic supramolecular, 56f
　supramolecular, 46–47
Polymethyl-methacrylate polymer surface, 184
Poly-paraphenylene-terephthalamide, 125
Polypeptide chain, 182
Polypeptide self-assembly, 162–167
　in bulk liquid phase, 167f
　surfaces in, 167–189
Polystyrene (PS), 170t
Polytetrafluoroethylene, 169t
Polyurethane, 212
Post hoc fine graining, 66
Proline (P), 98, 124, 126
Protease, 84
Protegrin-1, 103
Protein. *See also specific types*
　in nature, 97–100
　production in bacteria, 88
　self-assembling, 97–100
Protein recovery, 81–87
　purification and, 83–84
Protein-protein binding energy, 69
Proteolytic degradation, 143
PS. *See* Polystyrene
Pseudoproline, 166
PTAA. *See* Poly(thiophene acetic acid)
PurF, 107
Purification, protein recovery and, 83–84

Q
Quasi-atomistic simulations, 68

R
Rabies, 96
RAD16, 110, 200–201
Ralstonia eutropha, 89
Rate-determining steps, 59
Reactor platforms, 167
Receptor-ligand interactions, 163
Recombinant bioactive peptides, 102–108
  cathelicidins, 105–107
  cecropins, 103–105
  dermcidin, 105
  histonin, 107–108
  piscidins, 106–107
Recombinant expression
  bacteria and, 87–90
  host organisms for, 87–97
  of self-assembling peptides, 108–112
Recombinant peptide production, 100–108
Recombinant production
  of elastin, 98–100
  of self-assembling proteins, 97–100
  of silks, 97–98
Recombinant proteins, 85t–86t
  specific cleavage of, 87t
Recombinant spidroin, 137t, 138
Reconstruction, 67
Regenerated cellulose, 169t
Regenerated fibroin, 136, 137t
Regenerated silk fibroin (RSF), 140
Regenerated spidroin, 136, 137t
Relaxational kinetic equations, 58
Remineralization, 203
Reversible polymerization, 51, 70
Reversible scission, 57
RGD ligands, 199
RGD tripeptide motif, 197
Ribbons, 14–15, 22
  electron micrographs of, 21f
  P_{11}-2, 23
  phase diagram of, 16f
RMS. *See* Root mean squared
RNADraw, 83
Rod-coil conformations, 218–220
Root mean squared (RMS), 168
RSF. *See* Regenerated silk fibroin

S
Saccharomyces cerevisiae, 90, 91
SA-LBL. *See* Spin-assisted layer-by-layer
Salmonella, 104
Salt, 183
SAMs. *See* Self-assembled monolayers
Scanning electron microscopy, 181f, 188f
Seeding, 165–166
Self-assembled Ising chain, 62
Self-assembled monolayers (SAMs), 6
  fibrils, 173
Self-assembling peptide literature, 4–7
Self-assembling proteins, 97–100
Self-assembly, 1
  bioinspired protein-like, 12
  external trigger responsiveness and, 25–37
  Glu and, 25
  in nature, 80
  Orn and, 25
  P_{11}-1 and, 18f
  P_{11}-2, 22f
  P_{11}-4, 26f
  P_{11}-5, 26f
  P_{11}-9, 29f
  P_{11}-12, 32f
  polypeptide, 162–167
  surfaces in, 167–189
Self-catalyzed nucleated assembly, 53
Self-organization, 1, 4
Ser, 83
Sericin, 120, 121
Serine, 35, 122, 130
Shape, 4
Shear force, 134, 135f
Sheep Myeloid Antibacterial Peptide, 103, 108
β-sheets, 6, 13, 35, 66, 163
  barrel channel, 6
  fibrils, 192
  forming peptides, 111–112
  suprastructures, 217–218
Side chains, 32
Sigmoidal kinetics, 57, 59

Subject Index

Silk, 80, 148. *See also* Fibroin; Spidroin; Spinning
   artificial spinning of, 133–142
   biomineralization, 144–147
   dragline, 126, 134
   electrospinning, 140–142
   man-made, 127–128
   mechanical properties of, 125–132, 129t
   natural, 125–127
   natural fabrication of, 134–136
   proteins, 120
   recombinant production of, 97–98
   silica fusion proteins, 147
   solution spinning, 136–140
   structure and properties of, 120–132
   supramolecular structure in, 131f
Silkworms, 120–121
Silver ions, 190
SK-29, 105, 106
Sodium, 136
Solid-liquid interfaces, 167
   fibril formations in, 175
Solid-phase peptide synthesis, 214
Solution spinning, 136–140
Solvents in nucleated assembly, 55
Soybean, 92
Spiders, 121, 134. *See also* Silk; Spidroin
Spidroin, 120, 122t, 124f, 133, 145, 148
   morphologies of, 138f
   recombinant, 137t, 138
   regenerated, 136, 137t
Spinal cord injuries, 199
Spin-assisted layer-by-layer (SA-LBL), 132
Spinning. *See also* Silk
   artificial, 133–142
   dope, 134, 136
   dry-wet, 136
   electrospinning, 140–142
   solution, 136–140
   wet, 137
Stag-peptide, 106
Strategene, 83
Streptavidin-biotin interactions, 187–188
Structure, 4
Stupp, Sam, 5
Subcellular locations, 95
Substance P, 101

Sup35 fibrils, 187f, 188
Sup35 NM protein, 182f, 187f, 193
Supramolecular polymers, 46–47
Surface interactions following assembly, 182–189
Surface-based assembly, 168–175
Surface-directed assembly, 175–181
Surface-induced nucleation, 169t–172t
Surfaces
   in assembly, 167–189
   hydrophobic, 174
   mica, 178f
   peptide self-assembly, 37–39
Switch peptides, 35

T

Tandem repeat strategy, 101
Tape-forming peptides, 17–20
Target-coding regions, cloning, 82f
T-DNA, 93
TEM, 165, 182f
   Fibrils, 33
   P_{11}-9, 30f
   P_{11}-12, 33f
Templated assembly, 65
TEOS. *See* Tetraethoxysilane
Tetraethoxysilane (TEOS), 192
Tetrahydrofuran, 219
Tetramethylorthosilicate (TMOS), 220
TEV. *See* Tobacco etch virus
TFA. *See* Trifluoroacetate acid
Thermodynamic phase transitions, 57
Thermodynamic stability, 60
Thioester, 203
Thioflavin T, 165
Thioredoxin (Trx), 104
Thompson, D'Arcy, 4
Threonine, 35
TIRFM. *See* Total internal reflection fluorescence microscopy
TMOS. *See* Tetramethylorthosilicate
Tobacco, 92–93
Tobacco etch virus (TEV), 84
Tobacco mosaic virus, 63
   assembled state of, 70f
Total internal reflection fluorescence microscopy (TIRFM), 165

Toxicology, 197
Transcription, 81–82, 95
Transgene integration, 94
Transgenic plants, 92–97
Transient expression, 93
Translation, 82–83
Trifluoroacetate acid (TFA), 138–139
Trp (W), 83
Trx. See Thioredoxin
Tryptophan, 20, 215
TTR1, 197, 198
TTR105-115, 192
Tubulin, 50
Tyr, 122, 130, 215
(Fmoc)-tyrosine, 34

V
V. See Valine
Vaccinia, 195
Valine (V), 92, 98
van der Waals forces, 3, 20, 34, 182
Vesicle-forming peptides, 110–111
Viral RNA, 61

Viruses, 46–47
  detection, 195

W
W. See Trp
Water
  P_{11}-4 in, 27f
  pH in, 27f
Weight distribution, 48
Wet spinning, 137

X
X-ray diffraction, 189

Y
Yeasts, 80, 90–91
  advantages of, 90
  disadvantages of, 91
YIG, 201

Z
Zim-Bragg theory, 62
ZYB grade HOPG, 171t
ZYH grade HOPG, 171t

Contents of Volumes in This Serial

Volume 1 (1956)

J. W. Westwater, *Boiling of Liquids*
A. B. Metzner, *Non-Newtonian Technology: Fluid Mechanics, Mixing, and Heat Transfer*
R. Byron Bird, *Theory of Diffusion*
J. B. Opfell and B. H. Sage, *Turbulence in Thermal and Material Transport*
Robert E. Treybal, *Mechanically Aided Liquid Extraction*
Robert W. Schrage, *The Automatic Computer in the Control and Planning of Manufacturing Operations*
Ernest J. Henley and Nathaniel F. Barr, *Ionizing Radiation Applied to Chemical Processes and to Food and Drug Processing*

Volume 2 (1958)

J. W. Westwater, *Boiling of Liquids*
Ernest F. Johnson, *Automatic Process Control*
Bernard Manowitz, *Treatment and Disposal of Wastes in Nuclear Chemical Technology*
George A. Sofer and Harold C. Weingartner, *High Vacuum Technology*
Theodore Vermeulen, *Separation by Adsorption Methods*
Sherman S. Weidenbaum, *Mixing of Solids*

Volume 3 (1962)

C. S. Grove, Jr., Robert V. Jelinek, and Herbert M. Schoen, *Crystallization from Solution*
F. Alan Ferguson and Russell C. Phillips, *High Temperature Technology*
Daniel Hyman, *Mixing and Agitation*
John Beck, *Design of Packed Catalytic Reactors*
Douglass J. Wilde, *Optimization Methods*

Volume 4 (1964)

J. T. Davies, *Mass-Transfer and Inierfacial Phenomena*
R. C. Kintner, *Drop Phenomena Affecting Liquid Extraction*
Octave Levenspiel and Kenneth B. Bischoff, *Patterns of Flow in Chemical Process Vessels*
Donald S. Scott, *Properties of Concurrent Gas–Liquid Flow*
D. N. Hanson and G. F. Somerville, *A General Program for Computing Multistage Vapor–Liquid Processes*

Volume 5 (1964)

J. F. Wehner, *Flame Processes–Theoretical and Experimental*
J. H. Sinfelt, *Bifunctional Catalysts*
S. G. Bankoff, *Heat Conduction or Diffusion with Change of Phase*
George D. Fulford, *The Flow of Lktuids in Thin Films*
K. Rietema, *Segregation in Liquid–Liquid Dispersions and its Effects on Chemical Reactions*

Volume 6 (1966)

S. G. Bankoff, *Diffusion-Controlled Bubble Growth*
John C. Berg, Andreas Acrivos, and Michel Boudart, *Evaporation Convection*
H. M. Tsuchiya, A. G. Fredrickson, and R. Aris, *Dynamics of Microbial Cell Populations*
Samuel Sideman, *Direct Contact Heat Transfer between Immiscible Liquids*
Howard Brenner, *Hydrodynamic Resistance of Particles at Small Reynolds Numbers*

Volume 7 (1968)

Robert S. Brown, Ralph Anderson, and Larry J. Shannon, *Ignition and Combustion of Solid Rocket Propellants*
Knud Østergaard, *Gas–Liquid–Particle Operations in Chemical Reaction Engineering*
J. M. Prausnilz, *Thermodynamics of Fluid–Phase Equilibria at High Pressures*
Robert V. Macbeth, *The Burn–Out Phenomenon in Forced-Convection Boiling*
William Resnick and Benjamin Gal–Or, *Gas-Liquid Dispersions*

Volume 8 (1970)

C. E. Lapple, *Electrostatic Phenomena with Particulates*
J. R. Kittrell, *Mathematical Modeling of Chemical Reactions*
W. P. Ledet and D. M. Himmelblau, *Decomposition Procedures foe the Solving of Large Scale Systems*
R. Kumar and N. R. Kuloor, *The Formation of Bubbles and Drops*

Volume 9 (1974)

Renato G. Bautista, *Hydrometallurgy*
Kishan B. Mathur and Norman Epstein, *Dynamics of Spouted Beds*
W. C. Reynolds, *Recent Advances in the Computation of Turbulent Flows*
R. E. Peck and D. T. Wasan, *Drying of Solid Particles and Sheets*

Volume 10 (1978)

G. E. O'Connor and T. W. F. Russell, *Heat Transfer in Tubular Fluid–Fluid Systems*
P. C. Kapur, *Balling and Granulation*
Richard S. H. Mah and Mordechai Shacham, *Pipeline Network Design and Synthesis*
J. Robert Selman and Charles W. Tobias, *Mass-Transfer Measurements by the Limiting-Current Technique*

Volume 11 (1981)

Jean-Claude Charpentier, *Mass-Transfer Rates in Gas–Liquid Absorbers and Reactors*
Dee H. Barker and C. R. Mitra, *The Indian Chemical Industry—Its Development and Needs*
Lawrence L. Tavlarides and Michael Stamatoudis, *The Analysis of Interphase Reactions and Mass Transfer in Liquid–Liquid Dispersions*
Terukatsu Miyauchi, Shintaro Furusaki, Shigeharu Morooka, and Yoneichi Ikeda, *Transport Phenomena and Reaction in Fluidized Catalyst Beds*

Volume 12 (1983)

C. D. Prater, J, Wei, V. W. Weekman, Jr., and B. Gross, *A Reaction Engineering Case History: Coke Burning in Thermofor Catalytic Cracking Regenerators*
Costel D. Denson, *Stripping Operations in Polymer Processing*
Robert C. Reid, *Rapid Phase Transitions from Liquid to Vapor*
John H. Seinfeld, *Atmospheric Diffusion Theory*

Volume 13 (1987)

Edward G. Jefferson, *Future Opportunities in Chemical Engineering*
Eli Ruckenstein, *Analysis of Transport Phenomena Using Scaling and Physical Models*
Rohit Khanna and John H. Seinfeld, *Mathematical Modeling of Packed Bed Reactors: Numerical Solutions and Control Model Development*
Michael P. Ramage, Kenneth R. Graziano, Paul H. Schipper, Frederick J. Krambeck, and Byung C. Choi, *KINPTR (Mobil's Kinetic Reforming Model): A Review of Mobil's Industrial Process Modeling Philosophy*

Volume 14 (1988)

Richard D. Colberg and Manfred Morari, *Analysis and Synthesis of Resilient Heat Exchange Networks*
Richard J. Quann, Robert A. Ware, Chi-Wen Hung, and James Wei, *Catalytic Hydrometallation of Petroleum*
Kent David, *The Safety Matrix: People Applying Technology to Yield Safe Chemical Plants and Products*

Volume 15 (1990)

Pierre M. Adler, Ali Nadim, and Howard Brenner, *Rheological Models of Suspenions*
Stanley M. Englund, *Opportunities in the Design of Inherently Safer Chemical Plants*
H. J. Ploehn and W. B. Russel, *Interations between Colloidal Particles and Soluble Polymers*

Volume 16 (1991)

Perspectives in Chemical Engineering: Research and Education

Clark K. Colton, *Editor*

Historical Perspective and Overview

L. E. Scriven, *On the Emergence and Evolution of Chemical Engineering*
Ralph Landau, *Academic–industrial Interaction in the Early Development of Chemical Engineering*
James Wei, *Future Directions of Chemical Engineering*

Fluid Mechanics and Transport

L. G. Leal, *Challenges and Opportunities in Fluid Mechanics and Transport Phenomena*
William B. Russel, *Fluid Mechanics and Transport Research in Chemical Engineering*
J. R. A. Pearson, *Fluid Mechanics and Transport Phenomena*

Thermodynamics

Keith E. Gubbins, *Thermodynamics*
J. M. Prausnitz, *Chemical Engineering Thermodynamics: Continuity and Expanding Frontiers*
H. Ted Davis, *Future Opportunities in Thermodynamics*

Kinetics, Catalysis, and Reactor Engineering

Alexis T. Bell, *Reflections on the Current Status and Future Directions of Chemical Reaction Engineering*
James R. Katzer and S. S. Wong, *Frontiers in Chemical Reaction Engineering*
L. Louis Hegedus, *Catalyst Design*

Environmental Protection and Energy

John H. Seinfeld, *Environmental Chemical Engineering*
T. W. F. Russell, *Energy and Environmental Concerns*
Janos M. Beer, Jack B. Howard, John P. Longwell, and Adel F. Sarofim, *The Role of Chemical Engineering in Fuel Manufacture and Use of Fuels*

Polymers

Matthew Tirrell, *Polymer Science in Chemical Engineering*
Richard A. Register and Stuart L. Cooper, *Chemical Engineers in Polymer Science: The Need for an Interdisciplinary Approach*

Microelectronic and Optical Material

Larry F. Thompson, *Chemical Engineering Research Opportunities in Electronic and Optical Materials Research*
Klavs F. Jensen, *Chemical Engineering in the Processing of Electronic and Optical Materials: A Discussion Bioengineering*

Bioengineering

James E. Bailey, *Bioprocess Engineering*
Arthur E. Humphrey, *Some Unsolved Problems of Biotechnology*
Channing Robertson, *Chemical Engineering: Its Role in the Medical and Health Sciences*

Process Engineering

Arthur W. Westerberg, *Process Engineering*
Manfred Morari, *Process Control Theory: Reflections on the Past Decade and Goals for the Next*
James M. Douglas, *The Paradigm After Next*
George Stephanopoulos, *Symbolic Computing and Artificial Intelligence in Chemical Engineering: A New Challenge*

The Identity of Our Profession

Morton M. Denn, *The Identity of Our Profession*

Volume 17 (1991)

Y. T. Shah, *Design Parameters for Mechanically Agitated Reactors*
Mooson Kwauk, *Particulate Fluidization: An Overview*

Volume 18 (1992)

E. James Davis, *Microchemical Engineering: The Physics and Chemistry of the Microparticle*
Selim M. Senkan, *Detailed Chemical Kinetic Modeling: Chemical Reaction Engineering of the Future*
Lorenz T. Biegler, *Optimization Strategies for Complex Process Models*

Volume 19 (1994)

Robert Langer, *Polymer Systems for Controlled Release of Macromolecules, Immobilized Enzyme Medical Bioreactors, and Tissue Engineering*
J. J. Linderman, P. A. Mahama, K. E. Forsten, and D. A. Lauffenburger, *Diffusion and Probability in Receptor Binding and Signaling*
Rakesh K. Jain, *Transport Phenomena in Tumors*
R. Krishna, *A Systems Approach to Multiphase Reactor Selection*
David T. Allen, *Pollution Prevention: Engineering Design at Macro-, Meso-, and Microscales*
John H. Seinfeld, Jean M. Andino, Frank M. Bowman, Hali J. L. Forstner, and Spyros Pandis, *Tropospheric Chemistry*

Volume 20 (1994)

Arthur M. Squires, *Origins of the Fast Fluid Bed*
Yu Zhiqing, *Application Collocation*
Youchu Li, *Hydrodynamics*
Li Jinghai, *Modeling*
Yu Zhiqing and Jin Yong, *Heat and Mass Transfer*
Mooson Kwauk, *Powder Assessment*
Li Hongzhong, *Hardware Development*
Youchu Li and Xuyi Zhang, *Circulating Fluidized Bed Combustion*
Chen Junwu, Cao Hanchang, and Liu Taiji, *Catalyst Regeneration in Fluid Catalytic Cracking*

Volume 21 (1995)

Christopher J. Nagel, Chonghum Han, and George Stephanopoulos, *Modeling Languages: Declarative and Imperative Descriptions of Chemical Reactions and Processing Systems*
Chonghun Han, George Stephanopoulos, and James M. Douglas, *Automation in Design: The Conceptual Synthesis of Chemical Processing Schemes*
Michael L. Mavrovouniotis, *Symbolic and Quantitative Reasoning: Design of Reaction Pathways through Recursive Satisfaction of Constraints*
Christopher Nagel and George Stephanopoulos, *Inductive and Deductive Reasoning: The Case of Identifying Potential Hazards in Chemical Processes*
Keven G. Joback and George Stephanopoulos, *Searching Spaces of Discrete Solutions: The Design of Molecules Processing Desired Physical Properties*

Volume 22 (1995)

Chonghun Han, Ramachandran Lakshmanan, Bhavik Bakshi, and George Stephanopoulos, *Nonmonotonic Reasoning: The Synthesis of Operating Procedures in Chemical Plants*

Pedro M. Saraiva, *Inductive and Analogical Learning: Data-Driven Improvement of Process Operations*

Alexandros Koulouris, Bhavik R. Bakshi and George Stephanopoulos, *Empirical Learning through Neural Networks: The Wave-Net Solution*

Bhavik R. Bakshi and George Stephanopoulos, *Reasoning in Time: Modeling, Analysis, and Pattern Recognition of Temporal Process Trends*

Matthew J. Realff, *Intelligence in Numerical Computing: Improving Batch Scheduling Algorithms through Explanation-Based Learning*

Volume 23 (1996)

Jeffrey J. Siirola, *Industrial Applications of Chemical Process Synthesis*

Arthur W. Westerberg and Oliver Wahnschafft, *The Synthesis of Distillation-Based Separation Systems*

Ignacio E. Grossmann, *Mixed-Integer Optimization Techniques for Algorithmic Process Synthesis*

Subash Balakrishna and Lorenz T. Biegler, *Chemical Reactor Network Targeting and Integration: An Optimization Approach*

Steve Walsh and John Perkins, *Operability and Control inn Process Synthesis and Design*

Volume 24 (1998)

Raffaella Ocone and Gianni Astarita, *Kinetics and Thermodynamics in Multicomponent Mixtures*

Arvind Varma, Alexander S. Rogachev, Alexandra S. Mukasyan, and Stephen Hwang, *Combustion Synthesis of Advanced Materials: Principles and Applications*

J. A. M. Kuipers and W. P. Mo, van Swaaij, *Computational Fluid Dynamics Applied to Chemical Reaction Engineering*

Ronald E. Schmitt, Howard Klee, Debora M. Sparks, and Mahesh K. Podar, *Using Relative Risk Analysis to Set Priorities for Pollution Prevention at a Petroleum Refinery*

Volume 25 (1999)

J. F. Davis, M. J. Piovoso, K. A. Hoo, and B. R. Bakshi, *Process Data Analysis and Interpretation*

J. M. Ottino, P. DeRoussel, S., Hansen, and D. V. Khakhar, *Mixing and Dispersion of Viscous Liquids and Powdered Solids*

Peter L. Silverston, Li Chengyue, Yuan Wei-Kang, *Application of Periodic Operation to Sulfur Dioxide Oxidation*

Volume 26 (2001)

J. B. Joshi, N. S. Deshpande, M. Dinkar, and D. V. Phanikumar, *Hydrodynamic Stability of Multiphase Reactors*

Michael Nikolaou, *Model Predictive Controllers: A Critical Synthesis of Theory and Industrial Needs*

Volume 27 (2001)

William R. Moser, Josef Find, Sean C. Emerson, and Ivo M, Krausz, *Engineered Synthesis of Nanostructure Materials and Catalysts*
Bruce C. Gates, *Supported Nanostructured Catalysts: Metal Complexes and Metal Clusters*
Ralph T. Yang, *Nanostructured Absorbents*
Thomas J. Webster, *Nanophase Ceramics: The Future Orthopedic and Dental Implant Material*
Yu-Ming Lin, Mildred S. Dresselhaus, and Jackie Y. Ying, *Fabrication, Structure, and Transport Properties of Nanowires*

Volume 28 (2001)

Qiliang Yan and Juan J. DePablo, *Hyper-Parallel Tempering Monte Carlo and Its Applications*
Pablo G. Debenedetti, Frank H. Stillinger, Thomas M. Truskett, and Catherine P. Lewis, *Theory of Supercooled Liquids and Glasses: Energy Landscape and Statistical Geometry Perspectives*
Michael W. Deem, *A Statistical Mechanical Approach to Combinatorial Chemistry*
Venkat Ganesan and Glenn H. Fredrickson, *Fluctuation Effects in Microemulsion Reaction Media*
David B. Graves and Cameron F. Abrams, *Molecular Dynamics Simulations of Ion–Surface Interactions with Applications to Plasma Processing*
Christian M. Lastoskie and Keith E, Gubbins, *Characterization of Porous Materials Using Molecular Theory and Simulation*
Dimitrios Maroudas, *Modeling of Radical-Surface Interactions in the Plasma-Enhanced Chemical Vapor Deposition of Silicon Thin Films*
Sanat Kumar, M. Antonio Floriano, and Athanassiors Z. Panagiotopoulos, *Nanostructured Formation and Phase Separation in Surfactant Solutions*
Stanley I. Sandler, Amadeu K. Sum, and Shiang-Tai Lin, *Some Chemical Engineering Applications of Quantum Chemical Calculations*
Bernhardt L. Trout, *Car-Parrinello Methods in Chemical Engineering: Their Scope and potential*
R. A. van Santen and X. Rozanska, *Theory of Zeolite Catalysis*
Zhen-Gang Wang, *Morphology, Fluctuation, Metastability and Kinetics in Ordered Block Copolymers*

Volume 29 (2004)

Michael V. Sefton, *The New Biomaterials*
Kristi S. Anseth and Kristyn S. Masters, *Cell–Material Interactions*
Surya K. Mallapragada and Jennifer B. Recknor, *Polymeric Biomaterias for Nerve Regeneration*
Anthony M. Lowman, Thomas D. Dziubla, Petr Bures, and Nicholas A. Peppas, *Structural and Dynamic Response of Neutral and Intelligent Networks in Biomedical Environments*
F. Kurtis Kasper and Antonios G. Mikos, *Biomaterials and Gene Therapy*
Balaji Narasimhan and Matt J. Kipper, *Surface-Erodible Biomaterials for Drug Delivery*

Volume 30 (2005)

Dionisio Vlachos, *A Review of Multiscale Analysis: Examples from System Biology, Materials Engineering,and Other Fluids-Surface Interacting Systems*
Lynn F. Gladden, M.D. Mantle and A.J. Sederman, *Quantifying Physics and Chemistry at Multiple Length-Scales using Magnetic Resonance Techniques*
Juraj Kosek, Frantisek Steěpánek, and Miloš Marek, *Modelling of Transport and Transformation Processes in Porous and Multiphase Bodies*
Vemuri Balakotaiah and Saikat Chakraborty, *Spatially Averaged Multiscale Models for Chemical Reactors*

Volume 31 (2006)

Yang Ge and Liang-Shih Fan, *3-D Direct Numerical Simulation of Gas–Liquid and Gas–Liquid–Solid Flow Systems Using the Level-Set and Immersed-Boundary Methods*

M.A. van der Hoef, M. Ye, M. van Sint Annaland, A.T. Andrews IV, S. Sundaresan, and J.A.M. Kuipers, *Multiscale Modeling of Gas-Fluidized Beds*

Harry E.A. Van den Akker, *The Details of Turbulent Mixing Process and their Simulation*

Rodney O. Fox, *CFD Models for Analysis and Design of Chemical Reactors*

Anthony G. Dixon, Michiel Nijemeisland, and E. Hugh Stitt, *Packed Tubular Reactor Modeling and Catalyst Design Using Computational Fluid Dynamics*

Volume 32 (2007)

William H. Green, Jr., *Predictive Kinetics: A New Approach for the 21st Century*

Mario Dente, Giulia Bozzano, Tiziano Faravelli, Alessandro Marongiu, Sauro Pierucci and Eliseo Ranzi, *Kinetic Modelling of Pyrolysis Processes in Gas and Condensed Phase*

Mikhail Sinev, Vladimir Arutyunov and Andrey Romanets, *Kinetic Models of C1–C4 Alkane Oxidation as Applied to Processing of Hydrocarbon Gases: Principles, Approaches and Developments*

Pierre Galtier, *Kinetic Methods in Petroleum Process Engineering*

Volume 33 (2007)

Shinichi Matsumoto and Hirofumi Shinjoh, *Dynamic Behavior and Characterization of Automobile Catalysts*

Mehrdad Ahmadinejad, Maya R. Desai, Timothy C. Watling and Andrew P.E. York, *Simulation of Automotive Emission Control Systems*

Anke Güthenke, Daniel Chatterjee, Michel Weibel, Bernd Krutzsch, Petr Kočí, Miloš Marek, Isabella Nova and Enrico Tronconi, *Current Status of Modeling Lean Exhaust Gas Aftertreatment Catalysts*

Athanasios G. Konstandopoulos, Margaritis Kostoglou, Nickolas Vlachos and Evdoxia Kladopoulou, *Advances in the Science and Technology of Diesel Particulate Filter Simulation*

Volume 34 (2008)

C.J. van Duijn, Andro Mikelić, I.S. Pop, and Carole Rosier, *Effective Dispersion Equations for Reactive Flows with Dominant Péclet and Damkohler Numbers*

Mark Z. Lazman and Gregory S. Yablonsky, *Overall Reaction Rate Equation of Single-Route Complex Catalytic Reaction in Terms of Hypergeometric Series*

A.N. Gorban and O. Radulescu, *Dynamic and Static Limitation in Multiscale Reaction Networks, Revisited*

Liqiu Wang, Mingtian Xu, and Xiaohao Wei, *Multiscale Theorems*

Volume 35 (2009)

Rudy J. Koopmans and Anton P.J. Middelberg, *Engineering Materials from the Bottom Up – Overview*

Robert P.W. Davies, Amalia Aggeli, Neville Boden, Tom C.B. McLeish, Irena A. Nyrkova, and Alexander N. Semenov, *Mechanisms and Principles of 1D Self-Assembly of Peptides into β-Sheet Tapes*

Paul van der Schoot, *Nucleation and Co-Operativity in Supramolecular Polymers*
Michael J. McPherson, Kier James, Stuart Kyle, Stephen Parsons, and Jessica Riley, *Recombinant Production of Self-Assembling Peptides*
Boxun Leng, Lei Huang, and Zhengzhong Shao, *Inspiration from Natural Silks and Their Proteins*
Sally L. Gras, *Surface- and Solution-Based Assembly of Amyloid Fibrils for Biomedical and Nanotechnology Applications*
Conan J. Fee, *Hybrid Systems Engineering: Polymer–Peptide Conjugates*

DATE DUE